SCIENCE EDUCATION AND CULTURE

T0138164

Science Education and Culture

The Contribution of History and Philosophy of Science

Edited by

Fabio Bevilacqua

Pavia University, Italy

Enrico Giannetto

Pavia University, Italy

and

Michael R. Matthews

University of New South Wales, Australia

Selected papers from the "Science as Culture" conference held at Lake Como and Pavia, September 1999, and generously sponsored by the Volta Bicentenary Fund – Lombardy Region, Pavia University, and the Italian Research Council

KLUWER ACADEMIC PUBLISHERS

DORDRECHT / BOSTON / LONDON

A C.I.P. Catalogue record for this book is available from the Library of Congress.

ISBN 0-7923-6972-6 (HB)
ISBN 0-7923-6973-4 (PB)

Published by Kluwer Academic Publishers,
P.O. Box 17, 3300 AA Dordrecht, The Netherlands.

Sold and distributed in North, Central and South America
by Kluwer Academic Publishers,
101 Philip Drive, Norwell, MA 02061, U.S.A.

In all other countries, sold and distributed
by Kluwer Academic Publishers,
P.O. Box 322, 3300 AH Dordrecht, The Netherlands.

Printed on acid-free paper

Cover photo:
A vista of 18th century Pavia with Alessandro Volta

All Rights Reserved
© 2001 Kluwer Academic Publishers
No part of the material protected by this copyright notice may be reproduced or
utilized in any form or by any means, electronic or mechanical,
including photocopying, recording or by any information storage and
retrieval system, without written permission from the copyright owner.

Printed in the Netherlands.

Table of Contents

Editorial Introduction

This anthology contains selected papers from the 'Science as Culture' conference held at Lake Como, and Pavia University Italy, 15-19 September 1999. The conference, attended by about 220 individuals from thirty countries, was a joint venture of the International History, Philosophy and Science Teaching Group (its fifth conference) and the History of Physics and Physics Teaching Division of the European Physical Society (its eighth conference). The magnificient Villa Olmo, on the lakeshore, provided a memorable location for the presentors of the 160 papers and the audience that discussed them. The conference was part of local celebrations of the bicentenary of Alessandro Volta's creation of the battery in 1799. Volta was born in Como in 1745, and for forty years from 1778 he was professor of experimental physics at Pavia University. The conference was fortunate to have had the generous financial support of the Italian government's Volta Bicentenary Fund, Lombardy region, Pavia University, Italian Research Council, and Kluwer Academic Publishers.

The papers included here, have or will be, published in the journal *Science & Education*, the inaugural volume (1992) of which was a landmark in the history of science education publication, because it was the first journal in the field devoted to contributions from historical, philosophical and sociological scholarship. Clearly these 'foundational' disciplines inform numerous theoretical, curricular and pedagogical debates in science education.

Contemporary Concerns

The reseach promoted by the International and European Groups, and by the journal, is central to science education programmes in most areas of the world. Increasingly school science courses are being asked to address issues concerning the Nature of Science. Students are expected to gain a rudimentary understanding of the 'big picture' of science: its history, its philosophical assumptions and implications, its interaction with culture and society, and so on. It is increasingly expected that students will leave school with not just knowledge *of* science, but also with knowledge *about* science.

These 'liberal' curricular developments have occurred in the UK, US, Canada, Australia and Japan. In the US, aspects of the history and philosophy of science are written into the National Science Education Standards and are enthusiastically promoted by the AAAS's Project 2061. In England and Wales, these liberal goals have survived the vicissitudes of the unfolding National Curriculum (Donnelly 2001). In Greece, satisfactory completion of a final year course on the history of science is now a requirement for graduation from high school. In the Australian state of New South Wales, history of science and nature of science comprise two of six prescribed curriculum focus areas.

Even those with a more narrow and disciplinary focus maintain that if we want students to learn and become competent in science, then they must be taught something about the nature of science. For instance, Frederick Reif wrote:

> All too often introductory physics courses 'cover' numerous topics, but the knowledge actually acquired by students is often nominal rather than functional. If students are to acquire basic physics knowledge ... it is necessary to understand better the requisite thought processes and to teach these more explicitly ... if one wants to improve significantly students' learning of physics ... It is also necessary to modify students' naive notions about the nature of science. (Reif 1995, p. 281)

But it is not just curriculum injunctions that require teachers to address the nature of science: frequently they are called on to address issues raised by the so called 'Science Wars', by the Creationism controversy, by Feminist critiques of science, by multiculturalist claims for inclusion of indigenous science in the curriculum, and so forth. In many parts of the world, proponents of Islamic Science demand that the school science curriculum reflect the Koran's teaching on biological, cosomological and methodological matters. Intelligent discussion of all of these questions depends upon some grasp of the history, philosophy and sociology of science.

There are engaging tensions between science education and culture. Most people hope that children who learn science will be affected by what they learn; that lessons from the science classroom will flow on to children's ways of thinking about personal and social matters; that it will contribute to their world views, and ultimately to culture. If science teaching is done well, these flow-on effects are positive; if it is done badly, they are negative. But of course often there are no flow-on effects of either kind - some years ago it was famously shown that belief in astrology was unaffected by completion of a US science degree. A major issue in multicultural science debates is the extent to which there should be flow-ons from the science classroom to religious, social and traditional beliefs and practices.

Historical Background

Questions about the nature of science have long been of concern to science teachers and curriculum developers. It has been hoped that science education would fructify in society and have a beneficial impact on the quality of culture and public life in virtue of students appreciating something of the nature of science, internalising something of the scientific spirit, and developing a scientific frame of mind that might carry over into other spheres of life. John Dewey well expressed this Enlightenment hope for science education when he said:

> Our predilection for premature acceptance and assertion, our aversion to suspended judgment, are signs that we tend naturally to cut short the process of testing. We are satisified with superficial and immediate short-visioned applications. ... Science represents the safeguard of the race against these natural propensities and the evils which flow from them. ... It is artificial (an acquired art), not spontaneous; learned, not native. To this fact is due the unique, the invaluable place of science in education. (Dewey 1916, p. 189)

Such historical and philosophical contributions had been urged for well over a century. The Duke of Argyll in his 1856 Presidential Address to the British Association for the Advancement of Science challenged the meeting with the claim that: 'What we want in the teaching of the young, is, not so much the mere results, as the methods and, above all, the history of science ... that is what we ought to teach, if we desire to see education, well-conducted to the great ends in view'. Ernst Mach – who could be considered to have founded the discipline of science education when, in 1887, he published and edited the first issue of *Zeitschrift fur den Physikalischen und Chemischen Unterricht* (*Journal of Instruction in Physics and Chemistry*) – said that: 'The historical investigation of the development of a science is most needful, lest the principles treasured up in it become a system of half-understood prescripts, or worse, a system of prejudices' (Mach 1886/1943).

Other contributions to this liberal tradition included John Dewey's work at the turn of the century (Dewey 1910); F.W. Westaway's teacher preparation text of the 1920s (Westaway 1929); Joseph Schwab's writings in the 1940s and 1950s (Schwab 1945, 1958); the books of James Conant in the late 1940s and 1950s, especially his *On Understanding Science* (Conant 1947) and *Harvard Case Studies* (Conant 1948); Gerald Holton's writings in the 1950s (Holton 1952) and the Harvard Project Physics Course that he directed (Rutherford, Holton & Watson 1970); the books of Leo Klopfer, James Robinson and Arnold Arons in the 1960s (Klopfer 1969, Robinson 1968, Arons 1965); Martin Wagenschein's German work in the 1960s (Wagenschein 1962); the publications of Jim Rutherford and Michael Martin in the 1970s (Rutherford 1972, Martin 1972); and numerous articles that appeared through the 1980s urging the incorporation of history and philosophy of science into science education and into teacher education programmes (Jung 1983, Hodson 1986, 1988, Duschl 1985, Lederman 1986, Solomon 1989, and Matthews 1988).

Journal and Group History

Throughout the long history of liberal advocacy, there had been no central forum where the participants could keep abreast of the debate, and where curricular and classroom inovations might be advertised an appraisals of these read. The communities of philosophers, historians, sociologists, scientists, cognitive psychologists, and science educators were insular and seldom paid attention to issues outside their field. Certainly historians and philosophers of science paid notoriously little attention to pedagogical, curricular or theoretical issues in science education. Indeed one article appearing in the mid-1980s was titled 'Science Education and the History and Philosophy of Science: Twenty-five Years of Mutually Exclusive Development' (Duschl 1985); while another article of the period said that the number of philosophers of science who had bothered themselves with educational matters 'could be counted on the fingers of one hand' (Ennis 1979).

The appearance of *Science & Education* in 1992 began to change this culture of isolation. This in part has been due to the journal's association with the two organisations that sponsored the Como conference: the History of Physics and Physics Teaching Division of the European Physical Society, and the International History, Philosophy and Science Teaching Group.

The European group is the older, having its first conference in Pavia in September 1983, which was attended by about 90 scientists from 25 countries (Bevilacqua & Kennedy 1983). Among contributors to the conference were Walter Jung, Lewis Pyenson, Gerd Buchdahl, David Edge, John Heilbron, Samuel Goldberg, Anthony French, Stephen Shapin, Jurgen Teichmann, Harry Collins, Gianni Bonera and Salvo D'Agostino. Subsequent conferences, with published proceedings, were held in Munich (1986), Paris (1988), Cambridge (1990), Madrid (1992), Szombathely (1994), and Bratislava (1996). Fabio Bevilacqua was, and has remained, the foundation secretary of the group.

The International group coalesced around a series of journal special issues devoted to 'History, Philosophy and Science Teaching' that were published in the late 1980s. The seed was a special issue of *Synthese* journal (vol.80, no.1, 1989) that Michael Matthews was invited to guest edit during a sabbatical leave period in the Philosophy Department at Florida State University. The group became 'formalised' at its first conference held at Florida State University, Tallahassee, in 1989, a conference co-organised by David Gruender and Kenneth Tobin, and attended by about 200 teachers and researchers from about thirty countries (Herget 1989, 1990). Presentors included Arnold Arons, Derek Hodson, Joan Solomon, Alberto Cordero, James Wandersee, Jane Martin, Joseph Nussbaum, James Cushing, Joseph Pitt, Nancy Nersessian, Harvey Siegel, Jim Garrison, Ian Winchester and Ernst von Glasersfeld. Subsequent conferences have been held in Kingston, Ontario (1992), Minneapolis (1995), and Calgary (1997). The next conference of the International Group, which is being held in conjunction with the US History of Science Society, will be held in Denver Colorado, 7-11 November 2001. Michael Matthews was, and has remained, the foundation secretary of the group, details of which can be found at www.ihpst.org.

Since its inception, the journal has promoted wide-ranging historical, philosophical and sociological research in science education. This is reflected in the themes for the special issues that have been published: 'The Cultural Significance of Science' (vol.3 no.1, 1994); 'Hermeneutics and Science Education' (vol.4 no.2, 1995); 'Religion and Science Education' (vol.5 no.2, 1996); 'Philosophy and Constructivism in Science Education' (vol.6 nos.1-2, 1997); 'The Nature of Science and Science Education, Parts I, II' (vol.6 no.4, 1997, vol.7 no.6, 1998); 'Values in Science and in Science Education' (vol.8 no.1, 1999); 'Galileo and Science Education' (vol.8 no.2, 1999); 'What is This Thing Called Science?' (vol.8 no.4, 1999); 'Children's Theories and Scientific Theories' (vol.8 no.5, 1999); 'Science for Non-Majors' (vol.8 no.6, 1999); 'Thomas Kuhn and Science Education' (vol.9 nos.1-2, 2000); 'History and Philosophy in German Science Education' (vol.9 no.4, 2000); and 'Constructivism in Science and Mathematics Education' (vol.9 no.6, 2000).

The papers presented at the Como conference, and those selected for this anthology, range widely over the spectrum of theoretical, pedagogical and curricular issues in science education that can be illuminated by historical, philosophical and sociological scholarship. We trust that readers will benefit from this translocation of the papers, from the shore of beautiful Lake Como and the medieval courts of Pavia University, onto the wider international stage.

Professor Fabio Bevilacqua, Dipartimento di Fisica, 'A.Volta', Universita di Pavia, Via A.Bassi 6, 27100 Pavia, Italy.

Dr Enrico Giannetto, Dipartimento di Fisica, 'A.Volta', Universita di Pavia, Via A.Bassi 6, 27100 Pavia, Italy.

A/Professor Michael Matthews, School of Education, University of New South Wales, Sydney 2052, Australia.

REFERENCES

Arons, A.B.: 1965, *Development of Concepts of Physics*, Addison-Wesley, Reading MA.

Bevilacqua, F. & Kennedy, P.J. (eds.): 1983, *Proceedings of the International Conference on Using History of Physics in Innovatory Physics Education*, University of Pavia, Pavia.

Conant, J.B. (ed.): 1948, *Harvard Case Histories in Experimental Science*, 2 vols., Harvard University Press, Cambridge.

Conant, J.B.: 1947, *On Understanding Science*, Yale University Press, New Haven.

Dewey, J.: 1910, 'Science as Subject-Matter and as Method', *Science* **31**, 121-127. Reproduced in *Science & Education*, 1995, **4**(4), 391-398.

Dewey, J.: 1916, *Democracy and Education*, Macmillan Company, New York.

Donnelly, J.F.: 2001, 'Contested Terrain or Unified Project? "The Nature of Science" in the National Curriculum for England and Wales', *International Journal of Science Education* **23**(2), 181-195.

Duschl, R.A.: 1985, 'Science Education and Philosophy of Science Twenty-five, Years of Mutually Exclusive Development', *School Science and Mathematics* **87**(7), 541-555.

Ennis, R.H.: 1979, 'Research in Philosophy of Science Bearing on Science Education'. In P.D. Asquith & H.E. Kyburg (eds.) *Current Research in Philosophy of Science*, PSA, East Lansing, pp.138-170.

Herget, D.E. (ed.): 1989, *The History and Philosophy of Science in Science Teaching*, Florida State University, Tallahassee FL.

Herget, D.E. (ed.): 1990, *The History and Philosophy of Science in Science Teaching*, Florida State University, Tallahassee FL.

Hodson, D.: 1986, Philosophy of Science and the Science Curriculum', *Journal of Philosophy of Education* **20**, 241-251. Reprinted in M.R. Matthews (ed.), *History, Philosophy and Science Teaching: Selected Readings*, OISE Press, Toronto, 1991.

Hodson, D.: 1988, 'Toward a Philosophically More Valid Science Curriculum', *Science Education* **72**, 19-40.

Holton, G. et al.: 1967, 'Symposium on the Project Physics Course', *The Physics Teacher* **5**(5), 196-231.

Holton, G.: 1952, *Introduction to Concepts and Theories in Physical Science*, Princeton University Press, Princeton. Second edition (revised with S.G. Brush) 1985.

Jung, W.: 1983, 'Toward Preparing Students for Change: A Critical Discussion of the Contribution of the History of Physics to Physics Teaching'. In F. Bevilacqua & P.J. Kennedy (eds.) *Using History of Physics in Innovatory Physics Education*, Pavia University, pp.6-57. Reprinted in *Science & Education* 1994, **3**(2), 99-130.

Klopfer, L.E.: 1969, *Case Histories and Science Education*, Wadsworth Publishing Company, San Francisco.

Lederman, N.G.: 1986, 'Students' and Teachers' Understanding of the Nature of Science: A Reassessment', *School Science and Mathematics* **86**(2), 91-99.

Mach, E.: 1886/1986, 'On Instruction in the Classics and the Sciences'. In his *Popular Scientific Lectures*, Open Court Publishing Company, La Salle, pp. 338-374.

Martin, M.: 1972, *Concepts of Science Education*, Scott, Foresman & Co., New York (reprint, University Press of America, 1985)

Matthews, M.R.: 1988, 'A Role for History and Philosophy in Science Teaching', *Educational Philosophy and Theory* **20**(2), 67-81.

Reif, F.: 1995, 'Understanding and Teaching Important Scientific Thought Processes', *Journal of Science Education and Technology* **4**(4), 261-282.

Robinson, J.T.: 1968, *The Nature of Science and Science Teaching*, Wadsworth, Belmont CA.

Rutherford, F.J.: 1972, 'A Humanistic Approach to Science Teaching', *National Association of Secondary School Principals Bulletin* **56**(361), 53-63.

Rutherford, F.J., Holton, G. & Watson, F.G. (eds.): 1970, *The Project Physics Course: Text*, Holt, Rinehart, & Winston, New York.

Schwab, J.J.: 1945, 'The Nature of Scientific Knowledge as Related to Liberal Education', *Journal of General Education* **3**, 245-266. Reproduced in I. Westbury & N.J. Wilkof (eds.) *Joseph J. Schwab: Science, Curriclum, and Liberal Education*, University of Chicago Press, Chicago, 1978.

Schwab, J.J.: 1958, 'The Teaching of Science as Inquiry', *Bulletin of Atomic Scientists* **14**, 374-379. Reprinted in J.J. Schwab & P.F. Brandwein (eds.), *The Teaching of Science*, Harvard University Press, Cambridge MA.

Solomon, J.: 1989, 'Teaching the History of Science: Is Nothing Sacred?'. In M. Shortland & A. Warick (eds.), *Teaching the History of Science*, Basil Blackwell, Oxford, pp.42-53.

Wagenschein, M.: 1962, *Die Padagogische Dimension der Physik*, Westermann, Braunschweig.

Westaway, F.W.: 1929, *Science Teaching*, Blackie and Son, London.

PART ONE

History of Science, Education and Culture

Contributors to Part One of the anthology continue a long tradition of advocacy in science education. Ernst Mach at the end of the nineteenth century urged that:

> every young student could come into living contact with and pursue to their ultimate logical consequences merely a *few* mathematical or scientific discoveries. Such selections would be mainly and naturally associated with selections from the great scientific classics. A few powerful and lucid ideas could thus be made to take root in the mind and receive thorough elaboration. (Mach 1886/1986, p. 368).

Mach was enunciating the view that good science education has a role to play in *cultural* formation, as well as *technical* formation. Students being taught science should come to see and appreciate the role of science in the development of human social- and self-understanding. This engagement with some of the key ideas of science, and with the historical process by which the ideas emerged, were tested and came to be adopted - the methodology of science - should contribute to the student's rational development, and indeed to their character development. Rationality is a hard-won cultural and individual achievement Such educative growth flows through to the maintainence and development of culture.

These ideas have been an important part of the *Liberal* tradition in education. In England, Percy Nunn and F.W. Westaway articulated these ideas during the 1920s and 1930s. In the United States, John Dewey argued for the cultural importance of science education in the period between the wars. Joseph Schwab did the same in the 1940s with his paper 'The Nature of Scientific Knowledge as Related to Liberal Education' (Schwab 1945). James Conant famously argued for this view in the 1945 Harvard University Report titled *General Education in a Free Society* (Conant 1945). There he said that: 'The facts of science and the experience of the laboratory no longer can stand by themselves ... the facts must be learned in another context, cultural, historical and philosophical' (Conant 1945, p. 155). This orientation subsequently informed Conant's landmark *Harvard Case Studies in Experimental Science* (Conant 1957), and Leo Klopfer's adaptation of these for high schools (Klopfer 1969).

John Heilbron draws attention to the fact that the 'taken-for-granted' historical perspective of Mach, Westaway, Conant and others has disappeared from most modern textbooks, and he offers suggestions on how history of science might once again fructify the curriculum and texts. Heilbron cautions that the history of science should serve the purpose of having students better understand science, not having them better understand history. Yet he clearly rejects the 'history-as-sugar-coating' option that some teachers and textbook writers adopt. His examples of the use of history focus on the central methodological matter of how theory relates to evidence in science.

Alberto Cordero outlines the cognitive, social and personal values that inform the conduct of science - or what we might call the culture of science - and how these values articulate with the goals of education. But such an articulation requires that science be taught in such a way that students see and appreciate something of its 'big picture', including its history and epistemology.

Peter Machamer discusses a central part of the 'big picture', namely Galileo's innovative appeal to experimental evidence to justify knowledge claims. This is the beginning of the Galilean-Newtonian Paradigm in methodology that soon swept all others in natural science away, and indeed that profoundly influenced most other intellectual pursuits. It is the birth of intellectual modernity. Galileo wrestles with a new way of conceiving how experience which is personal and individual can nevertheless be a ground for objective, transpersonal, knowledge claims. Students can do worse than wrestle with the same issue.

James Rutherford - one of the directors of the influential Harvard Project Physics course of the 1970s, and former director of the American Association for the Advancement of Science Project 2061 - charts the recent efforts to incorporate history of science into US science programmes and curricula. More specifically he details the rationale and achievements of the 1970s Harvard Project Physics programme, and the 1990s Project 2061 programme.

Ron Good and James Shymansky address the core ambivalence about the Nature of Science that is found in the two major contemporary US science education documents - the AAAS's *Benchmarks for Science Literacy* and the *National Science Education Standards*. Both documents endorse core aspects of the liberal tradition in science education, yet there is an important tension about just what picture of science emerges from the documents - a traditional realist and universalist picture, or a postmodern relativist picture?

Robert Carson provides a very ambitious educational vision of the cultural purpose of science. He outlines, in considerable detail, how some major episodes in the history of science can inform not just a science curriculum, but a rich cross-curricular programme in middle (junior high) schools.

Hsingchi Wang and William Schmidt are ideally placed to provide just about the first perspective on the degree to which historical, philosophical and sociological aspects of science are included in curricula, and student learning, across the world. This perspective comes from their work with the TIMSS (Third International Maths and Science Study) project.

Conant, J.B.: 1945, *General Education in a Free Society: Report of the Harvard Committee*, Harvard University Press, Cambridge.

Conant, J.B. (ed.): 1957, *Harvard Case Histories in Experimental Science*, 2 vols., Harvard University Press, Cambridge (orig. 1948).

Klopfer, L.E.: 1969, *Case Histories and Science Education*, Wadsworth Publishing Company, San Francisco.

Mach, E.: 1886/1986, 'On Instruction in the Classics and the Sciences'. In his *Popular Scientific Lectures*, Open Court Publishing Company, La Salle, pp. 338-374.

Schwab, J.J.: 1945, 'The Nature of Scientific Knowledge as Related to Liberal Education', *Journal of General Education* **3**, 245-266.

History in Science Education, with Cautionary Tales about the Agreement of Measurement and Theory*

J.L. HEILBRON
Worcester College, Oxford, OX1 2HB, United Kingdom

We are gathered here because we think that the history of science has an important, even a fundamental role to play in science education. Unfortunately, its most conspicuous current use is as a sugar coating to the hard nuts of the real curriculum. We have failed to persuade textbook writers and science teachers that they have not done their job if they do not make the history of their discipline a significant part of their pedagogical work. One reason for this failure is that we do not offer what they need.

Another reason is that, in strong contrast to ourselves, textbook writers face a demanding and unforgiving market. If they omit customary topics in favor of what most of their colleagues consider a frill, they risk reputation and livelihood. The difficulty is significant. But it has not been fatal in the past and may not be so in the future if historians and philosophers of science provide appropriate pedagogical material.

I begin my talk by mentioning a few old but encouraging signs of the value of history for science education. Then, as examples of material that might be offered to textbook writers and classroom teachers, come brief renditions of the cautionary tales mentioned in my title. Next, I'll take the liberty of deviating from my main theme to discuss wider benefits that might ensue from bringing history into science. I end with suggestions about how to proceed. Unfortunately, my ignorance forces me to draw all my examples from physics and astronomy.

1. Hopeful Indications

A standard Christmas present for a student of physics 100 years ago was a copy of Ernst Mach's *Die Mechanik in ihrer Entwicklung, historisch-kritisch dargestellt* (1883). It has had at least nine German and six English editions, and several in French and Italian as well. It is not strictly a history, but a selection of problems important in the development of mechanics together with instructive analyses based on the solutions proposed at the time. Some very important people profited from

F. Bevilacqua et al. (eds.), Science Education and Culture, 5–15.
© 2001 *Kluwer Academic Publishers. Printed in the Netherlands.*

studying Mach's *Mechanik*, Planck and Einstein among them, though, to be sure, both later repudiated him. If critical history helped form Planck and Einstein, we are well advised to find a place for it, suitably updated, in our curricula.

An earlier example of the inspirational use of history in science is Joseph Priestley's *History and Present State of Electricity*, which had five English editions and several translations during his lifetime, when very few books on electricity appeared more than once in any language. Priestley's approach was entirely different from Mach's. Priestley prepared his readers not by incisive analysis of selected problems but by laying out the course of discovery of the entire inventory of electrical knowledge of the mid-18th century. The method was as suited to the rude state of the science of electricity in Priestley's day as Mach's treatment was to the advanced state of mechanics in his. Perhaps no more need be said here about the value of Priestley's history for students of science than that Volta gained much from it.

A third example comes from the teaching of modern physics. Max Born's book, *Die Relativitätstheorie Einstein's* (1920), still offers a royal road to neophytes via historical-critical accounts of a range of problems, such as the aberration of starlight and Fresnel drag, implicated in the thinking that eventuated in the theory of relativity. Born's book went through at least three German and two English editions and several reprintings. In some respects it is a supplement to Mach's *Mechanik*, which Born acknowledged as his principal reference. Born's recourse to history has counterparts in a few good introductory texts in quantum physics, which lends itself particularly well to historical treatment since it still employs the concepts it overcame, although in limited and often mysterious ways.

These modern texts make a different use of history from Priestley's and Mach's. Whereas Priestley presented his information historically so as not to lose information that was not yet known to be unimportant and to give credit to discoverers in fields just forming; and whereas Mach exploited historical examples to deepen and broaden principles already regarded as models of clarity and reason; Born and the others turned to history to explain why physicists had been forced to base their discipline on concepts that do not appeal to the intuition.

Further to this theme, I can offer my own experience that exposure to old astronomy can materially assist students of modern astronomy. That is because, as in the previous examples, the old science is not entirely outmoded: anyone who masters the full geocentric accounts of the motions of the sun, moon, and stars knows as much about their appearances as the naked-eye astronomer requires even now. The geocentric accounts relate immediately to the observable world – the changes of season, the times and places of sunrise and sunset, the lengths and directions of shadows, and so on. This sort of material is not always taught in astronomy departments, where students are thrown into black holes as soon as possible.

From these few indications let us take heart that historical materials can be useful, even indispensable, in science education provided – and this is a major qualification – provided that they are used to inculcate science, not history.

2. The Examples

I turn to my cases, in all of which quantitative agreement between theory and experiment is problematic. This common bond of course is an arbitrary principle of selection. I'll present the first case in detail because it is unfamiliar, peculiarly informative, and pictorially arresting.

2.1. THE OBLIQUITY OF THE ECLIPTIC

From ancient times astronomers have worried that the inclination of the earth's axis to the plane of the sun's apparent motion, the so-called obliquity of the ecliptic, is not constant. The fundamental data for answering the question are the sun's altitude at noon on the day of the summer solstice and the latitude of the place of observation. For well over two thousand years, until the 17th century, measurements of the obliquity even in one place varied so much that no unambiguous decision about the constancy of the obliquity could be made. One reason for the wide scatter in the data was that the instruments did not remain intact and in place; another was that astronomy then was ignorant about matters essential to a solution of the problem.

Around 1600, following the lead of Tycho and Kepler, astronomers routinely corrected their observations for atmospheric refraction and parallax. But since the observations contained these effects intertwined, and since the corrections required have opposite signs, the old astronomers could err greatly in their estimates of one effect and compensate by fiddling with the other. For example, Tycho made the solar distance grossly too small, by a factor of 20, and hence the solar parallax 20 times too large; an excess that had to be killed by exaggerating the amount of refraction. Since refraction is more serious for objects close to the horizon than for ones near the zenith, the corrections changed with altitude; and since the height of the midsummer sun is different from that of the stars used to determine latitude, the question of the change in the obliquity could not be resolved before both the parallax and the refraction were known separately as functions of altitude. The errors in these quantities, which often amounted to over two minutes of arc, vastly exceeded the size of the secular change in the obliquity, which, around 1700, amounted to around 45 arc seconds a century.

The establishment of accurate tables of refraction and parallax was the work of Gian Domenico Cassini, who used for the purpose the cathedral of San Petronio in Bologna. He converted it into a heliometer (or meridiana) by opening a hole in its roof and running a horizontal metal rod in the plane of the meridian containing the hole. The instrument measured the altitude of the sun at noon.

Cassini's successors at Bologna continued his observations for over half a century. After 1700 they could compare their results with ones obtained at a second church observatory built in Santa Maria degli Angeli in Rome, at the order and the expense of the pope. The churches spoke equivocally in favor of a negative change of minus one arc second ($-1''$) a year. (The negative sign means a diminishing obliquity, that is, a straightening up of the earth's axis; if the process existed and continued unabated, in only 1,000 centuries it would be perpetual spring.) Other observers with other instruments – lesser heliometers or telescopes fixed in the plane of the meridian – gave other results, ranging from zero to one degree a century.

In the mid 1750s a new church observatory came on line to settle the matter. Its builder, a Jesuit mathematician named Leonardo Ximenes, spared no pains in leveling the line, measuring the height of the hole, and laying out a scale against which to record the position of the sun's midday image. He had two precious advantages over previous measurers. For one, the church in which he laid out his instrument had a bronze plate in its floor on the exact spot on which the solstitial sun shone in 1510. Although many of you have been in Ximenes' church, you probably did not see the plate of 1510, which is kept covered to protect it from the feet of tourists. Ximenes had only to measure how far the midsummer sun fell from the image of 1510 to settle the rate of change of the obliquity.

The second advantage Ximenes had over his predecessors was the knowledge that previous measurements would not have been intercomparable even if they had been made with the accuracy – a second or two of arc – at which he worked. A decade or so before Ximenes began at Santa Maria del Fiore, the astronomer royal of England, James Bradley, had delivered the second of two body blows to the corpus of secure observations. The first blow was that all stars execute a little circle around their average position during the course of a year. The maximum value of this excursion from the average is about 10 arc seconds; hence observations of the same star, even if made with exquisite accuracy, can differ from time to time by as much as 20 seconds. The second blow was that, in addition to this annual dance, the stars oscillate up and down. The oscillations take 19 years to complete and have an amplitude of 9 seconds. The dance, called the aberration, is explained as the result of the earth's motion and the finite speed of light; the oscillation, called the nutation, arises from gravitational forces that cause the earth's axis to bob. The bobbing of course changes the obliquity; but that was not the long-term change astronomers had sought for a millenium.

To obtain comparable data, Ximenes had to correct both his and the 1510 measurement for refraction, parallax, aberration, and nutation; to be safe, he also corrected for the settling of the church, atmospheric conditions that might change the length of the meridiana, and other things truly negligible. After much massaging of the data, he could announce that the difference of four centimeters, which he had found between the images of 1510 and 1755, indicated a secular change in the obliquity of the ecliptic of $-30''$ seconds a century. He repeated his operation 19 years later, at the same phase in the nutation cycle, and got the same result. His measurement was in fact extremely good, about the best possible with the intrument he used.

That of course did not settle the matter. A meridiana set up in the church of Saint Sulpice in Paris obtained no change in obliquity. The best telescopic measurements made at the Paris Observatory when combined with what appeared to be the most reliable earlier observations gave values of 0″, 45″, 60″, and 100″, all values negative. The better the instruments, the more knowledgeable the users, the subtler the corrections, the worse the agreement.

At this point – we are now in the 1760s – the mathematicians began to speak. Without looking at the sun, Euler announced from his study that interplanetary gravitational forces caused a slow shift in the axes of rotation of all the planets. He made various estimates and settled on −45″/century for the shift for the earth. Laplace confirmed the result and also that the change is periodic. We cannot expect perpetual spring.

Did the measurements confirm the theory, or the theory the measurements, or neither the other? Ximenes thought that, since 45 does not equal 30 and since he had worked to sublime accuracy, taking all known disturbances into account, there must be something wrong with the gravitational theory or with the calculations. But Newton's laws had proved too successful to be upset by so subtle and uncertain a matter as a discrepancy of 50 or 100 percent in measurements of the change in the obliquity. Astronomers soon brought theory into agreement with experiment. They employed a technique often practiced in science. They changed instruments.

By 1750 the church heliometers were outmoded. Their advantages over telescopes – stability and size – were negated by improved mountings and single-metal constructions, achromatic lenses, and much improved graduation. Without the competition of the church observatories, astronomers armed with telescopes came closer to the mathematicians' value as observational protocols and instrument design improved and routinized. This convergence was no doubted assisted by the belief that English telescopes, especially instruments made by Jesse Ramsden, gave the best results.

There you have a story that delivers sound lessons in epistemology, measurement, and instrumentation and, at the same time, imparts important information about the universe, which, I must again insist, is not out of date. The imagery and the unexpected part played by cathedrals give openings for lessons of an entirely different kind, to which I shall return.

2.2. THE MOTION OF THE MOON

One reason that mathematicians declined to consider seriously Ximenes' suggestion that the discrepancy between his measurements and their calculations might be blamed on the theory of gravitation was that they had recently tried a similar move and had ended in fiasco. In 1747 Alexis Claude Clairaut, frustrated by the shortfall between his calculations of the moon's motions under the gravitational pulls of the earth and the sun, announced that Newton's law of the inverse square was not the entire story of gravity. To save the phenomena, Clairaut proposed to alter the law by adding a term involving the inverse cube of the distance. The effect of the new term was to cause the moon's orbit to precess in its plane. Clairaut calculated that the slow precession thus introduced would put the moon where it was observed to be and also clear up an apparent difficulty in the shape of the earth brough to light by then up-to-date geodetic surveys. Clairaut had the support of two other powerful reckoners, Jean le Rond d'Alembert and Leonhard Euler. It appeared that Newton had found only the first term in the gravitational force between mass points. It was just the possibility of this sort of finagling that made the reviewer of the *Principia* in the *Journal des sçavans* reject Newton's approach as unphysical; for how, he asked, could the underlying physics be found if the law could always be amended to cover apparent violations? The great defender of Newton in the Paris Académie des sciences, the Comte de Buffon, insisted on the inverse-square as the only reasonable and rational relationship. Clairaut rejected his opinion as ignorant and metaphysical.

The mathematicians made ready to amend the universal law of gravitation. But at the moment of truth, even Clairaut did not want to change Newton's law merely because it disagreed with the facts. So he returned to his computations. He had not made a mistake. But neither had he been right. He had stopped too soon in his approximations. When carried further, they accounted for the motions of the moon without invoking the inverse-cube precession. D'Alembert and Euler reached similar happy conclusions.

Once again the story carries methodological lessons – in this case the malleability of mathematical description, the trickiness of approximations, and the danger of premature defeatism (or, as the quantum physicists used to say, renunciation) – along with information about the moon and exercise in the three-body problem.

2.3. "RELATIVISTIC" FINE STRUCTURE

The spectrum of ionized helium differs from that of hydrogen in two respects that were of the first importance in the development of the quantum theory of the atom. For one, some lines in hydrogen fall very close to, but do not coincide with, corresponding lines in ionized helium. Bohr explained the disparity as a consequence of the greater mass of the helium nucleus. By a literal application of an elementary mechanical principle to a situation in which, as he said, mechanics does not apply, he calculated the displacement to many places of decimals. The perfect agreement with observation caused many important physicists, including Einstein, to take Bohr's quantized atomic model seriously.

The other significant difference between the spectra was that the helium lines had a fine structure while the hydrogen lines did not. (In fact, hydrogen lines have the same fine structure as ionized helium's, but spectroscopists had not resolved it when Bohr first promulgated his theory.) Bohr could not find the source of the helium fine structure. Arnold Sommerfeld came to the rescue with the observation that the radiating electron must travel at relativistic speeds. Analytically, the relativistic correction is equivalent to adding a little inverse cube to the Coulomb force, introducing the sort of precession invoked by the 18th-century moon men to control the lunar orbit. The frequency of the precession depends upon the eccentricity of the orbit, which is fixed by the so-called azimuthal quantum number, k. Applying his version of Bohr's theory, Sommerfeld could calculate the energies allowed precessing ellipses and thence the frequencies of the lines making up the fine structure. They agreed exactly with precision measurements of the satellites of ionized helium. Physicists rejoiced; the agreement reached to five or six places of decimals. That was a remarkable achievement since Sommerfeld's theory lacked what is now the essential factor in the analysis of the fine structure, the inner quantum number, j.

Using the same relativistic formulas, Sommerfeld calculated the energy difference between the two levels in the L region of the atom – that is, the second ring of electrons counting from the nucleus – allowed by his theory. This energy difference agreed perfectly with measurements of absorption edges of X-rays associated with the L region. It appeared therefore that the doublet – the two L edges – arose from orbits of different eccentricities, characterized by different values of k; to be explicit, from a circle ($k = 2$) and a highly eccentric ellipse ($k = 1$).

Alas, the L region has three, not two absorption edges. Sommerfeld's dynamical theory supplied only one circle and one ellipse. To label the third, he invoked j, which he had introduced, without dynamical significance, to classify the multiplets in optical spectra. Thus he had a "relativistic" doublet he could calculate arising from the difference in the precessional energy between a circular and an elliptical orbit, and a supernummerary level he could not explain arising also, somehow, from the ellipse.

To achieve his impressive agreement, however, Sommerfeld had to assume that the effective nuclear force in the $k = 2$ circle was exactly the same as that in the $k = 1$ ellipse. That did not seem plausible to those who took the orbital picture literally, since the circle lies entirely outside the innermost electron shell and the ellipse penetrates it. Also, certain analogies between optical and X-ray spectra suggested that Sommerfeld had got his attributions backward; the terms in the relativistic doublet should have the same k, but different j values. Those who preferred general qualitative analogies to isolated exact quantitative agreements were left without a way to reproduce Sommerfeld's remarkable achievement.

They were almost saved by the invention of the concept of electron spin, which allowed two energetically different orbits with the same value of k (and thus the same effective nuclear charge). Unfortunately, calculations of the difference in energy between the levels in these "spin doublets" continually differed from observation – and therefore from Sommerfeld's successful computation of the "relativity doublets" – by an apparently irreducible factor of two. Many physicists thought that a good ground for rejecting electron spin. As in the motion of the moon, however, those who pushed the established mechanical theory another step won out. A new relativistic effect connected with the spinning electron, the so-called Thomas precession, supplied just what was needed to bring the spin doublets into agreement with Sommerfeld's old calculations. The striking connection between spin and relativity thus revealed soon found its explanation in Dirac's theory of the electron.

Fundamental principles in relativistic, atomic, and quantum physics can be elucidated by this story, which, of course, also contains a warning against believing in a calculation just because it agrees with observation to seven places of decimals.

3. Outreach

3.1. CROSSING THE BRIDGE

Students who are not jaded or brain dead should develop a curiosity about the people, institutions, instruments, and other circumstances mentioned in the examples just given. The unexpected appearance of cathedrals as solar observatories may be a particularly good hook with which to draw science students across the cultural bridge that supposedly separates them from students of the humanities.

The science teacher can offer incentives to encourage crossing the bridge. For example, in the case of the obliquity, he or she can point to the apparent discrepancy between the Catholic church's condemnation of heliocentrism in 1633 and its initiation, some twenty years later and within its own cathedrals, of investigations into precisely the same question that had brought Galileo into trouble. Interests thus raised should not be pursued within the science curriculum, however, but in history courses. Similarly, interests in philosophy, art, architecture, and, maybe, history of science, awakened in science courses should be pursued on the other side of the bridge.

Crossing the bridge can bring substantial advantages to science students not only in personal cultivation but also in professional formation. The careers of scientists and engineers outside of academia suggest that replacement of a few technical courses by non-technical ones would not only be harmless, but even beneficial. People seldom get jobs that call for exactly what they studied during their professional training. They learn on the job, the more quickly and effectively the better they understand general principles and procedures. Those who rise the furthest tend to have a broader culture than those who remain at the bench.

Needless to add, if history courses dwelt more on the technologies that underpin the societies they describe they might help to push history students across the bridge into courses in engineering or science. But that is not the direction of travel of immediate interest to us.

3.2. GUIDING THE PERPLEXED

It would be ungenerous to omit scientists from the beneficiaries of our pedagogical services. Judged by their enthusiasm over anniversaries, scientists are among the most historically minded people on earth. From Copernicus to computers, any event, discovery, or invention of scientific interest is eligible for celebration provided only that it occurred a multiple of five or ten years before the present.

Scientists also make use of history didactically, again usually in a celebratory vein. In an effort to anchor their subject in time and space, or to find a place to begin, or to follow custom, writers of *Festschriften* and review articles and, of course, obituary notices, make use of historical material.

Now scientists, when they are being scientific, subject their methods, materials, and results to the most demanding scrutiny they can command, knowing that if they do not do so a colleague will perform the service for them. Fear of embarassment as well as faithfulness to the ideology of science keep up professional standards. But when they reminisce or write history, scientists do not fear to make themselves ridiculous. They do not approach the task scientifically. They need our help.

That brings us to a third use of history in science (the first two were celebratory and pedagogical). Geology, paleontology, evolutionary biology, and, increasingly, astrophysics and cosmology, have historical components. Even astronomy, the leading sector of science during the scientific revolution, the exemplar of the abstract and the mathematical, was, and still is, in part an historical science. Only by analyzing observations made over time could small long-period effects, like the ever instructive change in the obliquity, be determined.

4. Three Slight Suggestions

You will perhaps forgive an outsider's observation that too much of your time is spent discussing ideas that, whatever their merit, have no chance of being implemented in science courses in any plausible variant of our educational system. However, concern about instruction in science, particularly physical science, has made enough money available for curricular experiments that a well planned project to make appropriate use of historical materials may have a chance not only of securing funding, but also of making a difference. I propose for your consideration the sketch of an outline of such a project. It has three parts.

4.1. COMPENDIA

The first desideratum is for case studies of the sort I've outlined. They must be prepared so as to slip easily into courses where the scientific ideas they illustrate are discussed; thus they must be modular, that is, presentable in whole or in part as best fits the curriculum. They must convey useful scientific information beyond what the student would otherwise receive and do so in a way that strengthens the student's understanding of the principles presented in standard textbooks. This increase in understanding should be testable. Here we give ourselves a true challenge. If we meet it, we create a strong presumption in favor of our program. If we fail it, we may have to give it up.

One reason that appeal to history can aid understanding is that it offers examples of the difficulties that established scientists have had in constructing the concepts, and fitting the facts, that make up the theories the students are struggling to master. I've found that appropriate reference to Sommerfeld's troubles in classifying spectra and interpreting quantum numbers both geometrically and dynamically often clears up a cloud of misconceptions even though we now regard his entire enterprise as misdirected. In one case a student thus prepared underwent a quantum jump from the bottom to the top of his class.

Finally, wherever possible the case studies should carry epistemological or methodological lessons and dangle ties to humanistic subject matter. But never should the primary purpose of the cases be the teaching of history.

Part 1 of the project is the writing of case studies by teams of historians, philosophers, scientists, and teachers. The participation of the scientist-teacher is essential not only to the choice of the cases, but also to their acceptance for classroom use. The cases could be issued individually, through a journal, and in collections on kindred subjects. Several volumes will be required.

The participants should be paid for their work and time. The money should come largely from national and international science foundations and councils. Professional societies in the sciences as well as officials responsible for educational policy should be mobilized. The project should have as wide a representation and sponsorship as possible, although the working teams must be small enough to ensure easy communication and efficient performance.

4.2. GUIDES FOR CELEBRANTS

The second part of the project will offer help to scientists faced with the writing of obituary notices, anniversary discourses, or autobiographies, and demonstrate the everyday value of historians. The main product of this part might be a bibliography of good biographies and an annotated sampler of necrologies. This sampler would contain an analysis of the varieties of necrologies and their purposes: to extol or commemorate the dead, to enhance national prestige, to inspire the young, to recommend a style of science, to settle grudges, to whitewash reputations, and so on.

Help to speakers on anniversarial occasions may be given via a monograph that discusses the nature and objectives of commemorative speeches, indicates the styles and strategies that can be adopted, exhibits appropriate rhetoric, and gives examples of what is to be avoided. A close parallel would be a manual for sermonizers.

In both cases, the obituaries and the anniversaries, we should establish a consulting service. An obituarist needing a ghost writer or an anniversarialist needing a new slant should have access to a hot line.

4.3. BOOKS AND HOOKS

Part 3 of the project is the creation or documenting of literature for students and their teachers who are caught by the hooks to wider subjects buried in the case studies. Here we are generally in better shape than in the other parts of the project since quantities of good books already exist to satisfy most legitimate interests that the cases may arouse. For example, biographies of scientists, especially of the Galileos, Newtons, and Einsteins, written for students of all ages, abound in many languages. The project need only enlarge the bibliography proposed for Part 2 and classify it by level. Conspicuous gaps may come to light. The project can advertise them and try to commission new biographies, or arrange for the translation of older ones.

Another sort of literature to which scientists and science students wishing to build out from their established interests are drawn, concerns technological applications of science and their effects on the wider society. Accessible items in this genre are not as plentiful as they should be. Professional writing in the history of technology is too often confined to technical details of interest only to specialists. Part 3 of our project would achieve something if it prompted the writing of books showing how the ideas studied with the help of the materials in Part 1 have been translated into machines and devices that have enhanced and threatened civilized life. As for the reverse literature, explaining how social forces influence or determine the organization, conduct, and output of science, we may have enough of it already.

Note

* Text of a plenary address to the 5th International History, Philosophy and Science Teaching Conference, Pavia University, September 1999.

Further information about the cases may be obtained from my *The Sun in the Church: Cathedrals as Solar Observatories* (Cambridge: Harvard University Press, 1999), chap. 7; *Electricity in the 17th and 18th Centuries. A Study of Early Modern Physics* (Berkeley: University of California Press, 1979; 2nd edn, New York: Dover, 1999), pp. 59–61; and 'The Origins of the Exclusion Principle', *Historical Studies in the Physical Sciences* Vol. 19 (1982), pp. 261–310.

Scientific Culture and Public Education

ALBERTO CORDERO
Philosophy Department, Queens College & The Graduate Center, City University of New York, Flushing, NY 11367, USA

Abstract. Current science and science-friendly philosophy jointly yield a picture of the world and ourselves in it that is more substantial, detailed and coherent than any other produced before by natural philosophy. When carefully formulated this picture provides us with: (a) an unprecedentedly reliable representation of vast regions of the natural world; and (b) a non-arbitrary public framework for understanding and furthering important areas of public concern. This paper comments on the cultural and educational significance of this picture. Influential arguments against granting a privileged role to serious science on the basis of differential credibility are examined and found wanting. This result is then folded into an analysis of the significance of scientific thought and practice for a cautious conception of the goals and methods of public education.

1. Introduction

There is something worth calling a 'scientific picture of the world'. It is a joint contribution of science and science-friendly philosophy of science that tells us about the constitution and development of the universe, including a penetrating story about ourselves as natural entities. The picture is far from perfect, and it varies in reliable accuracy from area to area, and is manifestly incomplete in terms of coverage and depth (both diachronically and synchronically). Still, it seems far deeper, more detailed and coherent than anything offered before by natural philosophy. A scientific culture is correspondingly on the rise, centered on this picture and its attendant style of thinking. It draws from learned considerations of human nature and the world and presents itself as resourceful enough to lay down certain fundamental aspects of the good life for man. Arguably, no other contemporary framework is more rational than this for the public exploration of goals and valuations.

However, granting special cultural status to scientific thought is heavily resisted by many influential circles. Contemporary opinion is ambivalent about the cognitive status of science. We inhabit a social milieu that lives increasingly off the products of science and technology, yet simultaneously takes increasingly for granted the relativity of all knowledge.

F. Bevilacqua et al. (eds.), Science Education and Culture, 17–29.
© 2001 *Kluwer Academic Publishers. Printed in the Netherlands.*

2. Reason Denied

Opposition to according science special intellectual prominence is not in short supply. This point can be elaborated in too many directions, so let me just highlight two influential lines of thought which I think are emblematic of the present situation. One, from radical sociology of knowledge, is the reductionist doctrine that facts are merely social constructs. The other, linked to a more perennial form of romantic thoughtfulness, is a 'pluralist' critique of science. Both lines radically reject the notions of rationality and descriptive progress. Both deny coherence to the idea of reference invariance through conceptual change.

The first line is epitomized by the work of David Bloor (1976) and the so-called 'Strong Program' in sociology of knowledge, according to which the cognitive order is entirely dependent on the social order. From this perspective, only the social sciences can explain the development and nature of scientific knowledge: scientific beliefs are not primarily driven by the natural world but by the conflicting interests of competing groups. For any actual belief, partisans of this school maintain, causal laws can always be established that reveal how it was socially generated. Bloor's subsequent work (1983) has concentrated on a 'Wittgensteinian' outlook in which no aspect of knowledge transcends power relations, and language and belief are understood as reflections of power relations within society. It is not clear, however, how someone who is not already committed to the Strong Program can begin to consider it seriously. By its own lights, the proposal amounts to an idea that is 'driven by the conflicting interests of competing groups'. But, if so, why bother? People who have not given up the notion of 'approximately correct' belief would seem to have every reason for turning their back on this kind of radicalism. Even if this shortcoming is bracketed for the sake of argument, other serious difficulties are apparent.

The Strong Program proceeds as if the best contemporary scientific claims had the character or texture of conversions secured by the sword. They do not. It may be true that such claims express deep sociological forces; but merely voicing this as a possibility can hardly suffice. We need to be explained just how is one supposed to understand solid scientific claims in sociological terms. Think of such assertions as 'Mars is further away from the Sun than Mercury', 'whales descend from terrestrial mammals', 'protons are made of quarks", or 'the accessible universe has been expanding for more than 5 billion years'? Just how do claims like these relate causally to the conflicting interests of competing groups (as opposed to objective reason and the external world). And, to the extent that they are so related, what relevant light does the social causation involved cast on the acceptability of the claims in question? No minimally compelling sociological response to these questions is provided by sociologists, and none seems forthcoming.

At higher categorical levels the case for sociologism seems, if anything, worse. Bloor and his supporters have a vaporous conception of causation. Sociologists of knowledge typically try to flesh out their program by telling us about the ways in which different positions in the social structure correlate with different beliefs. That kind of finding can illuminate mob-belief formation. It is hard to appreciate, however, how mere social correlations could relevantly illuminate something like, for example, Dennis Sciama's reasoned acceptance in the late 1960s that the Big Bang model was probably right and the Steady-State model probably wrong – a conclusion Sciama maintains to have reached on the basis of detailed findings about so-called 'cosmic background radiation'. We need detailed models here, not promissory notes. Apart from not being causes, correlations are notoriously treacherous as indicators of underlying causal mechanisms. To do an explanatory job correlations need first to gain abductive weight as prospective fundamental facts about the world – i.e., they must reveal their worth in terms of 'unexpected' theoretical fruitfulness, differential predictive power, and so on. That is why, for example, some correlations are increasingly regarded as fundamental in quantum physics.[1]

No similarly compelling case exists for taking that way any of the correlations between cognition and social forces volunteered by radical sociologists so far. The Strong Program simply leaves epistemologically interesting phenomena in the dark. From the pragmatic point of view the Program's prospects seem equally dim. Freed from any responsibility of searching for the truth, of looking for objective reasons, arguments in the Strong Program tend to become simply rhetorical. Beliefs grow into expressions indistinguishable from interjections, in a power game in which the players portray themselves as bundles of mere causal effects. And so, one major problem with the Strong program is that it fails to motivate or render plausible its reductionistic approach. Standard scientific appeal to reason and external reference may be flawed, but this cannot be merely 'postulated'. In no obvious way are the mature sciences mere by-products of power games; nor is their manifest externality a mere 'illusion' in any obvious way. The world studied by science keeps disappointing its greatest models of it, showing how underdetermined by its representations it is.

In no way, therefore, does the Strong Program manage to block or render implausible epistemic accounts of scientific belief.[2] However, other more potentially illuminating denunciations of science arguably exist. Feyerabend's charges against the rationality and integrity of science from the 1970s are often understood to be one of them (Feyerabend 1975, 1978).

Science receives state and public support primarily because it achieves truths that deserve to be preserved and privileged. Feyerabend disagrees with this policy. It is a mistake, he thinks, because science is not the only enterprise delivering valuable results, and – its pretensions to the contrary notwithstanding – science lacks a special method for achieving such results. In Feyerabend's view nothing but political circumstance privileges what we currently take as the 'scientific truth' over claims from allegedly disreputable specialties like astrology, voodoo or magic. Build up the society he recommends, he says, and voodoo and astrology and the like will return in such splendor that you will have to work hard to maintain your own position and will perhaps be entirely unable to do so.

But, isn't science plainly better than those 'primitive practices'? According to Feyerabend, success is achieved by scientists only because of politics, rhetoric, and propaganda, rather than because of their advancement of our objective knowledge of the world – the absence of good reasons against current scientific lore simply being due to a historical accident. Therefore his view that a more tolerant examination of the 'material basis' of neglected specialties like the ones just mentioned could enrich, and perhaps even revise physiology.

Feyerabend's main charge against science, however, runs deeper. According to him, institutionalized science has come to encourage naivete and to inhibit freedom of thought. Even if there is truth to be found about the world and science succeeds in finding it, Feyerabend acidly observes, it is not true that we have to follow the truth – human life is guided by many ideas. Feyerabend's grand conclusion is that science should be regarded as just another religion. Science, he stresses, is just one of the many ideologies that propel society and it should be treated as such. To him this provides reason enough for demanding 'a formal separation between State and science just as there is now a formal separation between state and church'. In his view, science should be allowed to influence society but only to the extent to which any pressure group is permitted to do so. Scientific thought, in short, would not play any predominant role in the society Feyerabend envisages. In that society, scientists would be more than balanced by magicians, voodoo doctors, priests, and astrologers.

We thus get Feyerabend's liberal pronouncements on education. The purpose of education, he maintains, is to introduce the young into life, and so into the society where they are born and into the physical universe that surrounds it (Feyerabend 1978). Largely because there is a legitimate pedagogical need to simplify and idealize, the method of education typically consists in the teaching of some basic myth. This is acceptable, says Feyerabend, provided the chosen myth is deliberately counterbalanced by other myths, for we do need an education that makes people contrary, without making them incapable of devoting themselves to the elaboration of any single view. How can this be achieved? Feyerabend exhorts society to protect the naturally fertile imagination of children and to develop to the full the spirit of contradiction that exists in them, all of which, he thinks, would be greatly helped by granting equal classroom opportunity for science, voodoo, creationism, astrology, and the like.

Feyerabend's hunch-feelings against dogmatic scientism have some basis. Much in his good-natured libertarianism, critical reaction to naive scientific realism, and reasoned contextualism I, for one, find commendable. But the problems with his specific proposals are legion. Marred by overreaction to positivist models of science, Feyerabend misconstrues the character, scope and limits of contemporary scientific thought. As with the Strong Program, his views get their bite from a mischievous caricature of scientific theorizing. In the end, his skeptical points reveal themselves as toothless tigers when inspected using as background a more alert and nuanced account of science, or so I will argue in what the remaining sections.

This brings us back to the scientific picture of the world.

3. Science and the Public Picture of the World

The empirical sciences now provide us with an array of theories about the structure and development of the physical world, organic life and, to a lesser extent, the mind and human society. They yield demonstrably accurate representations of various aspects of these domains.

What these rich accounts amount to jointly, however, is made problematic by the lack of clear fit between many of the theoretical components involved. The textbook version of almost every theory is excessively sanguine about the actual range of its credible applicability. In physics, to mention one case, General Relativity and quantum theory are routinely presented as physical theories of universal applicability, even though they do not blend at all well. The two theories are marred by many as yet unresolved conceptual difficulties, and this limits their respective ranges of well-established applicability to a fraction of their intended domains. The quantum-theoretical picture, centered on a field that dwells in configuration space, reliably applies to material systems with degrees of freedom below those of found in medium-size organic molecules. The picture furnished by General Relativity, centered on the relationship between matter and spacetime structure, reliably applies to macroscopic material structures distributed over great distances. The situation is similar in other fields. Darwinian biology yields a superb explanatory model of the tree of life, from cellular creatures to the rudiments of mind; beyond these limits, however, it becomes increasingly speculative. Scientific activity continues well beyond the present ranges of safe representation, but in more controversial ways, involving engrossing disputes on issues as disparate as improving specific modeling, the theory's basic dynamical framework and even the theory's ontology.

There is thus a need for a sober counterpart to the blithe picture found in standard scientific textbooks. No reasonable case exists, however, for rejecting scientific representations altogether. On the contrary, science and philosophy of science seem to encourage at least one circumspect and credible version of the scientific picture in which deep descriptive success can be recognized as a complex but viable feat.

Successful description in science is a bit like map making. Looked at this way, each discipline provides us now with increasingly detailed and comprehensive maps of its intended territory or domain, drawn on increasingly complex 'conceptual hyper-surfaces'. The number, dimensionality and topology and texture of the resulting maps are dictated by the current conceptual structure of the discipline. The best representations thus produced work extremely well over regions which generally exceed those of the originally intended applications. As with the scientific theories mentioned, the area of high reliability is typically constrained by lower and upper bounds. However, many aspects of interest regarding the natural world remain in the dark. This notwithstanding, the current total picture does display a surprisingly high level of unity. There seems to be little question, for example, that the entire corpus of electromagnetic and transmutation phenomena in nature is governed by a common dynamics, or that all the living creatures on Earth are united by relations of descent, or that the domains of biology and mind are linked by relations of supervenience to the physical level.

The point is this. When soberly articulated, the current scientific picture presents a number of characteristics: it has depth (i.e., it goes well beyond the perceptual-observational level); it is uneven over the territory surveyed (i.e., the reliable representations are not equally detailed and comprehensive everywhere), it is pixel-like (no reliable representation is completely sharp). In this sense, its 'cartographic' success is best understood as being primarily 'local'.

Still, even with the more speculative parts of the map bracketed, the remaining 'cautious picture', circumspect and patchy as it is, constitutes an unprecedently deep picture of the natural world and our place in it. The picture is also robustly rational, supported by careful argumentation and stringent experimental testing, and endowed with more unity and pragmatic success than anything ever produced by natural philosophy. Its most prominent constituents include the following:[3]

– *Cautious Standard Quantum Theory*, which describes the behavior of material systems with less degrees of freedom than medium-size molecules. It includes the juxtaposition of formally similar theories of fundamental interactions known as the 'Standard Model of elementary particles'.
– *Cautious General Relativity*, which describes gravity as a metrical property of a spacetime continuum that gets curved in the neighborhood of matter. the range of high reliability for this component of the picture runs, at least, from small macroscopic regions to the largest scale in the accessible Universe.
– *Cautious Version of the Standard Cosmological Model* (Big Bang), which is a peculiar blend of core quantum theory and core general relativity, within which many models of natural history currently compete. Its range of high reliability runs from at least the first second through the next few billion years – with a number of gaps in the story (regarding, for instance, exact age of the universe and galaxy formation).
– *Cautious Description of Organic Life*, which presents living organisms on Earth as being united by a host of related common structures and mechanisms, conspicuously the genetic code at the synchronic level and genealogy at the diachronic level. This component explains how a primitive form of life present on Earth about 4 billion years ago could have produced so much diversity. The rise various degrees of consciousness and autonomy in animal species is deciphered in terms of the natural selection of differential advantages that improve the survival and reproductive capability of organisms in their local environments.

- *Cautious Naturalist Picture of the Mind and Culture*: The mind is an extreme case of the evolutionary path toward greater consciousness and autonomy mentioned above in connection with animal species. Its relation with underlying physical structure is one of supervenience rather than nomological reduction. As an organ that arose through natural evolution, the mind is presented as having the opportunistic constitution of a Swiss-Army knife, in particular one endowed with a collection of overlapping reasoning mechanisms rather than a single, all-purpose central problem-solving system. From this perspective, human behavior is largely determined by culture – itself largely autonomous system of symbols and values, growing from a biological base, but also growing indefinitely capable of moving away from it.

And much more.

Let us call the duly filled in version of this coherent (if limited) set of representations the 'Cautious Scientific Picture'. Its credibility and coherence are as remarkable as its gaps are considerable. The list of pending questions is very long, its most salient items including the following:

- A seemingly intractable tension between the most natural extensions (unrestricted versions) of General Relativity and basic Quantum Theory.
- Competing theories about origins at various levels, beginning with the evolution of the Universe prior to the state associated with the completion of its 'first second'. There are many other origin issues about which scientific consensus is presently poor. Three conspicuous ones concern the formation of galaxies, the origin of life on Earth, and the origin of mind.
- The overall picture rests on seriously unsettled conceptual foundations, which remain underdetermined at various levels of descriptive depth. This situation is vividly displayed by foundational studies that reveal the ways in which the facts of quantum theory allow for effective metaphysical proliferation.

And much more.

Many significant, vital, questions are left open by the current Cautious Scientific Picture. Indeed present knowledge gives reason to fear that quite a few of the pending questions may exceed our cognitive capacities. For instance, in the present search for a coherent general theory of spacetime and matter, there are grounds for fearing that we might never be able to determine points as basic as whether the world evolve deterministically or whether its history is punctuated by irreducible stochastic events (Cordero 1998).

These provisos notwithstanding, the Cautious Scientific Picture embodies centuries of education about how understand and fulfill our epistemic aims. Its reliability comes directly from its profound 'rational' character. Everything in this picture has been forged through a conscientious application of logic and learned methodological criteria guided by equally learned epistemic traits (i.e., ones that have been found to promote our epistemic aims) like internal consistency, explanatory and problem solving ability, quantitative precision, simplicity, and manifest fruitfulness, all working in conjunction with a distinctly modern emphasis on clear predictive power.

The result is a picture which is explicitly modest, recognizably incomplete, and open to the possibility of revision at every level, whose reliability rest on the nontrivial fusion of rationalism, empiricism and pragmatism implied by the above considerations. Though rationally underdetermined from certain levels of depth down, above these the overall picture enjoys a substantial degree of retention and supplementation that makes it as stable and robust as anything we have ever had.

4. Robustness and Partial Meaning Invariance

Epistemologies that conceive of description and representation as something that stands or falls with the entire conceptual network are suspect. They burden us with unrealistically ambitious and ultimately arbitrary metaphysical requirements on human cognition. The ancient Mesopotamians knew a great deal about dogs, even if almost everything pertaining to their biological evolution escaped them. Likewise, in nineteenth century science, J. C. Maxwell was able to study and discover crucial aspects of the physical process of the propagation of light without being initially committed to any physical quantity as constituting light. Science (and, for that matter, ordinary knowledge) in the 20th century is no different. Two relevant points here concern the partial invariance of the descriptions to conceptual change, and the actual depth of such invariance relative to the perceptual level. Holistic epistemologies thus miss a crucial old fact about our actual epistemic achievements: we do not need to know everything in order to know something.

In the natural sciences, especially the more mathematized ones, cases of learning through partial modelling and of partial theoretical invariance are routine. These two features can be said to have become characteristic of actual science. A typical case, studied by Psillos (Psillos 1995), is the transition from the ethereal wave theory to classical electromagnetic theory of light. Many physical properties that an ethereal wave was supposed to posses (transversality, ability to sustain potential and kinetic energy, finite velocity of propagation, etc) were kept as properties of an electromagnetic wave. Similar cases of learning through partial modelling and of invariance can be discerned in virtually every interesting case of conceptual change occurred in this century.[4]

The invariance part is particularly important here. Consider, for instance, the 'currently correct' claims about the reality of our genealogical connection with the rest of the biosphere, or about the Jupiter–Sun system's following Newtonian mechanics within 5% precision, or about the geometrical structure of benzene molecules within 5% precision, or about cockroaches sharing common ancestors with all of us. Claims like these don't just 'feel' correct. We have every reason to trust they will be seen as correct by our most critical descendants. Much about the subjects referred to by these claims escapes us as yet – are benzene Bohmian entities, or concentrated quantum fields, or 'many-world' partial fields, or something else?

We do not know, and may never know. Nevertheless, we seem to have learned a great deal about the nature of such systems – unless of course such knowledge is ruled out by definition.

The main points to which all this leads are, I suggest, two. First, wholesale skepticism about evolving theories is neither compelling nor particularly specific to science. Epistemologies that embody it are not just hard to express coherently (Nagel 1974); by tying reference to sense in a radical way, they spawn skeptical arguments that apply not only to science but to everything. Are we to believe, for example, that 'dogs-to-Descartes' and 'dogs-to-us' refer to completely different things or to nothing at all, simply because Descartes and us understood dogs differently? Skeptical intimations of this sort merely invite us to question the premisses underlying them; their greatest service is thus to show how not to do epistemology. This connects with my second point: a credible scientific picture of the world is now available, one which is more internally consistent, explanatory, problem-solving, precise, rich in predictive power, unified, and manifestly fruitful than anything made available by human reason before. Clear-headed philosophical awareness of the patchiness and limitation of this picture, of its passionate rational history, of the manifest externality of the world it purports to model, actually supplements and strengthens the picture, transforming it into a very resourceful philosophical framework.

With these considerations in mind, let us now return to the larger radical views that were left on hold at the end of Section 2.

5. Feyerabend's Fears Revisited

Of course science does not have a guaranteed method for completing its picture of the world. In addition, scientific knowledge is fallible, dependent on prior knowledge, and problematic with respect to truth. Last, but not least, it is always 'possible' that even the most reliable part of the present scientific picture might be completely wrong. These claims, however, are all compatible with legitimately granting special consideration to the Cautious Scientific Picture outlined in the previous section. The possibility that even the most reliable part of the current scientific picture might be completely wrong adds to the motivation and texture of the fallibilism embodied by scientific rationality. Beyond this, however, its status is that of mere possibility. To the extent that we live and believe by specific estimates of probability, mere possibilities license little horror by themselves no more than does, say, the continuous chance any of us (in this age of indiscriminate air transportation) have of being smashed by a falling elephant. Natural philosophers have insisted on this point repeatedly, from Christopher Clavius to Dudley Shapere: the mere possibility that a well-established claim might one day come to be doubted is not by itself a reason for doubting it now.[5]

What of the suggestion that the most solid parts of the scientific picture are held to be true only by mere accidents of history? The facts simply fail to support this. It is not just that plenty in the scientific picture seems as true as any empirical claim can be. The matter is one of differentials. Contrast the growing body of credible theoretical scientific claims with the theoretical descriptions issued by voodoo, telekinesis, alternative medicines, repressed memory therapies and similarly radical modes of knowing. For at least 30 years now such specialties have been granted every Feyerabendian courtesy in the USA and many European countries. What credible theoretical descriptions have they yielded? What critically acclaimable results have they produced? Their comparative poverty is glaring. Here the differences with science might be only ones of degree – but what a degree!

This is not to deny that, in principle, there might be some powerful nonscientific mode of knowing waiting to be discovered. We just have to keep an open-minded attitude about that, wait until somebody brings it forward, and then (and only then) react accordingly. If such a mode ever arises, it might represent calamity for our current cognitive views; or it might not. In the past, styles have added to styles, yielding a cumulative body of reasoning methods, with the body of best, most reliable information changing qualitatively and quantitatively along the way. The development of modern scientific rationality provides a good example here. Qualitatively, it has promoted a shift of rational belief away from foundationalism and illuminism, and toward probabilism. Quantitatively, it has led to an exponential growth of credible information in some fields (conspicuously the hard natural sciences), and to an exponential decrease of such information in many ancient fields (natural religion, astrology, uncheckable therapies).

So, science, it seems, is not just one of the many ideologies that currently propel society. In a growing number of fields, it is the most clearly and objectively reliable way of knowing at our disposal, and it should be treated as such.

With these considerations in mind, let me bring this exploration to a close with some thoughts on the place of scientific thinking in public education.

6. On Public Education

Let us agree with Feyerabend that the purpose of education is to introduce the young into life. If so, surely it is crucial for the young to learn and master the current public picture of the world. This means, firstly, to comprehend its intellectual underpinnings and to master the means of understanding it, which in turn requires being able to critically discern the views, models and methods incorporated into the picture; and being able to understand, historically and intellectually, how those elements came to gain acceptance. Secondly it means to appreciate the actual success and limitations of the prevailing public picture. Students need to critically appreciate its underlying tradition of rational thinking.

Teaching gets correspondingly burdened. Our epistemic success is not the same in all domains of human interest. Human knowledge is excellently rich, deep and robust in some areas. Simultaneously, it is poor, shallow and weak in many fields. There is much about which we have very little to go by in terms of publicly ascertainable knowledge. It would seem legitimate to demand, therefore, that public education be guided primarily by those views on 'what human beings and society are or might be' which are provided by the most reliable core of current public knowledge, which is now significantly influenced by scientific rationality. This core contains what I termed the Cautious Scientific Picture, as well as insight into the human condition from the most publicly credible part of views on human life and the world issued by the arts, literature, and traditional religions. Let us call this core 'Public Background Knowledge'. If this is accepted as reasonable, then many constraints follow. One, for instance, is that 'alternative' subjects cannot be imposed on the curriculum merely for the sake of diversity or conformity to pressure from private groups. Subjects presently lacking credibility by extant public standards do not belong in the public classroom but in extracurricular clubs.

Here is another implication. At the close of the century, we generally welcome the separation of church and state as a conscientious policy, largely because of the manifestly private character of religious belief and the recognized appropriateness of pursuing a maximum of liberty. Religious lore is amply tolerated, but it is not presently part of public knowledge. Accordingly, one important consequence of the previous reflections is that there should be also a formal separation between state and views that are not really part of the Public Background Knowledge. Such views may influence society, but only to the extent to which any pressure group is permitted to influence society. The pattern of recognition appropriate to forms of belief that lack public credibility would have to be similar to the one now agreed in the case of religion. A separation analogous to that between church and state seems plainly desirable for the relations between most nonstandard practices on view and the state. Claims issuing from currently private intellectual sources would be given every courtesy. They simply could not be imposed on anybody.

This way of turning Feyerabend's doctrine on its head is not a plea for one-sidedness in public education. On the contrary. The public core from which it springs embodies a deeply fallibilist view of knowledge. It is clear, therefore, that public education should actively encourage flexibility of mind, make students spontaneously contrary, counter-suggestive – as Feyerabend and so many revisionist educators say they want. Contrary to most radicals' expectations, however, that noble end is poorly served by simply ignoring the glaring differences of degree that exist between, say, scientific cosmology and creation myths, contemporary Western medicine and voodoo, etc. The bad educational effects of 'all-inclusive', 'multicultural', 'alternative' approaches are already abundant enough. They make people gullible and unskilled, intellectually frivolous and socially irresponsible. Not just students – university professors as well.

In a much discussed incident, Alan Sokal, a New York physicist, wrote a mischievous article about the implications and cosmological extensions of some post-modernist ideas under the title 'Transforming the Boundaries: Toward a Transformative Hermeneutics of Quantum Gravity'. The piece, written in the worst style of Derrida, Irigaray and Lacan, is a pornophonic parody of fashionable alternative modes of reflection on the contemporary situation. It was accepted without reservation by the influential and 'very demanding' journal *Social Text*.[6] To the shame and humiliation of the editors, however, Sokal simultaneously published a complete repudiation of the piece in another journal, along with an account of the chaotic and irresponsible way in which he produced his radical piece, and the reason for doing what he had done: he was alarmed by the state of the humanities in college campuses (Sokal and Bricmont 1998).

The faculty of imagination is always to be encouraged. But, for imagination to be of use, it needs to be developed by exercising it under tight prior constraints, be it in the arts, science or philosophy. Science, in particular, has an excellent record in this regard. It limits the imagination in a way that demonstrably forces it to get better. That is another reason why the achievements of the likes of Newton, Darwin, Einstein, and Crick and Watson should play a pivotal role in any public school curriculum.

Whatever else it may be, the cautious scientific picture sketched in Section 3 is not just one of the many ideologies that propel contemporary society. It is the most powerful engine of practical and intellectual advancement, and it should be treated as such.

Notes

[1] Quantum mechanical correlations are especially discussed in connection with the so-called 'Bell experiments'. See, for example, Cushing and McMullin (1989).

[2] For a discussion of Bloor's work, and more generally the Strong Programme in the Sociology of Scientific Knowledge, see the following contributions to *Science & Education*: Slezak (1994a, b), Suchting (1997) and Kragh (1998).

[3] See, for example, Shapere (1991), Cordero (1998), Dawkins (1995), Dennett (1991, 1995) and Cosmides and Tooby (1989).

[4] This trait not extends into very deep layers of the picture, enough to give substance to claims to knowledge on subjects seriously remote to the perceptual level – say, benzene molecules, protons, and the dynamical evolution of material systems with few degrees of freedom.

[5] Blake (1960) contains a relevant discussion of Clavius's epistemology. Shapere (1991) contains a perceptive discussion of the distinction between global and specific doubts in science.

[6] Published in the 1996 volume of *Social Text*, pp. 217–252.

References

Blake, R. M.: 1960, 'Theory of Hypothesis among Renaissance Astronomers', in R. M. Blake (ed.), *Theories of Scientific Method,* University of Washington Press, Seattle, pp. 22–49.

Bloor, D.: 1976, *Knowledge and Social Imagery,* Routledge & Kegan Paul, London.

Bloor, D.: 1983, *Wittgenstein: A Social Theory of Knowledge,* Columbia University Press, New York.

Cordero, A.: 1998, 'Physics and the Underdetermination Thesis: Some Lessons from Quantum Theory'. Forthcoming in S. Dawson (ed.), *Proceedings of the Twentieth World Congress of Philosophy* (Invited Papers), Federation Internationale de Societes de Philosophie & Boston University, Boston.

Cosmides, L. & Tooby, J.: 1989, 'Evolutionary Psychology and the Generation of Culture, II', *Ethology and Sociobiology* **10**, 51–97.

Cushing, J. T. & McMullin, E. (eds): 1989, *Philosophical Consequences of Quantum Theory*, University of Notre Dame Press, Notre Dame, IN.

Dawkins, R.: 1995, *River Out of Eden*, Basic Books, New York.

Dennett, D.C.: 1991, *Consciousness Explained*, Penguin Books, London.

Dennett, D.C.: 1995, *Darwin's Dangerous Idea*, Simon & Schuster, New York.

Feyerabend, P.K.: 1975, 'How to Defend Society Against Science', *Radical Philosophy* **11**, 3–8.

Feyerabend, P.K.: 1978, *Against Method*, Verso Press, London.

Kragh, H.: 1998, 'Social Constructivism, the Gospel of Science and the Teaching of Physics', *Science & Education* **7**(3), 231–243. Reprinted in M.R. Matthews (ed.), *Constructivism in Science Education: A Philosophical Examination*, Kluwer, Dordrecht, pp. 125–137.

Nagel, E.: 1974, *Teleology Revisited*, Columbia University Press, New York, pp. 95–113.

Psillos, S.: 1995, 'Is Structural Realism the Best of Both Worlds?' *Dialectica* **49**, 15–46.

Shapere, D.: 1991, 'The Universe of Modern Science and its Philosophical Exploration', in E. Agazzi & A. Cordero (eds), *Philosophy and the Origin and Evolution of the Universe*, Kluwer, Dordrecht, pp. 87–202.

Slezak, P.: 1994a, 'Sociology of Science and Science Education: Part I', *Science & Education* **3**(3), 265–294.

Slezak, P.: 1994b, 'Sociology of Science and Science Education: Part 11', *Science & Education* **3**(4), 329–356.

Sokal, A. & Bricmont, J.: 1998, *Fashionable Nonsense: Postmodern Intellectuals' Abuse of Science*, Picador, New York.

Suchting, W. A.: 1997, 'Reflections on Peter Slezak and the "Sociology of Scientific Knowledge"', *Science & Education* **6**(1–2), 151–195.

Worrall, J.: 1989, 'Structural Realism: The Best of Both Worlds', *Dialectica* **43**, 99–124.

Galileo and the Rhetoric of Relativity

PETER MACHAMER

History and Philosophy of Science Department, University of Pittsburgh, 1017 Cathedral of Learning, Pittsburgh, PA 15260, USA

ABSTRACT. This paper argues that Galileo well fitted in with the neo-Protagorian, person-relative framework that was emerging around him in the late sixteenth and early seventeenth centuries in western Europe. For Galileo all knowledge depended crucially and essentially on first person experience, and at the same time this knowledge was objective, not subjective. The paper develops this tension and concludes with some remarks on its educational implications.

That quite profound and remarkable changes occurred in the Sixteenth and Seventeenth Centuries hardly needs to be remarked. The Reformation and the Counter-reformation, the rise of capitalism and economic individualism, the dawn of the nation-state and the demise of divine rights, the social realization of personal privacy and the public display of this in architecture and, in literary form, the private diary, and the rise of the bourgeoisie and their literacy – all these and many more things were isolated or absent when the year 1500 appeared but were cultural institutions or practices by the time of the Enlightenment. During this period the human being lost a place in an ordered micro- and macro-Cosmos and culture but the individual took a place in a new world and social order.

Max Weber (1904–5), and after him many others, tried to develop a narrative interrelating some of these changes, and a few (like Robert Merton in 1938) tried to tie in the New Science. In an earlier paper I sketched a way of looking at the scientific revolution in terms of new ways of thought and justification that I called neo-Protagoreanism (Machamer 1991). The basic idea is one of epistemological individualism, wherein knowledge is gained, presented and justified in personal, human terms. Even the measures of mathematics and their use in the new science exhibit this fundamental first-person centered character.

However, this new way of knowledge – this way of ideas – brought forth a new problem: How could there be objective knowledge – knowledge good for all people and/or all times, if all knowledge was wholly based on what was inside individual person's minds (or bodies)? But must 'objectivity' mean for all people and all times, *sub specie aeternitatis*? Why should it? No one thinks science is good for all time, or presents timeless truths. Science advances, changes, develops. But then what can objectivity mean? This is the question that needs to be addressed in science classes, and one about which Galileo has something to teach us.

F. Bevilacqua et al. (eds.), Science Education and Culture, 31–40.
© 2001 *Kluwer Academic Publishers. Printed in the Netherlands.*

GALILEO AS A NEO-PROTAGORIAN

What I am about to show are examples of how it was that Galileo Galilei, called by some the 'father' of the new science, fits into the neo-Protagorian, person relative framework. What I will argue and what I hope you will understand is how Galileo's work well folds into the new pattern of culture established by the changes I mentioned above. Galileo was indeed a creature of this new age of individualism – though in many ways quite a remarkable one.

The easiest way to get a wide grasp of my thesis is to consider the claim that for Galileo all knowledge depended crucially and essentially on first person experience. It was his belief that the individual comes to know the world through perceptual and intellectual experience. Of course, experience is not uncritical; it is the interpreted experiences of the experts or geniuses that are important. Also, experience needs to be cast into certain forms that show its cogency, such as into mathematics (for clarity and intelligibility of inferences). This, of course, left him with the epistemological problem that would dominate Seventeenth Century philosophy and science until Newton: How can objectivity in science be obtained from an individual's subjective experience? Galileo did not solve this problem. However he did make many tries.

One quick caveat. In claiming that Galileo and indeed most everyone in the later Sixteenth and Seventeenth century are neo-Protagoreans, I am not claiming that everyone was an empiricist (as opposed, I guess, to a rationalist). The epistemology of empiricism, based solely on sensations and constructions from them, was not the way of Galileo, nor of Descartes, Arnauld, Huyghens, Hobbes, Boyle or others. Yet, on my view these men were all neo-Protagoreans, as were the empiricists, e.g., Locke and Berkeley.

THE *DIALOGUE*

Perhaps the best known place to begin with Galileo is in Day Two of *Dialogo*, where Galileo discusses the daily or diurnal rotation of the earth. After Salviati, who is the *persona* of Galileo, runs through his treatment and shows Simplicio, the authority and text ridden Aristotelian, and Sagredo, the critically wise man of common sense, how to properly interpret the experiences of balls falling from towers and ships' masts and how to figure out what happens to cannon balls shot from moving carriages, Salviati complains that

> . . . If I happened from time to time to meet anyone who held the Copernican opinion, I asked him whether he had always believed in it. Among the many whom I questioned, I found not a single one who did not tell me that he had long been of the contrary opinion, but had come over to this one, moved and persuaded by the force of its arguments. Examining them one by one then, to see how well they had mastered the arguments on

> the other side. I found them all to have these ready at hand.... On the other hand, so far as I questioned the Peripatetic and Ptolemaics... how much they had studied Copernicus' book, I found very few who had so much as seen it. (Galileo, 1632, p. 128)

Here we can see the argument that Galileo made again and again. It was a version of a typical Renaissance anti-Scholastic, anti-bad-authority argument, but with a neo-Protagorean twist. The Peripatetics et al. are bad guys, not just because they rely on authority or dogma, but because each one did not examine personally the theory of Copernicus. The Copernicans each individually went through such intellectual deliberations. It is in this spirit a few pages later (Galileo 1632, p. 132) that Salviati asserts that even Aristotle, if he were here, would be persuaded. The reason of course is that, if Aristotle were here, and if Aristotle were the smart man we know him to have been, he would examine the arguments for himself. In the balance he would be persuaded.

I choose this example as the first because it shows that the relevant first person experience for obtaining knowledge is not restricted to sensory experience. However arguments based on an individual's sensory experience are part of this view point. Later on in *Dialogo*, even Simplicio is forced to admit 'proof by the senses'.

Throughout Day Two of *Dialogo* Galileo employed the same strategy. A putative counter-example is brought forth against Copernicanism and the diurnal motion of the earth (balls falling, cannons booming), and there is a possible simple minded interpretation of it. Galileo contrasts that simple interpretation, likely to be given by an uncritical person or by one who is not intellectually open minded, i.e., by Simplicio the Aristotelian, with another interpretation that makes sense out of the example according to the Copernican theory. Indeed, Copernicus himself is praised because he did not let his common sense rule his interpretation. Rather he was a genius who went beyond the obvious common interpretation suggested at first blush by the senses.

Of course, Copernicus did not go beyond the senses in the way Galileo did. Copernicus did not develop the theory about the relativity of perceived motion (first given in Day Two of *Dialogo*, (Galileo 1632, p. 114)). For Galileo the key to interpreting these experiences having to do with the diurnal motion of the earth was based on the principle that motion in common among bodies is perceptually undetectable. What is seen as motion is relative to and independent of the motion of the perceiver. What is seen as motion depends on the point of view of the perceiver, in this case what motions the perceiver shares with other things that are in common relative to him.

In his most famous statement of this, Galileo writes:

> Salviati: For consider: Motion, in so far as it acts as motion, *to that extent exists relatively to things that lack it*; and among things which all share equally in any motion, it does not act, and is as if it did not exist. Thus the goods with which a ship is laden leaving Venice, pass by Corfu, by Crete, by Cyprus and go to Aleppo. Venice, Corfu, Crete, etc. stand still and do not move with the ship; but as to the sacks, boxes and bundles with which the

> boat is laden and with respect to the ship itself, the motion from Venice to Syria is as nothing, and in no way alters their relation among themselves. This is so because it is common to all of them and all share equally in it. If, from the cargo in the ship, a sack were shifted from a chest one single inch, this alone would be more movement for it than the two thousand mile journey made by all of them together. (Galileo 1632, p. 116)

Salviati then makes reference to Aristotle's definition, after Simplicio says this is just the Peripatetic doctrine:

> When he (Aristotle) wrote everything which is moved is moved *upon* something immovable, I think he only made equivocal the saying that whatever moves, moves *with respect to* something motionless. This proposition suffers no difficulties whereas the other has many. (Galileo 1632, p. 116)

The point is to show that it is important to use one's critical acumen to get beyond the appearances. The much vaunted drawing of the distinction between primary and secondary qualities (the tickles versus the bodies of *Il Saggiatore* (Galileo 1623) makes the same point. Secondary qualities are only appearances and the good scientist must see beyond them to what is real.

LETTER TO CHRISTINA

Let us now turn to a completely different kind of Galilean text, 'The Letter to the Grand Duchess Christina' (Galileo 1615). As is well known Galileo argued, in this letter, about how the Bible ought to be interpreted. He started in a familiar pattern by chastising those who condemned Copernicus without having ever read him, and proceeded to praise Copernicus for dealing with these matters of the heavens by 'astronomical and geometrical demonstrations founded upon sense experiences and very exact observations' (Galileo 1615, p. 179).

Galileo did not come out and say that if the Bible were to be rewritten now, it would be written from a Copernican point of view (or that the authors of the Old Testament would change their minds if they were here now.) Such dealing with a divinely inspired text would be too much even for Galileo. However, this clearly is his point. Galileo used the same interpretative ploy as he used later in *Dialogo*, arguing that the Bible was written for the common man in the language of its day, 'To avoid confusion in the common people' (Galileo 1615, p. 182). To interpret what is written in the Bible correctly and in an uncommon (read 'learned or expert') way takes an act of courage, a denial of what seems simple mindedly most obvious. Individuals who interpret and reason correctly will decide in favor of an interpretation that accords with reason and right thinking.

It is *the point of view* that needs to be taken into account. The point of view of the writers of the Bible was to convince people and bring them to the true faith, so 'even if the stability of the heaven and the motion of the earth should be more than certain in the minds of the wise, it would still be necessary to assert the contrary for the preservation of the belief among the all too numerous vulgar' (Galileo 1615, p. 200). The point of view of the wise person, such were each of the writers of the Bible, might be deviant, yet for the good of the faith they ignore their insights, all expression must be in the common language.

We have now seen two instances in different texts and contexts where Galileo brought up questions of interpretation and the point of view from which these interpretations are made. In both Galileo contrasts the point of view of the wise with that of the common people or the dogmatists.

THE *DISCOURSE*

Exactly the same tenor and point comes up at the beginning of *Discorsi* (Galileo 1638) where Galileo was setting a puzzle about the strength of materials: Why do long lengths of wood or marble crack when small, though proportional, ones do not? He contrasted the knowledge of expert artisans who work in the arsenal with those common opinions that are simple minded ways of extending of geometrical properties, viz. those who believe that all congruent shapes have the same material properties.

This way of putting the puzzle derives from Galileo's Neo-Protagorean individualistic theory, or, what we may now call Galileo's theory about the relativity of one's point of view. Though it really was not a 'theory', for it was never systematically elaborated or philosophically worked out. But it does form the basis for all of the remarks Galileo made about knowledge and the nature of good science.

This neo-Protagorean point of view probably came from his training with the artisans and practical mathematicians with whom he learned optics and geometry. These were part of his life-long beliefs or presuppositions which did not change even as he attempted to turn himself into a philosopher and become adept at philosophical language. This influence is well shown by Galileo's numerous illustrations of the method of science by invoking the method of the artisans, painters or sculptors. The sculptor works by genius to bring the sculpture out of the marble. Similarly the individual scientist works to discover and uncover the ways of nature (Galileo 1632, p. 109). Many times Galileo called upon the geniuses by name, Leonardo, Michelangelo, Rafaeleo, as though this would help persuade his audience of the aptness of the comparison.

THE BALANCE AS A MODEL FOR INTELLIGIBILITY

There are two other themes of this Galilean position to which I wish to draw your attention. First, is the relation of the principle of the relativity of observed motion to the basic principles of mechanics as presented first in *De Motu*, and finally in *Discorsi*, and how this relates to the kind of geometry that Galileo used. The second theme is how Galileo fought with the problem of objectivity in science.

Galileo in *De Motu* laid out a model for solving all problems of motion. He argued that the problems of floating bodies (with which he started his text) could all be reduced to problems of the Archimedian balance. He went on to show that all simple machines (the lever, the inclined plane and the pendulum) could be also reduced to balance problems. Free fall of bodies came to be an instance of floating bodies, or a balance that had no weight on the other side.

By the time of *Dialogo* the balance had become his metaphor for clear thought.

> So let us hear the rest of the arguments favorable to his [Aristotle's] opinion so that we may proceed with their testing, refining them in the crucible and weighing them in the assayer's balance. (Galileo 1632, p. 131)

This model readers of Galileo have seen before as it forms the whole of the image by which he judges Sarsi (Grassi) in *Il Saggiatore* (*The Assayer*) (Galileo 1623) which he labels in contrast to Grassi's tract, *The Astronomical and Philosophical Balance*. The contrasting, pregnantly ambiguous meanings of balance went from the balance scale (lances) to the alchemist's fire, the true tester (saggiatore) to wisdom (to the very notion of justice herself), Libra.

By the time of *Discorsi* his way of thinking about the world was set. He did not deal directly with the balance but being concerned with natural motion he did, after his definitions, begin with the equivalent Archimedian machine the inclined plane (*Galileo* 1638, p. 162) and then immediately changes that into a pendulum problem. From there he moved directly to free fall, or a vertical machine problem. All these types of problems were called 'mechanical conclusions' (Galileo 1638, p. 171). In this section of *Discorsi* he shifted from the inclined plane right back to talk about the balance and talk about equilibrium of weights. But equilibrium proofs are relativistic proofs. Weights on a balance are equal relative to each other.

Now the balance as the model for what is intelligible during this period in history had a great and convoluted history. The balance was physically and metaphorically the model of intelligibility of the age (for further detail see Machamer and Woody 1994). It was clearly observable when the balance was in equilibrium, when the weights and arms were equalized. Any individual could judge when a problem had been solved, when 'things were right'. And it was a concept of correctness or proof that could be easily taught. It was a way of interpreting phenomena that anybody could learn, and the standard for success was patent. There was no question of whether you had a proof or not; it was easily seen. Those who would not accept this model of intelligibility would not open their eyes. Personal ambition (such as claiming priority over Galileo) or dogmatism or authority blinded them.

Now this is an important aspect of Galileo's vision. While it was true that only the genius or true scientist had the critical powers to see things as they really are, these things after they were seen could be taught to others who did not have the original insight. This is a democratization of knowledge. Anyone, within limits, can be taught the proper principles of science, and when the method is learned, then they too are able to see clearly and correctly when things are right. This is as far as universality gets in Galileo, but it is one source of his objectivity.

In the late Sixteenth Century with rising capitalism and the introduction of concept of the nation state, the balance become the model for bourgeoisie book-keeping with its 'balanced accounts', for international commerce and its 'balance of trade', and for the relation between nations with 'balance of payments'. Truly the model of Archimedes had permeated the whole of the fabric of society and social relations (see Mirowski 1989). Later, with the idea of the contract as an equilibrium among individuals established by mutual agreement, social relations will come to have a new footing, and government itself will have a new legitimation. The legitimation is taken from the mathematics of mechanics, but this science itself took it from the mathematics and the ideas of commerce and trade for which that math was first used.

A note about Galileo's geometry is needed here. Euclid, Archimedes, and, following them, Galileo used geometry itself as a comparative, relativizing model for understanding. Nowhere in the Galilean corpus does he attempt to ascertain real values for any physical constant. Nowhere does Galileo attempt to find out, for example, what the real speed or weight of anything is. This proportional geometry is inherently comparative and relational. It measures one thing by showing its relation to another, one thing relative to another, which is conceived of as some arbitrarily or conveniently intelligible standard. In this sort of geometry there are no absolute values, no physical constants, which serve as the touchstone for certainty or objectivity. The standards by which the proof is measured are set by the person. This is the same relativity that we have seen before; one thing is judged relative to another.

This proportional geometry made it easy to think in terms of relative motion. Set the point of view or standard by which motion was to be judged and what counts as equilibrium or equality is determined thereby. In Galilean relative motion than standard was not arbitrary, but it was person relative. It was not until Huyghens that the arbitrariness of the standard for judging equilibrium became clear.

KNOWLEDGE AND OBJECTIVITY

Yet in this person-relative way of looking at the world and judging when proofs had been successful, Galileo recognized a problem. The balance model of intelligibility and the equilibrium model of proof demanded intersubjectivity. But even this was insufficient and objectivity was needed. In Day One of *Dialogo* Galileo contrasted God's extensive knowledge with the human being's intensive knowledge. When a person posed a puzzle in the language of proportional geometry where solutions were recognized by seeing that equilibrium was achieved, then the person was God-like in insight and understanding of the case at hand. By contrast, God sees all of the infinite cases. God's extensive knowledge of all cases only contrasted with human certainty in its intensive, particular mode of operation. The individual could be assured of his certainty intensively by using a proper method of proof.

It was in this way that the *more geometrico* provided the model of intelligibility and proof for science. The geometry involved was not a pure geometry but a physical geometry of the mixed sciences. It was the geometry of Archimedes, the geometry of proportions and of the properties of machines considered relative (or in relation to) one another. The visual paradigm of equilibrium proof for the balance brought together the Galilean tenets of experiment, long observation and rigorous demonstration (for a different take on this see Wallace 1992).

It is with this background in mind that one must understand Galileo's famous dictum about the book of nature. I shall quote this passage at length for it exhibits the themes of teaching and learning, of anti-authority, and of getting at the underlying truth by first person experience and the use of geometry. In *Il Saggiatore*, disputing the ideas and method of his opponent Sarsi, Galileo writes:

> It seems to me that I discern in Sarsi a firm belief that in philosophizing it is essential to support oneself upon the opinion of some celebrated author, as if when our minds are not wedded to the reasoning of some other person they ought to remain completely barren and sterile. Possibly he thinks that philosophy is a book of fiction created by some man, like the Iliad or Orlando Furioso – books in which the least important thing is whether what is written is true. Well, Sig. Sarsi, that is not the way matters stand. Philosophy is written in this grand book – I mean the universe – which stands continually open to our gaze, but cannot be understood unless one first learns to comprehend the language and interpret the characters in which it is written. It is written in the language of mathematics, and its characters are triangles, circles, and other geometrical figures, without which it is humanly impossible to understand a single word of it; without these one is wandering in a dark labyrinth. (Galileo 1623, pps. 183–4)

It was this geometrical model of equilibrium, when extended to colliding bodies by Descartes, and later, springs by Hooke, that set the standard for demonstration of the model of intelligibility for the science of motion. Indeed this equilibrium balance was the model for all natural science until Newton changed the ground rules with algebra replacing proportional geometry, absolute space replacing relational place, true motion replacing relative motion, and God becoming an active intervener in the world. The method of understanding the world by relating one thing to another in human terms became the problem of solving the equation to find a real number value, a universal constant. The world of science in the Eighteenth

Century became a world of absolutes. These changes gave science a new model of intelligibility and success. Science gained yet another new agenda, and the problem of objectivity in science was taken to be solved by finding absolute values.

Now it is historically interesting to note that this change occurred only in physics. In all other areas of human intellectual endeavor the model of mechanically balancing continued to be taken as the method of determining what is intelligible. So in biology, chemistry, psychology, government, and economics mechanical equilibrium models persisted, and still (to a large degree) persist today. This type of model, as we have seen, measured things in pragmatic ways by their relations to and usefulness for other things, especially their relevance to humans. This was a mechanical model that sought mechanisms through which things could be understood, and then changed or controlled.

EDUCATIONAL IMPLICATIONS

Now what is important for students of science to note is that relativity in judging motion or in judging arguments does not mean lack of objectivity. A scientist or any person investigating a problem needs to have standards as to what counts as a proof or what counts as a reasonable argument. The standards are in part given by how things cohere with the world, as found out by observation and experiment, and displayed in the evidence. The other part of objectivity depends on how clear and coherent are the demonstrations and proofs. How well these demonstrations lay out their assumptions and show the 'rules' by which they correlate the evidcnce with the other taken-for-granted (at least for the sake of this argument) assumptions.

The moral for teachers of science students is to get them to recognize relativity, and interpretation, but also to provide them with models for gathering and assessing evidence, and well as 'rules' and strategies for critically uncovering their assumptions, and for showing connections among parts of an argument.

REFERENCES

Galileo: 1590, *De Motu* (*On Motion*), Translated by Stillman Drake, Madison, WI.: The University of Wisconsin Press, 1960.

Galileo: 1615, 'Letter to the Grand Duchess Christina', translated by Stillman Drake, in Stillman Drake (ed.), *Discoveries and Opinions of Galileo*, New York: Doubleday Anchor 1957.

Galileo: 1623, *Il Saggiatore*, *The Assayer*, Translated by Stillman Drake in *The Controversy of the Comets of 1618*, Philadelphia: The University of Pennsylvania Press 1960.

Galileo: 1632, *Dialogo*, Dialogues Concerning the Two Chief World Systems, Translated by Stillman Drake, Berkeley, CA.: The University of California Press, 1967.

Galileo: 1640, *Discorsi, Discourses on the Two New Sciences*, Translated by Stillman Drake, Madison, WI.: The University of Wisconsin Press, 1974.

Machamer, Peter: 1991, 'The Person Centered Rhetoric of the 17th Century', in M. Pera and W. Shea (eds.), *Persuading Science: The Art of Scientific Rhetoric*, Canton, MA: Science History Publications.

Machamer, Peter and Woody, Andrea: 1994, 'A Model of Intelligiblity in Science: Using Galileo's Balance as a Model for Understanding the Motion of Bodies', *Science and Education*, 3, 215–244.

Merton, Robert: 1938, *Science, Technology and Society in Seventeenth Century England*, New York: Harper and Row 1970.

Mirowski, Philip: 1989, *More Heat Than Light: Economics as Social Physics, Physics as Nature's Economics*, Cambridge University Press.

Wallace, William A.: 1992, *Galileo's Logic of Discovery and Proof*, Dordrecht, Holland: Kluwer.

Weber, Max: 1904–5, *The Protestant Ethic and the Spirit of Capitalism*, New York: Charles Scribner's Sons, 1958.

Fostering the History of Science in American Science Education

F. JAMES RUTHERFORD
American Association for the Advancement of Science, 1200 New York Avenue, Washington, D.C. 20005, U.S.A., E-mail: jrutherf@aaas. org

1. Introduction

Given the expanding role of science and technology in the modern world, it becomes ever more urgent for general education – the education, that is, of all students, not exclusively those headed toward college and technical careers – to produce scientifically literate graduates. Not withstanding its desirability, that grand aim has proven elusive.

Perhaps this should not be surprising, given that making progress toward widespread science literacy is difficult for many reasons. Curricula are tradition bound, the leaders of each generation of adults believing that schooling should be pretty much as they had experienced it. But even when change is welcome, reaching agreement on precisely what those changes should be and on how best to go about effecting them is not easily achieved. And of course there simply are no quick and easy solutions to be had when it comes to dealing with such complex institutions as school systems. Progress depends on recasting learning goals, curricula, teacher education, instructional materials, assessment and teaching practices, educational policies, support systems, and more, and doing so in concert. Moreover, school systems vary widely from country to country and often from region to region within a country, and hence reforms that work well in one place may not do so in another. This is exaggerated in the United States by the extreme fragmentation and dilution of authority in its system of education.

In the face of such realities, this paper offers no comprehensive strategy for fostering science literacy everywhere. Instead, it takes the position that introducing the history and philosophy of science into school science curricula is a necessary, though not sufficient, part of any serious school effort to promote science literacy. The paper begins with a brief rationale for that position, and then presents two quite different efforts to that end taken in the United States: The Project Physics Course and Project 2061.

F. Bevilacqua et al. (eds.), Science Education and Culture, 41–52.
© 2001 *Kluwer Academic Publishers. Printed in the Netherlands.*

2. Rationale

In the United States there is substantial – though not nearly unanimous – agreement among scientists and educators that in addition to learning some of the important facts, concepts, and principles of science, all students should learn how science works. This position is clearly evident in the three key national reports of the last decade dealing with the content of K-12 science curriculum in American schools: *Science for All Americans* (American Association for the Advancement of Science,1989), *Benchmarks for Science Literacy* (American Association for the Advancement of Science,1993), and the *National Science Education Standards* (National Academy Press. All three emphasize learning goals dealing with the nature and history of science.

In those reports, arguments for the inclusion of the history of science in secondary-school science education are of two kinds. One argument has to do with the usefulness of the history of science in teaching science. The other argument has to do with the intrinsic value of some knowledge of history of science itself. These can be thought of, respectively, as instrumental and cultural arguments. In other words, the history of science is both a *tool* for teaching science well and a *part of the substance* of science literacy.

The utilitarian argument is set out in *Science for All Americans* as follows:

> ...generalizations about how the scientific enterprise operates would be empty without concrete examples. Consider, for example, the proposition that new ideas are limited by the context in which they are conceived; are often rejected by the scientific establishment; sometimes spring from unexpected findings; and usually grow slowly through contributions from many different investigators. Without historical examples, these generalizations would be no more than slogans, however well they might be remembered. (American Association for the Advancement of Science, 1989, p. 145)

It then exemplifies the cultural argument in this way:

> ...some episodes in the history of the scientific endeavor are of surpassing significance to our cultural heritage. Such episodes certainly include Galileo's role in changing our perception of our place in the universe; Newton's demonstration that the same laws apply to motion in the heavens and on earth; Darwin's long observations of the variety and relatedness of life forms that led to his postulating a mechanism for how the came about; Lyell's careful documentation of the unbelievable age of the earth; and Pasteur's identification of infectious disease with tiny organisms that could be seen only with a microscope. These stories stand among the milestones of the development of all thought in Western civilization. (American Association for the Advancement of Science, 1989, p. 145)

The case for introducing the history of science into secondary school science is all the stronger when the focus of science education is more on science literacy for all students than on the preparation of students for technical careers. As used in the two American Association for the Advancement of Science reports (AAAS) cited above, the 'scientific endeavor' encompasses the natural and social sciences, mathematics, and technology, the interdependencies among them, and the links between them and society. Without the help of history to illustrate actual instances, it is hard to imagine how students can come to understand such vital relationships.

But in spite of such arguments, the history of science has little presence in American schools. History courses typically avoid the history of science altogether and even, for the most part, the history of technology. And, for their part, science courses generally do little more than bow to the history of science by having students memorize some names and dates to be forgotten after the examination – in no important sense history.

3. Strategies for Change

Two reasons for the insignificance of the history of science in science teaching in American secondary schools are immediately obvious. One is that high school science teachers lack suitable preparation in the history of science, and the other is that high school science textbooks and other instructional materials mostly ignore the history of science. Neither teachers nor textbook publishers seemed disturbed by this fact. And there is a third reason, one that is more subtle, perhaps, but no less significant than the others. It is that the school system authorities who establish content requirements for the curriculum are generally unaware of the value of the history of science in science education (or in history education, for that matter), and therefore do not set appropriate learning goals.

If that assessment is correct, it follows that changing high school science curricula to include the history of science effectively will take reforming teacher education, creating new kinds of science courses and teaching materials, and establishing revised learning goals for all high school students. Moreover, even as those changes are underway, school systems will need help in revising their science curricula more broadly, since attaining lasting and significant change involves far more than simply injecting some history of science into already over-stuffed and somewhat incoherent science curricula.

In this paper, it is not possible to delve into the crucial matter of teacher education, even though, as just indicated, the prospect for changing the curriculum is bleak without dramatic reform in the professional preparation of teachers. Few undergraduates in American colleges and universities – not excluding those who will become science teachers – receiving anything like a serious introduction to the history of science. But reforming teacher preparation in science in the American system is itself a complex undertaking and involves, as argued in the Project 2061 report *Blueprints for Reform*, changing higher education more generally (American Association for the Advancement of Science, 1998).

Though not sufficient in itself, the development of new science courses and course materials that incorporate the history and philosophy of science is a necessary step toward curriculum reform in science education. It is also a relatively less complex reform undertaking than the equally necessary one of changing undergraduate education in hundreds of different colleges and universities. It is surprisingly, therefore, that so few attempts have been made to create science courses for the secondary schools that incorporate historical and philosophical perspectives, treating science as culture. In the United States, the Project Physics Course, developed at Harvard University during the period 1964–70, stands pretty much alone in that regard (Rutherford, F.J., Holton, G., and Watson, F.W. 1970).

Project 2061 was initiated by the American Association for the Advancement of Science on the premise that significant and lasting reform of science education in the schools of the United States could not be achieved by piecemeal efforts dealing with only this or that aspect of the system at any one time. Finding an appropriate place for the history and philosophy of science in the curricula would have to proceed hand in hand with other needed changes. The AAAS believed then (and continues to do so now) that a scientific society such as itself, while having no direct authority over school systems, and seeking none, can nevertheless contribute to systemic reforms in important ways. With regard to the history of science in particular, Project 2061 set out to (1) define science literacy in a way that incorporates the history and philosophy of science, (2) promote the adoption of learning goals that flow from that conception of science literacy, and (3) create tools for school systems, colleges and universities, publishers and funding agencies, and federal and state education agencies to use in carrying out science education reform efforts that focus on those learning goals. How this has played out is outlined following the comments on the Project Physics Course.

4. An Early Model: The Project Physics Course

I hesitate to discuss the Project Physics Course here for several reasons. One is that it did not, in the end, have the impact on American science education intended for it – though it is still in use in some schools in the United States and has been translated and adapted for use in several other countries, including Italy (Bastai, 1986). Another is that the course shows its age, having been developed at Harvard during the years 1964–70, since which time much has happened in science and education. A more positive reason is that we should turn our attention to Pavia Project Physics, a much more modern and educationally sophisticated course being developed by Professor Fabio Belivacqua and his colleagues at the University of Pavia, as an up-to-date model for what science courses that take the history of science seriously should be like in terms of content and instructional properties.

Nevertheless, I believe that it is worth a brief look at the Project Physics Course as a pioneering effort the features of which others may wish to consider in creating secondary school science courses that draw on the history of science. These include:

1. Rather than presenting history continuously throughout the student textbook – risking having it become a history text rather than a science one – history is presented in concentrated episodes from time to time. Thus, for example, the chapter 'The Language of Motion' is developed in a straightforward manner using real world phenomena, and is followed by 'Galileo Describes Motion', a chapter that contrasts Galileo's approach to the subject with Aristotle's.
2. In addition, the textbook includes throughout carefully selected quotations, timelines, and art to illustrate the way in which scientific knowledge is arrived at over time due to the contributions of many different investigators in many different places.
3. In parallel with the textbook are seven volumes of readings containing articles and chapters from books. Many of the selections are historical in nature, taken either from original sources (in translation, if necessary) or from writings of historians and scientists. Some typical examples are: 'Preface to *De Revolutionibus*' by Nicolaus Copernicus; 'On the Method of Theoretical Physics' by Albert Einstein; and 'Some Personal Notes on the Search for the Neutron' by Sir James Chadwick.
4. Some of the student laboratory activities deliberately and openly reproduce aspects of classical experiments, the purpose being to help students see what the idea behind the experiment was, the kind of data that accrue, and line of reasoning used. Examples include: 'A Seventeenth Century Experiment' an adaptation of the inclined plane discussed by Galileo in the *Two New Sciences*' using scantlings and student-made water clocks; and 'The Charge-to-Mass Ratio for an Electron', a version of J. J. Thomson's measurements on cathode rays (using student-made electron-beam tubes and coils for varying the deflecting magnetic field) to demonstrate their particle nature.
5. The Project Physics Course exploited the multimedia of the day – overhead transparencies, filmstrips, 8-mm film loops, 16-mm films, and broadcast television – and the history of science had a presence in them. Examples: a filmstrip made from three sequences of photographs taken at irregular intervals allow student to see Mars in retrograde motion and to determine the angular size and duration of its retrograde motion; a measurement film loop on Galilean relativity reproducing Galileo's thought experiment of a ball dropped from the mast of a boat by actually photographing such an event from different frames of reference in stop-motion; and 'The World of Enrico Fermi', a 16-mm documentary film – the world premier of which took place at the Accademia dei Lincei in 1970 at the Villa Farnesina in Rome.
6. Recognizing that what many students bother to learn is strongly influenced by what they will be examined on, the course provides teachers with sample tests that include questions relating to the history of science.

7. An extensive Resource Book for teachers provides content and teaching background information on many aspects of the Project Physics Course, including help with regard to understanding and using the history of science in teaching the course.

Even if those attributes of the Project Physics Course model are still valid in general, their use in the design of new science courses and science course materials must take into account new circumstances. Chief among these are the growth of scientific knowledge, the growth of knowledge about learning, and the rapidly increasing power of information and communications technologies. And while that is being done, it is well to keep in mind this lesson from the Project Physics Course experience: No matter how good they may be, new courses cannot by themselves do the reform job that needs doing. The system in which they are to be imbedded also needs to be reformed.

5. Influencing the System: Project 2061

Project 2061produces science education reform tools. These are publications – print, CD-ROM, and Web – designed to help educators carry out the kinds of reform measures that will result ultimately in graduates who are science literate. But Project 2061 does not attempt to develop materials for *student* instruction and assessment, to create curricula for the elementary and secondary schools, to design programs for teacher education, or to formulate school policy; instead it tries to influence and provide support for those who do.

In a sense, Project 2061 is a work in progress. Since it began in 1985, the project has developed four distinct but interrelated products and four others are in the pipeline (and others are under consideration). Each of the products – reform tools, as Project 2061 sees them – will be revised periodically to take into account feedback from users and to reflect the growth of knowledge. At the same time, the products, already highly interrelated, are gradually being consolidated to form a coherent whole. This integrated system of support for science education reform should be in place by 2005 and have reached its full potential by 2010, thereafter to be maintained by the AAAS and continuously modified as necessary to take advantage of new knowledge and technological advances.

In addition to its R&D functions, Project 2061 is engaged in professional development. With the publication of *Science for All Americans* in 1989, the project began to provide workshops to help educators learn how to use the Project 2061 products effectively. As each new product emerges, the project's professional development activities expand accordingly, and will continue into the foreseeable future as long as the need for them exists. However, the emphasis in this paper is confined to the development component of the Project 2061 initiative.

Since the publication of *Science for All Americans*, Project 2061 typically has several products under development at any one time. For many reasons, they are not released at uniform intervals, nor in the order one might prefer. In the following discussion, they are presented by function. The purpose of the first three listed publications is to set out and clarify science literacy learning goals for elementary and secondary schools; the purpose of the next five publications is to help educators carry out reform measures to enable students to reach those goals. The eventual computer-based, Internet-accessible unified assembly of these is called *System 61*. The set (which does not contain other products under consideration for development) includes the following:

Science for All Americans
Benchmarks for Science Literacy
Atlas of Science Literacy
Resources for Science Literacy: Professional Development
Resources for Science Literacy: Curriculum Materials Evaluation
Resources for Science Literacy: Assessment
Designs for Science Literacy
Blueprints for Reform
System 61

Each of these is briefly described, below, with some attention to the history of science.

5.1. SCIENCE FOR ALL AMERICANS

The terms and circumstances of human existence can be expected to change as much and as unpredictably from 1986 to 2061 – a human lifespan and coincidentally the current cycle of Comet Halley – as they did from 1910 to 1986 or from 1935 to 1910. Science and technology have been and will continue to be at the epicenter of change, causing it, shaping it, responding to it. Learning more about science and technology becomes, therefore, ever more central to education for life and living. Just what is the content and character of such learning? That is what Project 2061 first set out to answer.

Because of the large number of scientists, educators, and others involved (including historians and philosophers of science), and because of the care taken to formulate a description of science literacy that could serve as a lasting intellectual foundation for all that would follow, it took three years to arrive at a satisfactory answer to that question. The validity of *Science for All Americans* derives from a combination of the appropriateness and stature of the participants, the thoroughness of the process, the clarity and good sense of its recommendations, and the quality of the arguments and documentation used to support them. It defines science literacy broadly, emphasizing the connections among science, mathematics, and technology, and recommends what all students should learn in the following chapters:

'The Nature of Science' discusses the scientific world view, scientific methods of inquiry, and the nature of the scientific enterprise.

'The Nature of Mathematics' describes the creative process involved in both theoretical and applied mathematics.

'The Nature of Technology' examines how technology extends our abilities to change the world and the tradeoffs necessarily involved.

'The Physical Setting' lays out basic ideas about the content and structure of the universe (on astronomical, terrestrial, and sub- microscopic levels) and the physical principles on which it seems to run.

'The Living Environment' delineates basic facts and scientific concepts about how living things function and how they interact with one another and their environment.

'The Human Organism' discusses human biology as exemplary of biological systems.

'Human Society' considers individual and group behavior, social organization, and the process of social change.

'The Designed World' reviews principles of how people shape and control the world through some key areas of technology.

'The Mathematics World' gives presents basic mathematical ideas, especially those with practical application, that together play a important role in almost all human endeavors.

'Historical Perspectives' illustrates the science endeavor with ten episodes of exceptional significance in the development of science and in shaping human understanding of ourselves, our origins, and our universe. They are: displacing the earth from the center of the universe; uniting the heavens and earth; relating matter & energy and time & space; extending time; moving the continents; understanding fire; explaining the diversity of life; discovering germs; and harnessing power.

'Common Themes' presents general concepts, such as systems, models, scale, and change, that cut across science, mathematics, and technology.

'Habits of Mind' identifies attitudes, skills, and ways of thinking that are essential to science literacy.

In addition, *Science for All Americans* contains a chapter setting out a few of the most important principles for effective learning and teaching. Among them is one calling for students to encounter many scientific ideas in historical context, in the process becoming aware of the interdependencies among science, mathematics, and technology, the influence of society on the development of science and technology, and the impact of science and technology on society.

6. Benchmarks for Science Literacy

Over a period of four years, Project 2061 worked with six geographically and demographically diverse school-district teams (each consisting 25 of teachers and administrators backed up by university faculty members) to pin down a common set of specific learning goals for grade ranges K-2, 3-5, 6-8, and 9-12 based on *Science for All Americans*. The result provides a coherent set of benchmarks that reflect a logical progression of ideas for any given topic, with early-grade benchmarks anticipating the more advance benchmarks for later grades.

The grade-range recommendations in *Benchmarks for Science Literacy* were derived from *Science for All Americans*, and indeed are organized into the same chapter and section sequences. Accordingly, specific grade-related learning goals are spelled out for the nature and history of the scientific endeavor. The benchmarks are supplemented with integrative essays, cross references to related benchmarks in other sections, and relevant cognitive research, and they emphasize the interconnectedness of benchmarks within and between science, mathematics, and technology.

6.1. ATLAS OF SCIENCE LITERACY

As they were developing Benchmarks for Science Literacy, teachers and scientists found it helpful to construct growth-of-understanding maps showing how students might progress from an early understanding of a given concept at the K-2 level to a more sophisticated understanding at the 9–12 level. Subsequently, the maps were found by others to useful in a variety of ways, including particularly professional development (deepening their understanding of science) and curriculum change. A collection of about 50 of these maps will form the first volume of the Atlas, to be followed in about a year by a second volume of about the same number. A CD-ROM will accompany the Atlas, enabling users to access a great deal of information associated with each map and to manipulate the maps themselves. Some maps are on historical topics, and many of the other maps incorporate some historical and nature of science benchmarks.

6.2. RESOURCES FOR SCIENCE LITERACY: PROFESSIONAL DEVELOPMENT

This is the first of three print/CD-ROM products that will eventually merge into one. It is a tool to help educators gain the knowledge and skills necessary to understand and effectively work together toward science literacy in their schools and classrooms. It can be used by higher education faculty in teacher preparation programs, by school districts in designing in-service staff development programs, and by teachers for self-guided study of *Science for All Americans* and *Benchmarks for Science Literacy*.

Along with the full text of *Science for All Americans*, the CD-ROM includes a descriptive database of highly recommended books on science, mathematics, and technology (including the history of science) written for general audiences; citations of current cognitive research about how students learn concepts and skills that are important for science literacy; descriptions of undergraduate programs in different universities designed especially the understandings in *Science for All Americans*; and a workshop guide which includes a variety of presentation scripts, transparency masters, and background materials that can be used to create and conduct Project 2061 workshops.

6.3. RESOURCES FOR SCIENCE LITERACY: CURRICULUM MATERIALS EVALUATION

As educators become more informed about science literacy and its implications for teaching and learning, they will become more aware of the need for appropriate curriculum materials. To serve that need, Project 2061 worked with hundreds of teachers and scientists to develop a reliable procedure for evaluating curriculum materials for their alignment to learning goals such as benchmarks. This new tool – a print/CD-ROM combination containing the full text of *Benchmarks for Science Literacy* – can be used by school-district adoption committees to improve the selection of textbooks, by publishers to develop new textbooks and other instructional materials, and by teachers to revise or enhance their current materials. Of course part of this will be to evaluate the content and pedagogy of materials for their bearing on the history of science.

6.4. RESOURCES FOR SCIENCE LITERACY: ASSESSMENT

Now under development, this third component of the Project 2061 Resources series will try to do for assessment materials what the former one does for curriculum materials. A print/CD-ROM product, it will describe how to analyze student examinations (local, state, national, international) in terms of how well they target specific science literacy learning goals. The disk will provide utilities to help educators carry out the analysis. Needless to say, if the history of science does not have an presence in assessments of learning, little can be expected in the way of student understanding of the history of science, no matter what else happens.

6.5. DESIGNS FOR SCIENCE LITERACY

Significant and lasting improvement of science education in the United States requires the design, development, and implementation of new curricula. *Designs for Science Literacy* (a print/CD-ROM product now in press) is a guide for educators engaged in reforming curricula. It lays out some basic design principles, brings them to bear on the design of K-12 curricula, and provides examples of how these principles might be used to improve today's curricula and contribute to more comprehensive changes in curricula of the future.

Designs also contains suggestions for how to create coherent professional development programs that focus on the science literacy goals of *Science for All Americans* and *Benchmarks for Science Literacy*, reduce the core content of overstuffed curricula, and increase curriculum connections across grades and subjects. It emphasizes that the history of science is especially good for studying connections among science, mathematics, and technology and between them and the arts and humanities. The CD-ROM contains databases, templates, utilities, and other interactive features to help educators think about and keep track of their curriculum reform efforts.

6.6. BLUEPRINTS FOR REFORM

Project 2061 recognizes that significant and lasting reform in one area of the education system cannot take place in isolation from changes in other parts of the system. To explore the complex interactions of all parts of the education system in the light of the science literacy goals set out in *Science for All Americans*, Project 2061 commissioned a dozen concept papers from twelve expert panels, and then submitted the draft papers to extensive review by educators and education policy makers in every region and kind of school system. The resulting papers examine equity, policy, research, finance, school organization, curriculum connections, assessment, teaching materials and technologies, teacher education, higher education, family and community, and business and industry.

To back up the print version of the papers, Project 2061 has published, on its World Wide Web, summaries of the twelve papers, a descriptive database of exemplary systemic reform programs, bibliographic references, and information for contacting selected agencies and organizations for further information on the topics. Forums and other interactive exchanges on the Web site will be used to encourage more extensive debate of the important issues raised in the *Blueprints* papers.

6.7. SYSTEM 61

As indicated earlier, Project 2061 intends to bring the various CD-ROM and Web products together to form a computer-based integrated whole that will be both more powerful, more flexible, and easier to use than the separate products. Progress toward this end is already underway as the various disks referred to above contain overlapping material and are being fitted with a common electronic interface. For the foreseeable future, however, successive editions of the individual print/CD-ROM products will still be available.

By the time it reaches maturity, *System 61* may also incorporate other components now being considered. *Family Guide to Science Literacy* will connect homes to the system, or parts of the system. SCIENCE *for Teachers* will provide teachers using the system with weekly information on what's happening in science by selecting, rewriting, and supplementing material from the AAAS journal SCIENCE and connecting it to the topics in *Science for All Americans*. Finally, *Insights on Science* will provide multimedia articles for teachers keyed to the topics in *Science for All Americans*, especially for topics which teachers are less familiar with, such as the history of science, the nature of science, mathematics, and technology, and cross-cutting concepts.

7. Looking Ahead

How will this extraordinary effort play out? Will it, among other things, result in a greater presence for the history and philosophy of science in science education in American schools? Even after more than a decade of effort, it is still too early to predict what the lasting effects of the project will be in detail. Separate studies conducted shortly after the release of *Benchmarks for Science Literacy* by the Organization for Economic Cooperation and Development and Stanford Research International both concluded that *Science for All Americans* and *Benchmarks* had had considerable influence in general on federal and state education agencies, but less so at the classroom level. While this is to be expected, given the Project 2061 long-term strategy, it leaves unanswered such detailed questions as to what degree the project is successfully fostering the use of the history of science.

In any case, the Project 2061 is trying to provide the vision and tools to enable educators to expedite reform toward science literacy. That vision and those tools give a prominent place to the history of science, but actual reform in that direction will take the sustained, creative effort of many others – to redesign curricula, to create appropriate student learning materials, to devise assessment materials closely linked to learning goals, to prepare future teachers to be able to use the history of science in their teaching. By fostering a desire for reform in such a direction and helping to build a capacity for carrying it out, Project 2061 is at least increasing the likelihood that it will occur.

References

American Association for the Advancement of Science: 1989, *Science for All Americans*, Oxford University Press, New York.

American Association for the Advancement of Science: 1993, *Benchmarks for Science Literacy*, Oxford University Press, New York.

American Association for the Advancement of Science: 1997, *Resources for Science Literacy: Professional Development*, Oxford University Press, New York.

American Association for the Advancement of Science: 1998, *Blueprints for Reform*, Oxford University Press, New York.

American Association for the Advancement of Science: in press, *Atlas of Science Literacy*, Author, Washington, D.C.

American Association for the Advancement of Science: in press, *Designs for Science Literacy*, Oxford University Press, New York.

Bastai, A.P. et al: 1986, *Progetto Fisica*, second edition, Zanichelli, Bologna.

National Research Council: 1996, *National Science Education Standards*, National Academy Press, Washington, D.C.

Rutherford, F.J., Holton, G. & Watson, F.W.: *The Project Physics Course*, Holt, Rinehart & Winston, New York.

Nature-of-Science Literacy in *Benchmarks* and *Standards*: Post-Modern/Relativist or Modern/Realist?

RON GOOD[1] and JAMES SHYMANSKY[2]
[1] *Curriculum & Instruction, Physics Departments, Louisiana State University, Baton Rouge, LA 70803, USA; E-mail: rgood@lsu.edu;* [2] *Curriculum and Instruction Department, University of Missouri at St. Louis, St. Louis, MO 63121, USA; E-mail: edujshym@jinx.umsl.edu*

Abstract. The complexity of science is described in the two major science education reform documents in the US: *Benchmarks for Science Literacy* (1993) and *National Science Education Standards* (1996). Some have seen them as too 'postmodern' while others have charged they are too 'modern' in their descriptions of the nature of science. An analysis of the documents shows how each charge might arise. Science's complexity requires one to say that 'scientific knowledge is tentative or subject to change' and 'scientific knowledge is stable'; that 'change is a persistent feature of science' and 'continuity is a persistent feature of science'; that 'it is normal for scientists to differ with one another' and 'scientists work toward consensus'. Both *Benchmarks* and *Standards* describe science in terms that sometimes seem to emphasize tentative, local knowledge while at other times emphasizing stable, universal knowledge. Although the overall picture of science presented by each document appears to be one of modern realism, it is not difficult to see how the postmodern relativist could select statements that paint science as epistemically equivalent to the social sciences or even the arts and humanities. Implications for science education are discussed.

Introduction

Both *Benchmarks for Science Literacy* (1993) and *National Science Education Standards* (1996) have had considerable impact on science education policy and practice in the US. Most of the 50 states have revised their own science education frameworks or standards to be compatible with *Benchmarks* and/or *Standards*, including nature-of-science (NOS) standards. A November 1992 preliminary draft of *Standards*, sponsored by the United States National Academy of Sciences, contained the statement 'The National Science Education Standards are based on the postmodern view of the nature of science' (p. A-2). The scientific community in particular was upset at that statement and at other content in *Standards* that some thought indicated a postmodern/relativist view of science. The final version of *Standards* does not contain the term 'postmodern' and generally the document seems to portray science as a complex, rather variable process that nevertheless

F. Bevilacqua et al. (eds.), Science Education and Culture, 53–65.
© 2001 *Kluwer Academic Publishers. Printed in the Netherlands.*

yields knowledge about our world that is stable and progressively more accurate, in terms of learning Nature's fundamental secrets.

Benchmarks never contained a statement about being based on the postmodern view of the nature of science and has not suffered the same kind of criticism from the scientific community. Both documents, however, have been criticized by various academics as being too 'postmodern' or too 'modern', depending on the position of the critic. How can the same document be seen so differently? It is this question that we try to answer in this paper.

The Postmodern/Relativist View

Postmodernism is a term that seems more at home in the arts and humanities than in the natural sciences and Rosenau (1992) traces it to French postmodernists such as Jacques Derrida, Michel Foucault, and Paul Sartre and to German philosophers Martin Heidegger and Friedrich Nietzsche. It refers to many things, including (1) rejection of universals or metanarratives, (2) rejection of any final meaning, (3) rejection of the idea of progress, (4) disillusionment with science, (5) pessimism about our future, (6) rejection of logic and reason, (7) abandonment of objectivity, (8) a fascination with mystical, new-age ideas that challenge established 'reality', and (9) Feyerabend's (1975) motto of 'anything goes'.

Postmodern critics Paul Gross and Norman Levitt provide a stinging critique of post-modernism as a point of view that constantly borders on nihilism:

> Contrasted to the Enlightenment ideal of a unified epistemology that discovers the foundational truths of physical and biological phenomena and unites them with an accurate understanding of humanity in its psychological, social, political, and aesthetic aspects, postmodern skepticism rejects the possibility of enduring universal knowledge in any area. It holds that all knowledge is local or 'situated', the product of interaction of a social class, rigidly circumscribed by its interests and prejudices, with the historical conditions of its existence. There is no knowledge, then; there are merely stories, 'narratives', devised to satisfy the human need to make some sense on the world. In so doing, they track in unacknowledged ways the interests, prejudices, and conceits of their devisers. On this view, all knowledge projects are, like war, politics by other means. (Gross and Levitt 1994, p. 72)

If there is a method associated with postmodernism it is deconstruction. Deconstruction identifies inconsistencies or tensions but does not try to reconstruct or suggest alternatives. Rosenau identifies eight underlying principles of deconstruction:

1. Find an exception to a generalization in the text and push it to the limit so that this generalization appears absurd.
2. Interpret the arguments in a text being deconstructed in their most extreme form.

3. Avoid absolute statements in deconstructing a text, but cultivate a sense of intellectual excitement by making statements that are both startling and sensational.
4. Deny the legitimacy of all dichotomies because there are always a few exceptions to any generalization based on bipolar terms, and these can be used to undermine them.
5. Nothing is to be accepted; nothing is to be rejected. It is extremely difficult to criticize a deconstructive argument if no clear viewpoint is expressed.
6. Write so as to permit the greatest number of interpretations possible.
7. Employ new and unusual terminology.
8. Never consent to a change in terminology. (Rosenau 1992, p. 121)

Text is used by postmodernists to mean anything and everything and Culler (1982) asserts that deconstruction is not concerned with what a text means. By now it should be clear that postmodernism, although difficult to define, is very different than science in almost every way. Post-epistemological postmodernism abandons shared inquiry, with its communally-agreed upon methods, for standardless, criteria-free, individual perceptions.

The Modern/Realist View

Rather than seeing 'reality' as a social construct, scientists and most modern philosophers of science see Nature as real, existing independently of humans and their various 'philosophical' theories. Physicist Roger Newton puts it this way:

> It is difficult to imagine a scientist who doubts that a real world exists independently of ourselves. We measure its properties, we observe its changes, we try to understand it, and sometimes it astonishes us. 'The belief in an external world, independent of the perceiving subject, lies at the basis of all natural science', Einstein insisted. (Newton 1997, p. 160)

Philosopher W. H. Newton-Smith's view of science supports physicist Newton's ideas:

> The realist tradition in the philosophy of science is an optimistic one. Realists do not think merely that we have in principle the power specified in the epistemological ingredient. They take it that we have been able to exercise that power successfully so as to achieve progress in science. (Newton-Smith 1981, p. 39)

Both scientist and philosopher agree that modern science is a part of the modern/realist tradition and that progress is a hallmark of the tradition. John Dewey, America's most influential philosopher-educator, described science as our most effective operation of intelligence:

> The function which science has to perform in the curriculum is that which it has performed for the race: emancipation from local and temporary incid-

ents of experience, and the opening of intellectual vistas unobscured by the accidents of personal habits and prediliction. (Dewey 1916, p. 270)

We turn now to the U.S. reform documents *Benchmarks for Science Literacy* (1993) and *National Science Education Standards* (1996) to see how the nature of science is defined and how the postmodern/relativist version of science might be seen in these documents by those with such an agenda.

1. Science as Portrayed in *Benchmarks* and *Standards*

Behind *Benchmarks* and *Standards* are two of the largest and most influential scientific societies in the world: The American Association for the Advancement of Science (AAAS) for *Benchmarks* and The National Academy of Sciences (NAS) for *Standards*. Each document was developed through a consensus-building process that involved a large number of scientists, educators, and others throughout the U.S. Although both science education reform documents have been very influential, they are not without controversy, especially the *Standards*. As mentioned earlier in our Introduction, an early draft of the *Standards* contained the statement 'The National Science Education Standards are based on the postmodern view of the nature of science' (p, A-2). This statement does not appear in the final document because of considerable criticism from the scientific community and others who see postmodernism as a 'decidedly antiscience' (Nicholson 1993, p. 268) movement that has the potential to impair the status of science by blaming it for the ills of the world. However, as Michael Matthews (2000) notes in his recent book, *Time for Science Education*, the constructivist content of the *Standards* was not rejected, merely relocated.

The complexity of science requires that it be described in ways that capture its many- faceted nature; however this is seen by some (e.g., Harding 1994; Hodson 1994; Jegede 1989; Latour and Woolgar 1986; Ogawa 1989; Pomeroy 1992; Stanley and Brickhouse 1994) as an invitation to claim that modern science should be placed in the same category as folklore and local knowledge of indigenous people. To say that 'scientific ideas are tentative and open to change', a statement that accurately portrays one facet of the nature of science, fails to consider the fact that 'most scientific ideas are not likely to change greatly in the future'. Each statement accurately describes the nature of science, but it is the latter statement that is more characteristic of scientific knowledge when compared to other kinds of knowledge. How science can be described in post-modern/relativist terms by persons with such agendas is the focus of the next two sections.

CONTRASTING NOS STATEMENTS IN *STANDARDS*

A perusal of *Standards* shows that many statements about the nature of science (NOS) seem to describe science in contrasting ways. Depending on one's agenda science can be shown by *Standards* to be postmodern/relativist in nature or modern/realist in nature. Some examples are listed here, with the letter 'a' by statements more likely to be used by the nonscience, postmodern crowd and the letter 'b' by statements more likely to be used by modern/realist scientists to characterize the nature of science.

1a. Scientific ideas are tentative and open to change. (p. 171)
1b. Most scientific ideas are not likely to change greatly in the future. (p. 171)
2a. It is normal for scientists to differ with one another about ideas and evidence. (p. 171)
2b. Scientists work toward finding evidence that resolves disagreements. (p. 171)
3a. Scientists are influenced by societal, cultural, and personal beliefs, and ways of viewing the world. (p. 201)
3b. Explanations on how the natural world changes based on myths, personal beliefs, religious values, or authority are not scientific. (p. 201)
4a. All scientific knowledge is subject to change. (p. 201)
4b. The core ideas of science are unlikely to change. (p. 201)

Each of these statements accurately describes one of the many facets of the nature of science. However, if we choose to concentrate on 1a, 2a, 3a, and 4a rather than 1b, 2b, 3b, and 4b it is clear that a picture of science emerges that is more consistent with postmodern/ relativist ideas than modern/realist ideas. A perusal of *Benchmarks* results in a similar picture.

CONTRASTING NOS STATEMENTS IN *BENCHMARKS*

Benchmarks (1993) is based on the earlier reform document *Science for All Americans* (1989) and both begin with a chapter on the nature of science. In many ways the nature of science is described in more detail in these sources than in *Standards* so more contrasting examples are provided here than in the previous section. Again, 'a' precedes statements more likely to be used by the nonscience, postmodern folks and 'b' precedes statements more likely to be used by modern/realist scientists to characterize the nature of science.

1a. Scientific knowledge is subject to change. (p. 5)
2b. Scientific knowledge is stable. (p. 5)
3a. Results of similar scientific investigations seldom turn out exactly the same. (p. 6)
3b. Science investigations generally work the same in different places. (p. 6)
4a. Radical changes in science sometimes result from the appearance of new information. (p. 7)
4b. Usually the changes that take place in scientific knowledge are small modifications of prior knowledge. (p. 8)
5a. Change is a persistent feature of science. (p. 8)
5b. Continuity is a persistent feature of science. (p. 8)

6a. Sometimes scientists have different explanations for the same set of observations. (p. 11)
6b. That usually leads to their making more observations to resolve the differences. (p. 11)
7a. There is no fixed set of steps that all scientists follow. (p. 12)
7b. Scientific investigations usually involve the collection of relevant evidence, the use of logical reasoning, and the application of imagination in devising hypotheses and explanations to make sense of the collected evidence. (p. 12)
8a. What people expect to observe often affects what they actually do observe. (p. 12)
8b. Independent studies of the same question control for personal bias. (p. 12)
9a. There are different traditions in science about what is investigated and how. (p. 13)
9b. All scientists share basic beliefs about the value of evidence, logic, and good arguments. (p. 13)
10a. Scientists in any one research group tend to see things alike. (p. 13)
10b. Scientists check other's results to guard against bias. (p. 13)
11a. Science disciplines differ from one another in what is studied, techniques used, and outcomes sought. (p. 19)
11b. All science disciplines share a common purpose and are part of the same scientific enterprise. (p. 19)

This longer list of NOS statements from *Benchmarks* supports and extends the previous list from *Standards.* Lump all the 'a' statements together and you describe science as a changing, uncertain, inconsistent, variable, biased enterprise that differs little from other ways of knowing. Lump all the 'b' statements together and the picture is completely different. In the remainder of the paper we look more carefully at these differences, how they might cause confusion among science teachers and others in the science education community, and how the recent 'science wars' have focused more attention on the nature-of-science issue.

Science Wars and Science Education

Who should define science for the science education community? We have seen that the AAAS (*Benchmarks*) and the NAS (*Standards*) have described science in ways that can be categorized as modern/realist or, as postmodern/relativist, depending on one's motives. The scientific community sees science in modern/realist terms and certain others, including some sociologists of science, science educators, and curriculum theorists see science in post-modern/relativist terms. The struggle to define science for public school teachers, their students, and the general public, has come to be known, at least in some academic circles, as the 'science wars'. Although *Science for All Americans* (1989) appeared a decade ago and can be considered the precursor of both *Benchmarks* and *Standards*, a more recent event has focused much public attention on the question of who defines science for the science education community and the public.

Fashionable Nonsense: Postmodern Intellectuals' Abuse of Science (Sokal and Bricmont 1998) follows other critiques (e.g., Holton 1993; Gross and Levitt 1994; Gross et al. 1995) of postmodern 'myths' about science. The most recent book devoted to a similar critique is edited by philosopher Noretta Koertge – *A House Built on Sand: Exposing Postmodern Myths about Science* (1998). Physicist Alan Sokal started what is now called 'science wars' by writing a parody of postmodern science that was published in the spring/summer 1996 issue of the fashionable, cultural-studies journal *Social Text*. His paper, 'Transgressing the Boundaries: Toward a Transformative Hermeneutics of Quantum Gravity', is full of 'scientific' nonsense, from the scientist's viewpoint, but very flattering of postmodern critics of science such as Aronowitz (1988), Harding (1991), Latour (1987), and Woolgar (1988). Sokal's parody of postmodern accounts of the nature of science was exposed by him shortly after his paper was published in *Social Text*, triggering a great deal of discussion from both sides, postmodern/relativist and modern/realist.

Placing everyone in either the postmodern/relativist camp or the modern/realist camp greatly over-simplifies the situation, but for ease of argument and simplification of description of the situation we proceed using this dichotomy. Few science critics like the idea of being called a relativist, but their stated positions on issues such as the contrasting statements from *Standards* and *Benchmarks*, make it difficult to argue otherwise. For example, science's core assumption of universalism is rejected by postmodernism in favor of local, equally viable 'truths'. Both *Benchmarks* and *Standards* underline the importance of universalism to all of science: 'Science also assumes that the universe is, as its name implies, a vast single system in which the basic rules are everywhere the same' (AAAS 1989, p. 3) and 'Science assumes that the behavior of the universe is not capricious, that nature is the same everywhere, and that it is understandable and predictable' (NAS 1996, p. 116). These statements make it clear that the basic assumption of science is universalism, what Michael Matthews calls the core universalist idea: 'The core universalist idea is that the material world ultimately judges the adequacy of our accounts of it. Scientists propose, but ultimately, after debate, negotiation and all the rest, it is the world that disposes' (Matthews 1994, p. 182).

A leading spokesperson for the postmodern view of multiple, equally-valid sciences is feminist philosopher Sandra Harding. In various publications Harding raises the question – Is science multicultural? – with the answer, '... there could be many universally valid but culturally distinctive sciences' (Harding 1994, p. 320). To give some idea of how Harding reaches such a conclusion, here is a quote that she thinks provides 'evidence' for the claim:

> If we were to picture physical reality as a large blackboard, and the branches and shoots of the knowledge tree as markings in white chalk on this blackboard, it becomes clear that the yet unmarked and unexplored parts occupy a considerably greater space than that covered by the chalk tracks. The so-

> cially structured knowledge tree has thus explored only certain partial aspects of physical reality, explorations that correspond to the particular historical unfoldings of the civilization within which the knowledge tree emerged.
>
> Thus entirely different knowledge systems corresponding to different historical unfoldings in different civilizational settings become possible. This raises the possibility that in different historical situations and contexts sciences very different from the European tradition could emerge. Thus an entirely new set of 'universal' but socially determined natural science laws are possible. (From Goonatilake 1984, pp. 229–230)

Here it seems that the universalism of Nature is reduced to playing a minor role while different cultural/historical 'unfoldings' assume center stage. As Gross and Levitt (1994), Matthews (1994), Rosenau (1992), Slezak (1994a, b) and others have observed, postmodernists attempt to show that science is basically a political struggle, leaving little room for the role of Nature in science (see Good and Demastes 1995 for more on the diminished role of Nature in postmodernism).

In science education the science wars are played out in the names of multiculturalism and constructivism, two movements that are highly visible in today's educational scene. The nature of science is argued by proponents and critics of both movements, with little apparent agreement on either side. Those in favor of a multicultural science curriculum argue that we should include the knowledge that indigenous people have of their environment while critics argue for exclusion because accurate observation without universally agreed upon natural explanation is not science. Proponents want to include all ideas to make everyone feel good while critics want to include only the best ideas regardless of how some person or group might feel. Good et al. (1999) assert that proponents of the postmodern/relativist position actually engage in a form of censorship, similar to the efforts of the religious fundamentalists who try to suppress the teaching of Darwinian theory of evolution of life.

Regarding constructivism, proponents want all students to be able to construct their own ideas about science and the critics see science as mainly counterintuitive and beyond students' comprehension without a great deal of specific, directive teaching. The wars go on, perhaps because neither side can agree on the nature of science; or perhaps because appeal to logic and evidence is the central agenda on one side while a different agenda drives the other.

Science as Culture: Revolutionary and Counterintuitive

Multiculturalism tries to honor and conserve local ideas and customs while science is indifferent to them. Constructivism stresses an individual's unique outlook while science seeks consensus within the scientific community. In these ways multiculturalism and constructivism are more consistent with a postmodern/relativist worldview while science is more consistent with a modern/realist worldview. Although over-simplified, these contrasting positions will be used to describe the nature of science as a revolutionary and counterintuitive culture.

In *The Unnatural Nature of Science* (1993) British biologist Lewis Wolpert describes science as counterintuitive, even unnatural. Any science teacher sensitive to the multitude of prescientific conceptions students bring to the classroom, knows what Wolpert means and teachers know also that it is often very difficult to help students understand the scientific conception. Most key scientific ideas are truly counterintuitive; they run counter to the ideas about our world that we develop naturally as a result of growing up from child to adult.

Revolutionary is used here in the sense that displacing the Earth from the center of the universe was very disruptive to people following Copernicus and inserting natural selection as the origin of species was disruptive following Darwin. Ontological and epistemological ruptures in the sense used by Gaston Bachelard (1934) are revolutionary to people's thought and to their culture. The authors of *Science for All Americans* devote a chapter of their book to 'Habits of Mind', saying that:

> The revolutions that we associate with Newton, Darwin, and Lyell have had as much to do with our sense of humanity as they do with our knowledge of the earth and its inhabitants. Moreover, scientific knowledge can surprise us, even trouble us, especially when we discover that our world is not as we perceive it or would like it to be. (AAAS 1989, p. 172)

Conserving local cultural practice or protecting individuals' prescientific ideas about physical causality are not priorities for science. Religious practice and related ideas about our world have conflicted with science since Copernicus, Galileo, Newton, and Darwin introduced their ideas about the natural world and the two cultures continue to clash today. In a special issue (April 1996) of *Science & Education* on religion and science education, Martin Mahner and Mario Bunge make the case that science and religion are incompatible:

> Science and religion are not only methodologically different but incompatible. The same holds for the metaphysics and the ethos of science and religion. Finally, insofar as religion makes some cognitive statements about the world, there will also remain doctrinal incompatibilities between religion and science. (Mahner and Bunge 1996, p. 115)

They go on to talk of implications for a sound science education:

> Since science education as well as moral and religious education are supposed not only to convey propositional knowledge but also to elicit and develop a certain attitude or mentality in our children (see, e.g., Bunge 1989; Martin 1990), we come to the conclusion that, regarding the incompatibility of the scientific and religious attitude, a religious education, particularly at an early age, is a most effective obstacle to the development of a scientific mentality. (Mahner and Bunge 1996, p. 119)

Taking the clash of the two cultures, scientific and religious, to the level described by Mahner and Bunge ensures vigorous debate and disagreement among nearly everyone who feels strongly about these two areas of our culture. If Mahner and Bunge are correct about the potentially negative impact of early religious training on a later scientific education, the science education community has a significant challenge before it. In the case of evolution education, it is clear that that certain forms of religious training have a negative impact on students' later learning of the modern Darwinian theory of evolution of life. Scientific habits of mind such as curiosity, openness to new ideas, and skepticism may actually be inhibited by early religious training unless a way is found to confront the problem that Mahner and Bunge think is unavoidable.

In the remainder of the paper we return to the postmodern/relativist vs. the modern/realist dichotomy, as portrayed in *Benchmarks* and *Standards*, and suggest ways that science teachers and their students can deal more accurately with the complex nature of science while avoiding the confusing, misleading issues described earlier as the 'science wars'.

Implications for Science Education

How can the complexity of science be taught and learned without degenerating into the postmodern/relativist world? Our earlier analysis of *Benchmarks* and *Standards* shows that for many modern/realist statements on the nature of science there are corresponding statements that paint a more complex, tentative picture and yet each of these reform documents describes the overall enterprise of science in generally modern, realist, rational terms, especially when compared to other ways of knowing. The early phase of science of developing new ideas that better explain Nature, what Gerald Holton (1988) called the private, speculative aspect of science, is as creative as developing new art, poetry, or architecture. This phase of science changes as the scientific community then analyzes the new idea to see whether it really contributes anything of value to our understanding of Nature. This later phase, Holton's public aspect of science, may take many months, years or decades before there is a general consensus among the relevant members of the scientific community regarding the value of the new idea.

When the nature of science (NOS) is taught, as suggested in *Benchmarks* and *Standards*, it should be compared to other ways of knowing and believing. Philosophy of science tends to emphasize the stable, rational, progressive, universal, consensus nature of science while history of science tends to point out the unique, personal, variable, complex, local side of science and, of course, both sides or viewpoints are correct. However, when compared to other ways of knowing or believing, modern science is by far the most progressive, stable, and rational way of knowing yet devised by humans and it is this side (modern/realist) rather than the other (postmodern/relativist) that better characterizes the enterprise of science. It is only when the natural sciences are compared to the social sciences and the arts and humanities that the differences become so apparent, and it is the differences not the similarities that are more useful in introducing the beginning student or the lay public to the nature of science. The philosophers, historians, scientists, and other academics who understand the complexity of the many facets of the enterprise of science know the difference between Holton's 'private' and 'public' science, but novices do not and should not be expected to grasp the complexities before they understand what sets science apart from other ways of knowing; that is, its essentially rational, progressive, universal nature.

In his most recent book, *Consilience: The Unity of Knowledge*, evolutionary biologist Edward Wilson argues that only when all knowledge is grounded on the firm foundation of the natural sciences will we have the best chance of coping successfully with our existence:

> The legacy of the Enlightenment is the belief that entirely on our own we can know, and in knowing, understand, and in understanding, choose wisely. That self-confidence has risen with the exponential growth of scientific knowledge, which is being woven into an increasingly full explanatory web of cause and effect. In the course of the enterprise, we have learned a great deal about ourselves as a species. We now better understand where humanity came from, and what it is. *Homo sapiens*, like the rest of life, was self-assembled. So here we are, no one having guided us to this condition, no one looking over our shoulder, out future entirely up to us. Human autonomy having thus been recognized, we should now feel more disposed to reflect on where we wish to go. (Wilson 1998, p. 297)

Wilson argues convincingly that modern/realist science, the legacy of the Enlightenment, should become the foundation of the social sciences and even ethics. Such a unified, universally-shared conception of knowledge will allow our species to make decisions that increase the chances of a better life for all species on our planet.

The vision of science that Wilson uses in *Consilience* is clearly modern/realist rather than postmodern/relativist. He understands the complexity of the enterprise of science and, even more than the authors of *Benchmarks* and *Standards*, he emphasizes its modern/realist nature. Compared to other ways of knowing and believing, the natural sciences provide, by far, the best example of rational, progressive, universal knowledge and it is these aspects that should be learned by novices. Taken to its logical conclusion, as Wilson does in *Consilience*, this view of science can become the basis of a new conception of what it means 'to know'. The many nuances and complexities become apparent as one studies the history of science, but the essentially rational, progressive, universal nature of science is the more accurate picture that science teachers should help students understand.

The broader definition of science literacy found in *Benchmarks* relates the natural sciences to mathematics, technology, and the social sciences. Edward Wilson's ideas in *Consilience* are very compatible with this broader definition of science literacy and science teachers could use his ideas to help students understand the nature of science as compared to various other ways of knowing. Many years ago French philosopher Gaston Bachelard (1934) said science would show philosophy the way. At the close of this century Edward Wilson is saying that science will show all disciplines, not just philosophy, the way toward more valid and reliable knowledge. Science educators should pay close attention to Wilson's ideas as they search for ways to help students better understand the nature of science.

References

American Association for the Advancement of Science: 1989, *Science for All Americans*, AAAS Press, Washington, DC.

American Association for the Advancement of Science: 1993, *Benchmarks for Science Literacy*, AAAS Press, Washington, DC.

Aronowitz, A.: 1988, *Science as Power: Discourse and Ideology in Modern Society*, University of Minnesota Press, Minneapolis, MN.

Bachelard, G.: 1934, *The New Scientific Spirit*, Beacon Press, Boston, MA.

Bunge, M.: 1989, *Ethics – The Good and the Right*, D. Reidel, Dordrecht.

Culler, J.: 1982, *On Deconstruction: Theory and Criticism after Structuralism*, Cornell University Press, Ithaca, NY.

Dewey, J.: 1916, *Democracy and Education*, Macmillan, New York, NY.

Feyerabend, P.: 1975, *Against Method: Outline of an Anarchistic Theory of Knowledge*, New Left Books, London.

Good, R. & Demastes, S.: 1995, 'The Diminished Role of Nature in Postmodern Views of Science and Science Education', in F. Finley et al. (eds), *Proceedings, Vol. 1, Third International History, Philosophy, and Science Teaching Conference*, Minneapolis, MN.

Good, R., Shymansky, J. & Yore, L.: 1999, 'Censorship in Science and Education', in E. Brinkley (ed), *Caught Off Guard: Teachers Rethinking Censorship and Controversy*, Allyn & Bacon, New York, NY.

Goonatilake, S.: 1984, *Aborted Discovery: Science and Creativity in the Third World*, Zed Press, London.

Gross, P. & Levitt, N.: 1994, *Higher Superstition: The Academic Left and Its Quarrels with Science*, Johns Hopkins University Press, Baltimore, MD.

Gross, P., Levitt, N. & Lewis, M. (eds): 1995, *The Flight from Science and Reason*, New York Academy of Sciences/Johns Hopkins University Press, Baltimore, MD.

Harding, S.: 1991, *Whose Science? Whose Knowledge? Thinking From Women's Lives*, Cornell University Press, Ithaca, NY.

Harding, S.: 1994, 'Is Science Multicultural? Challenges, Resources, Opportunities, Uncertainties', *Configurations* **2**, 301–330.

Hodson, D.: 1993, 'In Search of a Rationale for Multicultural Science Education', *Science Education* **77**, 685–711.

Holton, G.: 1988, *Thematic Origins of Scientific Thought: Kepler to Einstein*, Harvard University Press, Cambridge, MA.

Holton, G.: 1993, *Science and Anti-Science*, Harvard University Press, Cambridge, MA.

Jegede, O.: 1989, 'Toward a Philosophical Basis for Science Education in the 1990s: An African Viewpoint', in D. Herget (ed.), *The History and Philosophy of Science in Science Teaching*, Florida State University, Tallahassee, FL.

Koertge, N. (ed.): 1998, *A House Built on Sand: Exposing Postmodern Myths about Science*, Oxford University Press, New York, NY.

Latour, B.: 1987, *Science in Action: How to Follow Scientists and Engineers Through Society*, Harvard University Press, Cambridge, MA.

Latour, B. & Woolgar, S.: *Laboratory Life: The Social Construction of Scientific Facts*, Sage, London.

Mahner, M. & Bunge, M.: 1996,'Is Religious Education Compatible with Science Education?' *Science & Education* **5**, 101–123.

Martin, M.: 1990, *Atheism – A Philosophical Justification*, Temple University Press, Philadelphia, PA.

Matthews, M.: 1994, *Science Teaching: The Role of History and Philosophy of Science*, Routledge, New York, NY.

Matthews, M.: 2000, *Time for Science Education: How Teaching the History and Philosophy of Pendulum Motion Can Contribute to Science Literacy*, Plenum Press, New York, NY.

National Academy of Sciences: 1996, *National Science Education Standards*, NAS, Washington, DC.

Newton, R.: 1997, *The Truth of Science: Physical Theories and Reality*, Harvard University Press, Cambridge, MA.

Newton-Smith, W.: 1981, *The Rationality of Science*, Routledge, New York, NY.

Nicholson, R.: 1993, 'Postmodernism', *Science* **277**(July 9), 268.

Ogawa, M.: 1989, 'Beyond the Tacit Framework of Science and Science Education among Science Educators', *International Journal of Science Education* **11**, 247–250.

Pomeroy, D.: 1992, 'Science Across Cultures: Building Bridges Between Traditional Western and Alaskan Native Cultures', in S. Hills (ed.), *History and Philosophy of Science in Science Education*, Vol. 2, Queens University, Kingston, ON.

Rosenau, P.: 1992, *Post-Modernism and the Social Sciences*, Princeton University Press, Princeton, NJ.

Slezak, P.: 1994a, 'Sociology of Scientific Knowledge and Scientific Education: Part I', *Science & Education* **3**, 265–294.

Slezak, P.: 1994b, 'Sociology of Scientific Knowledge and Scientific Education: Part II: Laboratory Life Under the Microscope', *Science & Education* **3**, 329– 355.

Sokal, A. & Bricmont, J.: 1998, *Fashionable Nonsense: Postmodern Intellectuals' Abuse of Science*, Picador, New York, NY.

Stanley, W. & Brickhouse, N.: 1994, 'Multiculturalism, Universalism, and Science Education', *Science Education* **28**, 387–398.

Wilson, E.: 1998, *Consilience: The Unity of Knowledge*, Knopf, New York, NY.

Wolpert, L.: 1993, *The Unnatural Nature of Science*, Harvard University Press, Cambridge, MA.

Woolgar, S. (ed.): 1988, *Knowledge and Reflexivity: New Frontiers in the Sociology of Knowledge*, Sage, London.

The Epic Narrative of Intellectual Culture as a Framework for Curricular Coherence

ROBERT N. CARSON
115 Reid Hall, Montana State University, 59717, USA
E-mail: uedrc@montana.edu

Abstract. This paper describes a proposed middle school curriculum designed to coordinate the major subject areas around a single coherent story line, and to tell the epic tale of the development of formal intellectual culture from its distant origins to the present day. *Ourstory* explores the history of scientific culture from the perspective of foundational disciplines (history, philosophy, sociology, psychology, anthropology). It examines the growth of scientific culture against the backdrop of the world's traditional cultures, and balances the role of the sciences against the role of the arts in their respective contributions to the life of the mind.

1. Introduction

For the past three years I have addressed an array of pedagogical, philosophical, and psychological topics in my educational psychology courses by arranging them around a hypothetical curriculum design project. This curriculum, called *Ourstory*, was designed as a heuristic device, a way of keeping a number of pedagogical and disciplinary conversations clustered and related to one another as we try to envision a curriculum that overcomes the reductionism, the fragmentation, and the aesthetic and conceptual sterility of a typical school curriculum. *Ourstory* recognizes the virtues of a multi-cultural perspective and a postmodern social ethos, but it is built upon a conceptual framework that borrows distinct features from an older classical liberal tradition.

Few people today mourn the passing of the classical curriculum, based as it was upon mastery of the Latin and Greek languages and cultures, but there is a sense in which that old tradition of schooling enjoyed some significant advantages that disappeared along with it. History served a central role, providing an organizational framework for all of the knowledge, events, information, and ideas contained in that curriculum, providing a mechanism of coherence and order, and inducing much of the study to take the form of a story, built around human perspectives.

In the following scenario I will argue for the use of history in the teaching of science, but I will argue for a different use of history than has been customary among advocates of the history and philosophy of science (HPS) community. I will enter a plea that we stop thinking only in the limited framework of science

F. Bevilacqua et al. (eds.), Science Education and Culture, 67–82.
© 2001 *Kluwer Academic Publishers. Printed in the Netherlands.*

education and try to recognize that the whole of scientific culture forms an indivisible mass that must be taught altogether if it is to make sense to students as a culture, as a world view, and as a way of life. Turning our backs on the humanities and the arts and adopting a posture of contempt toward traditional cultures, while insisting on greater progress in science education, bespeaks a kind of parochialism that diminishes us all.

2. *Ourstory* - Structuring Education for Meaning

In a paper delivered to the HPS group in Calgary, June 1997, I raised the suggestion that we broaden the scope of our concerns beyond the teaching of science *per se* and begin to consider how best to teach the whole of scientific culture (Carson, 1997a). Science did not develop independently of the other formal disciplines, nor should the teaching of science be considered independently from the teaching of the arts, literature, history, and so on. Knowing how the main branches of intellectual culture interact, and knowing how prominent ideas obtain different modes of expression throughout the various disciplines is, in a word, necessary. The compartmentalization of culture into isolated disciplines and subdisciplines, which takes place routinely in institutions of formal education, produces a peculiar misrepresentation of the complex and dynamic relationships between the arts and sciences, and between humankind and its various cultural systems.

My suggestion is that we consider ways, early in the students' schooling, to help them understand how mathematics, science, literature, history, art, social sciences, technology, and so forth, emerged from the ferment of humanity's long social and cultural struggles. *Ourstory* uses as a model the notion of an epic tale. It recognizes the power of these narratives to bring together an audience, to create a shared experience, to connect individuals with their social histories, and to embed massive amounts of information into an easily accessible, memorable, and enjoyable form.

Consider a simple experiment. Take a thousand page novel. Identify every idea, description, event, and piece of information contained in it. Write each of these on a separate index card. Then shuffle the stack of cards (several tens of thousands, no doubt), and see how easy it is for someone unfamiliar with the original story to remember or make sense of all that information. Disconnected, out of sequence, and unrelated to one another, each particle of the story becomes an isolated learning task, much like the content of the typical school curriculum. Gestalt psychologists established over fifty years ago that the mind is among other things a pattern-seeking and a pattern-making mechanism (see Hunt, 1993, for a good popular history of these developments). Since then, cognitive psychologists (e.g., Ausubel, 1963, 1968; Novak, 1998) have demonstrated in a variety of ways that knowledge is more easily understood, learned, and remembered when it is situated within meaningful, organizing frameworks. Concept mapping, advance organizers, narrative knowledge structure, and other devices are all techniques designed to relate pieces of information to one another within a larger structure.

The plan of *Ourstory* is to combine the temporal frame of history, the spatial frame of world geography, and the conceptual frame of philosophy to tell the story of the world's major cultural developments. In this plan, history would serve as the main integrative framework. Told as a series of cultural episodes, this story attempts to move the learner through each historical moment of change, whether a discovery, an innovation, or establishment of a new cultural convention. Something becomes 'meaningful' because it is connected to other things the learner already knows (Novak, 1998). If we are seeking authentic, integrating frameworks, we must look to the nature of knowledge, culture, and human activity as they exist in the world 'out there'.

Science has never grown or thrived in isolation from the arts: 'Bluntly stated, the goal *per se* is not to teach science. The goal is to teach scientific culture. All of it. Science is one of the definitive branches. Wise policy will serve the entire culture, and all of its parts. We cannot be indifferent to the whole of our intellectual culture' (Carson, 1997a). When we teach science in isolation from the larger social, cultural, historical and philosophical contexts within which its growth has been hosted and nourished, it becomes unnecessarily cryptic. We lose sight of why knowledge is framed the way it is, and why it gets represented as it does. Often, there is a story behind the conventions that seem otherwise so peculiar. Students have trouble seeing it as a human activity, thus they have trouble seeing themselves as scientists, or being sympathetic to the ways in which scientists investigate phenomena and crystalize their resultant knowledge. There is a whole, gestalt-like, intuitive feel for the nature of the discipline that eventually 'clicks' with those who finally succeed at it. Sometimes that feel is there early on, in which case the learner never does understand why others have so much trouble with it. For some it develops after struggling with enough information. Most never get past the desperate strategies of rote learning and uncomprehending reliance on algorithms. They never get the pieces into an accurate structural alignment, nor understand clearly which aspects are empirical in nature, and which derive from convention and from human imagination. If the learner could go back to the beginning and see how the discipline evolved in the first place, how the knowledge was uncovered, organized, formalized, and shared, then a better intuitive feel for the nature of the enterprise would be possible.

The 'intellectual fragmentation' Matthews (1994) laments is not just within disciplines. It is, importantly, between disciplines as well. The project this article describes is an attempt to find the modern equivalent of an integrative liberal education, centered more fully than traditional liberal education around the historiography of science and technology, but mindful nevertheless of the crucial roles played by the arts and the humanities. When we stand back and look at those historic epochs in which the global project of science was advancing vigorously (Greek classical civilization, the Enlightenment, and the age in which we currently live, for example) we quickly recognize that the other major disciplines were also expanding, changing, and contributing to that progress. Synoptic histories of culture, such as Janik and Toulmin's (1973) marvelous account of 1920s Vienna, demonstrate the point well.

As currently envisioned, *Ourstory* would serve to orchestrate the whole three year curriculum at the middle school level. (The term 'middle school' in the U.S. refers to junior high schools that have been reconfigured to create more intimate learning communities and to shift the pedagogical strategies toward social processes and constructivist learning orientations). Students enter middle school around the fifth or sixth grade (at approximately eleven or twelve years of age) just as an adult-like consciousness is beginning to emerge. Most stage theorists recognize that this is an age in which maturation takes a profound step, physically and emotionally, as well as morally and intellectually. Learners become capable of addressing relatively complex networks of ideas and topics, but they are just beginning to gain competence at formal abstract operations. They have a keen interest in human stories, personal dramas, the complexities of human life. They hunger for philosophical insights, drama, the intrigue of ethical dilemmas, exposure to the world's wealth of poetic beauty, wisdom, experience, and romantic engagement. Most are not ready for the austere precision of a formal discipline.

A curriculum framework like *Ourstory* would suit the middle school level well in part because this is the first age group capable of receiving it. In pilot studies we found middle school students to be highly receptive to the use of narrative histories of pivotal cultural events. And as these histories formed a sequence, students quickly made the necessary connections. If they can enter into the 'problem space' of a cultural advance and understand what the original problems and conditions were, they can do a creditable job of seeing a range of possible approaches and solutions, and that means they are also capable of understanding at some level the solution humankind generated under those conditions. At this age level, the history does not have to be precise, though of course it should be accurate.

There is another reason for locating *Ourstory* at the middle school level. In terms of realpolitik in the educational community, many middle school faculty and administrators are already determined to create an educational experience that is thematic, interdisciplinary, generalizing rather than overly specific, and based upon social processes and dialogue. They are more likely to consider a historically based model such as *Ourstory*. High schools, by contrast, are institutionally more rigid, and more inclined to model themselves after colleges. They focus on the content of the disciplines, and they teach each subject in relative isolation from one another. Coordination of any kind across disciplinary boundaries is notoriously difficult, as it is in universities.

Middle schools typically assign several cohorts of approximately twenty five students to a team of three or four teachers, who then rotate these cohorts among themselves throughout the day. One of these teachers (probably the history/social studies teacher) would take primary responsibility for teaching the main story line that *Ourstory* is framed around. That teacher would actually conduct the first lessons in the sequence that would provide the conceptual ramp into the associated topics in mathematics, or art, or science, and the story line would then be picked up by those teachers as each line of discussion condensed into those particular areas of specialization. The social studies teacher might begin to portray the life of Thales, his immigration to Egypt, the conversations he supposedly had with geometers there. She might even provide the first lesson or two in the sequence as the mathematical conversation begins to yield the beginnings of classical geometry, or she may co-teach a few of these lessons with the mathematics teacher. When the mathematics teacher then takes over the mathematical part of the story, the social studies teacher would return to the main story line, which in turn would begin to produce additional leads out into literature, into art and architecture, into science, and so forth. These leads would be picked up and developed in those other classes. From the students' perspective, this would be a sequential voyage through the main developmental moments of civilizations, from ancient to modern, built up out of re-creations of the most culturally significant events.

All of these teachers, including the history/social sciences teacher, would still teach the usual material that is taught without *Ourstory. Ourstory* is not being conceived of as a whole curriculum, but rather as a curricular framework with just enough added material to produce this central story line and to structure a meta-discourse on the nature of knowledge and cultures. It would require about two or three hours out of each week. While this story line itself begins at the end of the last ice age and gradually makes its way to the present, it does not require the entire curriculum to dwell in the past. Nothing is covered in *Ourstory* that does not have significant implications for the present. And in all cases, the purpose is to explain the way things are today, by means of their antecedents. In the case of science education, contemporary topics would still be the main venue, but students would also experience re-creations of the cultural commitments, the main discoveries, and the evolution of investigative techniques that account for the transformation of natural philosophy and metaphysics into modern science over the course of twenty five centuries. They would visit with the pre-Socratics, who first began to outline the logical possibilities of natural philosophy. They would visit with the mathematicians who developed rational thought and who contemplated the logical structure of the physical world. They would enter into the presence of Aristotle, the great collector of all things human and natural, who organized and catalogued thousands of objects into orderly taxonomies. Along subsequent travels through history they would meet up with Archimedes, Galileo, Newton, Bacon, and others. It is not just the specific scientific discoveries that need to be learned. Perhaps more importantly, it is the grappling with investigative strategies and other procedural matters, even social matters. Robert Boyle's address to the Royal Society (1661) contains more than the fruitful suggestion to view as elements any substance that cannot be further reduced. It also contains a blueprint for the social protocols for contesting ideas, theories and points of view without rancor or personal invective. It explains why scientists insist that claims be presented in a manner that others can reproduce.

3. Practical Considerations

The resources for this curriculum project could be web based. They could be organized using a simple grid (see Figure 1). The horizontal axis along the top identifies conventional historical epochs of the kind Van Doren (1991), the Durants (1935/1975), and others customarily refer to. The vertical axis lists the cultural systems that constitute the main disciplinary venues in schools. This grid could be used as the main index for organizing a collection of web-based resources with extensive links to other sites. In order to ensure coherence a minimal story line would be obligatory, but this rich collection of resources could be selected from at the discretion of the teachers to determine how far to go into secondary and tertiary topics.

Historical epochs are addressed in chronological order, and the developments during each epoch in each of the disciplinary categories are examined in relation to one another. This approach constitutes an interdisciplinary, multi-cultural, multimedia approach to the study of mathematics, science, art, architecture, music, history, geography, natural language, literature, and other formal disciplines. It examines each discipline from a foundational perspective, providing a sense of coherence by exploring the social, philosophical, historical, and cultural dimensions of the development of these various disciplines within the context of the world's evolving scientific culture. These historic developments are seen in relief against the broader picture of the world's traditional cultures. The history of the relationship between traditional cultures and scientific culture is also explored.

The title of this project is meant to serve as a gentle reminder that the rise of science, and its influence on all the other formal disciplines and upon all traditional cultures, while a complex and often troubled story, belongs to all of humankind. The advent of scientific culture has drawn its inspiration from numerous cultures and at the same time has had a profound influence on every society on earth. While the project of science gained significant advances in Europe, it did not originate exclusively in Europe, and in the twentieth century its modern impetus moved beyond the borders of Europe to become a truly global phenomenon. Its development is tied up with the painful history of colonialism and other sorrows. It cannot be presented merely as 'subject matter' in schools while ignoring its deep historical and cultural significance. It is our commitment to tell this story with as much integrity and intellectual grace as possible.

The title *Ourstory* also serves as a reminder that, in all societies, the first obligation of education has always been to present 'our' epic tale, to tell the story of who we are, where we came from, what we as a people have come to believe, and so on. Use of the story-form, where appropriate, serves to restore a much needed coherence (Egan, 1986) and to address the adolescent's need for rich human perspectives.

But who are 'we'? In the late twentieth century, the possibility of a single grand narrative broke down (Lyotard, 1984). The post-modern condition is often

Ourstory

A History of Formal Intellectual Culture as Seen Against the Backdrop of the World's Cultural Traditions

Historical Epochs / School Subjects	Early Cultures & Societies. (12^{th}m. - 8^{th}c. BC) *The Agricultural Revolution*	Classical Civilizations Greece (600-320BC). *Literacy, Reason & Democracy*	The European Middle Ages (410AD-1300) *Christianity and Islam*	The Renaissance (1300'S - 1600) *Rebirth of Classical Learning*	The Enlightenment (1660-1800) *Science and Romanticism*	Modernity (1800-1920) *Technology. Imperialism*	20th & 21st Centuries *Post-Modernism. Cultural Pluralism*
Social Studies, History & Geography	A1	A2	A3	A4	A5	A6	A7
Music, Art & Architecture	B1	B2	B3	B4	B5	B6	B7
Mathematics & Logic	C1	C2	C3	C4	C5	C6	C7
Science & Technology	D1	D2	D3	D4	D5	D6	D7
Literature & Language	E1	E2	E3	E4	E5	E6	E7
Traditional Cultures	F1 *Egypt, Africa Mesopotamia*	F2 *China, India, Israel, Rome.*	F3 *Europe; Arabic civilization*	F4 *American Indians; Inca, Aztec, Mayan*	F5 *Africa Revisited: Tribal structures vs Colonialism*	F6 *Clash of cultures: Traditional vs Scientific. Asia revisited.*	*Special topic: Philosophy of Language, Culture & Mind*

Figure 1.

described now as a decentralized mosaic of localized discourses with no central account even possible. Any narrative is political, any curriculum an indoctrination. Any single narrative is a chimera. Our story is a narrative of many voices. Like the old travelers' tales of the late middle ages, the only thing we have in common is the fact that all of us are on a similar journey. It is the form of the story, more than the specific events of each narrative, that forms the common bond.

Does *Ourstory* privilege Western civilization? Inevitably, perhaps, yes. But by their very nature schools do so anyway and in a far more insidious manner than would *Ourstory*. Schools perform many functions, worldwide, but in the end they are designed to teach those complicated formalized disciplines that cannot be learned by more natural modes of cultural apprenticeship. Schools are artificial environments developed for specific purposes, and those purposes generally take us into the cultural contributions associated with western civilization. Having said that, though, most critics of this project have been sympathetic to the argument that it is more honest to frame the discussion in terms of 'cultural systems', which every people on earth can lay claim to in one form or another, than it is to ignore the world's great wealth of cultural systems and to engage in an uncritical indoctrination into school subjects, and thus scientism. While *Ourstory* is attempting to focus on those specific developments that led to modern scientific culture, it certainly does not preclude teaching parallel developments in other cultures, or adapting the content to the cultural backgrounds of the students in any given educational setting. *Ourstory* is intended to dignify human ingenuity and variability in all its richness, but it also recognizes that the emancipatory function of education depends in large part on procuring for all students mastery of those domains of learning that are generally recognized as undergirding the scientific and technological culture that now pervades the earth.

As traditional cultures reassert their legitimacy, and as their members figure out how to negotiate co-residency in both a traditional and scientific culture, a new ideal of the educated individual will likely emerge. An educated and worldly person will be one who is comfortably situated within an ancient cultural tradition as well as competent in those domains of learning that will constitute a world wide scientific culture. One does not have to give up Judaism, Catholicism, or allegiance to Lakota culture to be a physicist. One may wish to recognize though that the austere logic and materialism of positive science simply cannot fulfill the human needs that gave rise to cultural traditions in the first place. Those traditions are ubiquitous for a reason. Scientific culture is a conceptual tool kit, but it is not a spiritual culture. Those who reject traditional cultures and attach their allegiance solely to a scientific worldview often make science over into a quasi- tradition, called scientism, and they risk becoming just as dogmatic as any tribal member toward his or her ancient ways of knowing.

The notion of cultural systems, like the notion of political economy, is deliberately broad and inclusive. It is a way of legitimating traditional cultures in the same way that we legitimate formal intellectual disciplines, as ways of knowing that satisfy human needs and desires. Each formal discipline taught in the schools is regarded in *Ourstory* as a cultural system. So too, each traditional culture may be regarded as a cultural system. If one can immerse oneself in it as a way of knowing, then it may be seen as a cultural system. 'The concept of culture', says Geertz, 'is essentially a semiotic one. Believing, with Max Weber, that man is an animal suspended in webs of significance he himself has spun, I take culture to be those webs ...' (1973, p. 5).

Although academic disciplines, like traditional cultures, tend to interact, they also tend to retain distinct identities. And, importantly, they tend to be incommensurable, one with another (Carson, 1997b). The conclusions of literature or of art simply are not of the same conceptual coinage as the conclusions of physics, any more than the conclusions of science can be reconciled with the world view of an America Indian culture. Different systems have different modes of investigation, different subject matters, different underlying assumptions, different standards of validity, different goals and purposes. They constitute different world views. All are sustained within different symbolic systems which, in turn, enable different views of reality. Whether ambiguity is a fringe phenomenon or the very essence of reality or of perception is unresolved. Ourstory takes a modest, 'trivial' position on the matter by simply recognizing that there are different, incommensurable cultural systems, which schools are expected to teach and which human beings can expect to encounter.

The notion of a curriculum based upon the exploration of cultural systems is not unique. Similar perspectives arise in the work of Aikenhead (1992, 1996a,b). He too identifies science as a cultural enterprise, a position that is not without controversy, especially among those who see science as the victor in a cultural-evolutionary struggle for superiority over tradition-based cultures, or who privilege it because of its putative universality. He recognizes that people participate in numerous cultures and subcultures, groups that share coherent yet distinct world views and perspectives, and that those cultures satisfy real needs. The passage into another culture is referred to by Aikenhead (1996a,b) as 'border crossing'. The difficulty of the passage into science depends upon the learner's existing cultural background and the degree to which the learner considers mastery of science necessary to future plans. He uses the five categories established by Costa (1995) to propose variations on the strategy of teaching, and even ponders for a moment whether five different curricula might be needed. Learners described as 'Potential Scientists' and 'Other Smart Kids' make the transition into scientific culture far more readily than those defined variously as '*I-Don't-Know* Students', 'Outsiders', or 'Inside Outers' (those who want to learn but are kept outside by prejudice or other institutional barriers). Using work from studies that have considered the problems faced by non-western students crossing the cultural borders into western science including Jegede (1994, 1995) and Jegede and Okebukola (1990, 1991), Aikenhead recognizes that even students of European heritage suffer similar kinds of disjunction if their own sociocultural backgrounds and aspirations do not happen to align with the world view characterized in science. In a private letter comparing our respective points of view before the current project was developed, Aikenhead warned me that '... the relevance agenda defined by students often interferes with our rational agenda for teaching science in interesting ways' (1997). Trying to create access to scientific culture by a series of historical narratives may not work if students cannot identify somehow with the people depicted in those stories, or never develop empathy for the various problem spaces those individuals found so fascinating, or simply do not care. Clearly, much work remains ahead.

4. Units and Lessons

Let us now consider how *Ourstory* would organize the episodes of cultural development it is based upon. A sparse, central narrative could be provided students either in written form or taught less formally by one of the teachers on the middle school team, in the tradition of oral story telling. That teacher would coordinate the related modules taught by other teachers so that coherence and continuity are maintained. The telling of the story, spread out over a three year period, handled by different teams of teachers, would occur in relatively concise sessions, which would serve as advance organizers, perhaps for a week's work. The student's engagement in this story would then consist of scripted exercises that would take the student working in small groups into the problem space of an event that was represented in that central story line.

The narrative on Thales for example could be followed by an experience using ropes and wooden stakes in which basic geometrical formulations are represented. Problems presented in concrete form, as the Egyptians knew them originally, become problems seen with the mind's eye. They are drawn on paper, and then these representations are taken to be representations not of ropes and wooden stakes but of lines and points. Theoretical entities thus emerge from these activities, and we begin to face the same ontological and epistemic questions that led Thales, Pythagoras and others to develop geometry as an abstract science, and led Plato to contemplate the ontological status of pure ideas.

Resources for teaching these episodes could be catalogued into a web-based collection using the grid described earlier. Each cell in the grid represents a discipline, or a cluster of associated disciplines, as it lines up under the heading of a particular historical epoch. Clicking on a cell brings the reader to that respective historical epoch and discipline. The resources contained within that section are limited and carefully selected. We are not pretending to provide a comprehensive history of the particular epoch or discipline. Rather, we seek to identify the most significant events which promoted the growth of those disciplines toward their current state, and those events that spilled out from their incubation zone in one discipline to affect the general course of intellectual culture. The selection process and orchestration of a story line involves careful concept mapping (Novak, 1998), for it is the connectedness of significant cultural developments that establishes much of their meaningfulness. It also involves an understanding of the role cultural tools play in the life of the mind, how specific developmental advances articulate from each new set of instruments developed for the mind, and how learning involves a re-capitulation of these historical developments in the development of the individual, as Vygotsky recognized. (For a discussion of these points, see Scribner, 1995).

One example would be in cell E1 (Literature & Language; Early Cultures & Societies) where we would find a unit on the advent of writing, the development of early alphabets, and the first democratization of literacy among the Hebrews and the Greeks (Jean, 1994). The advent of an easily mastered writing system and its diffusion into a whole society alters that society profoundly.

A second example, in cell C3 (Mathematics & Logic; The European Middle Ages), occurs when the combined use of place value and base ten spread from their incubation zone in India through the Arabic lands and into Europe. The system we use today makes use of Arabic numerals and symbols for zero, for addition and subtraction, multiplication and division, and during the nineteenth century was added a symbol to replace the words 'is equal to'. Our unit on these events would explore the greater efficiency, the additional capabilities, the aesthetic beauty and simplicity, the conceptual empowerment, and the historical significance of these brilliant contributions to intellectual culture. Nothing learned early by children should be taken for granted forever, but should be revisited when appreciation becomes possible for the learner.

A third example might be found in B7 (Music, Art & Architecture; Post-modernism) when innovators in all three of these artistic disciplines break entirely free of neo-classical conceptions of art and beauty, and create works which challenge the very definitions that these disciplines have taken as axiomatic since classical times (cf. Stangos, 1994, especially pp. 6; 110–134; 256–290). The notion that there is an eternal, permanent standard of beauty in the universe produces a profoundly different consciousness (and suggests an entirely different approach to education) from the notion that beauty is purely an individual preference and prerogative. Does this shift in the theory of aesthetics constitute a great liberation from the constraints of convention, or a demolition of timeless values? Students would have the opportunity to see how philosophically divergent viewpoints play out in the theatre of formal (or is it post-formal) art. In so doing, they will be challenged to think more deeply about what art is, and what role it plays in any society. Because all of the disciplines are being examined by historical epochs, they would also see that in an age when physicists are recognizing that there are 'no privileged frames of reference' in time and space, a similar notion has invaded the realms of ethics, aesthetics, literature, historiography, and so on. Powerful ideas define entire cultural epochs.

A fourth example might be found in D5 (Science & Technology; The Enlightenment) when Lavoisier and his associates created the new nomenclature for elements and compounds, a change not only in the language, but in the conceptualization of matter. Lavoisier's original introduction to *The Elements of Chemistry*

(1965/1789, pp. xiii–xxxvii) draws the reader into reflection on the relationship between language and thought, a topic still of keen interest two centuries later.

Pivotal events like these are connected to larger historical and cultural trends, always. While the student learns in science class about the contributions Lavoisier made to the origins of chemical science, she will also learn in her history class about the Revolution that cost him his life, and she will see how the shift in power from social elites to ordinary people found expression in the arts, as in the transition Mozart and others made from classical to Romantic music. In such a context, the art itself begins to make more sense. So do the styles in which it is created, and the modes of thought it represents.

When students study the Enlightenment in *Ourstory*, they would examine contemporary developments taking place in music, in art, in literature, in the political discourse, as well as in science, technology, and mathematics, and they would look for thematic connections and that metaphorical resonance of ideas across disciplinary boundaries that tends to occur in any culturally robust epoch. In this context, the students immersed in the culture of the Enlightenment would encounter the beginnings of analytical geometry, chemistry, physics, classical and romantic music, neo-classical architecture, the political essays underpinning liberal democracy, the beginnings of the romantic protest against science, the first machines, laissez-faire capitalism and its discontents, and so forth.

Ourstory requires significant collaboration by teachers. But it repays the effort by creating a coherent discourse for teachers and students alike. And it creates additional perspectives on those disciplines that may otherwise have lapsed into sterile entombment as 'school subjects'. '*History* is not a distinctive subject-matter to be inquired into. It is rather at once a trait of all subject-matters, something to be discovered and understood about each of them; and a distinctive way of inquiring into any subject matter' (Randall (1962), quoted in Scribner, 1995).

5. The Curriculum as Epic Tale

Ourstory moves history from the fringes of the curriculum to the very center, constrains it in this use to the history of formal intellectual culture, and then arranges the approach to all of the other disciplines as branches off from this main trunk. The primary focus would be on events that had lasting significance for the mental landscape we now inhabit as participants in modern scientific culture. Yet it recognizes a world made up of many cultures and a standard of liberal education in which mastery of different cultural systems is the key.

Students often complain that the subject matter they are taught in school is irrelevant and disconnected, that they are unable to see why these various subjects need to be learned. Why study mathematics, or science, or art, or history, or literature? There is no mechanism in the present curriculum for addressing these questions, other than a rather crass and superficial examination of how skill in math or science can lead to employment opportunities, higher salaries, and more commodities. Teachers who try to provide a deeper response quickly realize that the explanation needed is too extensive to produce *ex tempore*. It needs to be built into the entire curriculum.

This is a human story. It is about us. All of us. It tells us about how humans have responded to various challenges, and about the consequences of their various discoveries, innovations, and decisions. The full account of this story, even in the most telegraphic outline form, takes time to tell. It also takes time to construct the explanation that formalized intellectual disciplines, languages, and cultures, are the very stuff that mind is made of. They are the matrices in which formal cognition is manifested. Without the language and the cultural icons and the disciplined ways of thinking and seeing, our cognition reverts to unreflective awareness. But with these disciplines, we gain control of our minds, we extend the range of ideas we are able to entertain, and we deepen our understanding of the world around us. New languages, new semiotic systems, new concepts, formulas, theories, ideas, or works of art are the substrate through which the mind gains extension (Hirst, 1973). We cannot expect students to be motivated to learn unless they have some deep prescience of the benefits that will obtain from such demanding work.

As we mentioned at the beginning of this article, one advantage classical liberal education enjoyed over our present set of specialized and disconnected offerings was a kind of historically based coherence. Mathematics, philosophy, ethics, literature, art and other disciplines resided within a kind of storyline generated by the history of two civilizations that had completed their life cycles long ago. It is that kind of coherence, adopted to modern conditions, that this project seeks to emulate.

We have located *Ourstory* as early as possible in the schooling process, in part because we doubt if high schools or universities can (or even should) attempt a similar approach. This project is an attempt to shift the framework and foundation of a liberal arts and sciences education down to the middle school level. If done successfully, it should become easier for high schools and universities to engage in the more specialized study characteristic of these institutions without students feeling the kind of disconnection that comes from studying an abstract discipline out of context and without adequate background. Typically, we do not lose the student halfway through the year; we lose her in the first few weeks, such that she never feels at home within the symbols and processes of the conceptual game, be it calculus, physics, or history. Understanding how a discipline began, how it evolved, and how its early pioneers came to cherish it is part of the human interface that helps to personalize the entry into one of these formalized cultural disciplines. If we can relive those moments, then we should also be able to acquire the excitement and interest that attended them.

The approach is not without its legitimate cautions and criticisms. It does entail, almost of necessity, a rather superficial treatment of the historical dimensions of any of the subject areas, including science. Matthews (1992) points out the typical pitfalls of lacing the teaching of science with quasi-history and pseudo-history to spice it up or to enhance interest. Such history tends to be bent to the pedagogical intent. It tends to be superficial. Our approach does not pretend to provide a comprehensive history of each discipline, or an exhaustive treatment of specific historical events. It seeks to provide students with an organizing framework that allows them to see formal disciplines as products of human activity and human society. Major advances in one domain of human learning will tend to produce effects in other domains. Not only do mathematics, science and technology interact, but advances in these disciplines have tended to produce new challenges, new purposes, and new ideas in arts and letters as well. Being able to see them as connected helps students acquire the conviction that a fuller understanding of the world will require some level of proficiency in each of the major domains of learning.

Finally, it should be stated that this curriculum is not intended to substitute the history of a subject for the subject itself. It does not replace physics with physics-for-poets. Emphatically, it is designed to produce enough interest and a clear enough initial orientation in each of these human enterprises that students will be attracted to them and want to participate in the benefits these disciplines historically have bestowed upon humankind. The intent is to produce support for a rigorous, demanding curriculum.

6. Summary

Every traditional society on earth has its epic tale. These complex narratives answer to fundamental human needs which have become generally ignored in the specialization and compartmentalization of our own advancing intellectual culture. Cognitive science suggests a mind very different from the one behaviorists subscribed to, a mechanism that seeks and makes patterns, that copes with detail by relating it to larger organizing structures, and that sees the external world through the lens of personal frameworks (Caine & Caine, 1991). The combined use of narrative knowledge structures and history as organizational schemata is beginning to look more respectable than at any time since the collapse of classical liberal education a century ago. We learn the various disciplines in school because they empower the human mind. This is nowhere seen more dramatically than in the historical record where the collective empowerment of humankind is writ large. Efficient new instruments for the human intellect contribute to the cycle of development, enabling new cultural expressions, which in turn empower the mind with additional bases of thought, hence Vygotsky's views about the co-evolution of mind and culture (Vygotsky, 1978; Wertsch, 1985, 1995). A cultural education does in a sense recapitulate this cycle of development within the individual (Egan, 1997, pp. 26–32). Narrative knowledge structures are present in the epic tale of traditional cultures; they aid memory and comprehension. They were present in the classical curriculum several generations ago for the same reason, and they should contribute to the curriculum of today.

Speaking at the Harvard tercentenary celebration early in the twentieth century, president James B. Conant said: 'The older educational discipline, whether we like it or not, was disrupted before any of us were born. It was based on the study of the classics and mathematics; it provided a common background which steadied the thinking of all educated men. We can not bring back this system even if we would, but we must find its modern equivalent' (McCord, 1936, p. 213). In this project, we are seeking a modest step in that direction.

References

Aikenhead, G.S.: 1992, 'How to Teach the Epistemology and Sociology of Science in a Historical Context', in G.L.C. Hills (ed.), *The History and Philosophy of Science in Science Education*, Vol. 1, Queen's University, Kingston, Ontario, pp. 22–34.

Aikenhead, G.S.: 1996a, 'Science Education: Border Crossing into the Subculture of Science', *Studies in Science Education* **27**, 1–52.

Aikenhead, G.S.: 1996b, 'Border Crossing: Culture, School Science, Assimilation of Students', in D.A. Roberts & L. Ostman (eds.), *The Multiple Meanings of a School Subject: Essays on Science and the School Curriculum*, Teachers College Press, New York.

Aikenhead, G.S.: 1997, Personal correspondence dated July 13, 1997.

Ausubel, D.P.: 1963, *The Psychology of Meaningful Verbal Learning*, Grune & Stratton, New York.

Ausubel, D.P.: 1968, *Educational Psychology: A Cognitive View*, Holt, Rinehart & Winston, New York.

Boyle, R.: 1911/1661, *The Sceptical Chymist*, Everyman's Library, Dent, London.

Caine, R. & Caine, G.: 1991, *Making Connections – Teaching and the Human Brain*, Association for Supervision and Curriculum Development, Alexandria, VA

Carson, R.: 1997a, 'Why Science Education Alone is Not Enough', *Interchange* **28**(2&3), 109–120.

Carson, R.: 1997b, 'Science and the Ideals of Liberal Education', *Science & Education* **6**, 225–238.

Costa, V.B.: 1995, 'When Science is "Another World": Relationships between Worlds of Family, Friends, School, and Science', *Science Education* **79**(3), 313–333.

Durant, W. & Durant, A.: 1935/1975, *The Story of Civilization* (11 volumes published between 1935 and 1975), Simon & Schuster, New York.

Egan, K.: 1986, *Teaching as Story Telling – An Alternative Approach to Teaching and Curriculum in the Elementary School*, University of Chicago Press, Chicago.

Egan, K.: 1997, *The Educated Mind – How Cognitive Tools Shape Our Understanding*, University of Chicago Press, Chicago.

Geertz, C.: 1973, *The Interpretation of Cultures*, Basic Books, New York.

Hirst, P.H.: 1973, 'Liberal Education and the Nature of Knowledge', in R.S. Peters (ed.), *The Philosophy of Education*, Oxford University Press, Oxford, pp. 87–111.

Hunt, M.: 1993, *The Story of Psychology*, Anchor, New York, pp. 280–310, 435–477, 511–558.

Janik, A. & Toulmin, S.: 1973, *Wittgenstein's Vienna*, Touchstone, New York.

Jean, G.: 1994, *Writing – The Story of Alphabets and Scripts*, Thames and Hudson, London.

Jegede, O.: 1994, 'African Cultural Perspective and the Teaching of Science', in J. Solomon & G. Aikenhead (eds.), *STS Education: International Perspectives on Reform*, Teachers College Press, New York, pp. 120–130.

Jegede, O.: 1995, 'Collateral Learning and the Eco-Cultural Paradigm in Science and Mathematics Education in Africa', *Studies in Science Education* **25**, 97–137.

Jegede, O. & Okebukola, P.A.: 1990, 'The Relationship between African Traditional Cosmology and Students' Acquisition of a Science Process Skill', *International Journal of Science Education* **12**(1), 37–47.

Jegede, O. & Okebukola, P.A.: 1991, 'The Effect of Instruction on Socio-cultural Beliefs Hindering the Learning of Science', *Journal of Research in Science Teaching* **28**(3), 275–285.

Lavoisier, A.: 1965/1789, *The Elements of Chemistry*, Dover, New York.

Lyotard, J.-F.: 1984, *The Post-Modern Condition: A Report on Knowledge* (Bennington & Massumi, trans.), University of Minnesota Press, Minneapolis.

Matthews, M.: 1992, 'History, Philosophy, and Science Teaching: The Present Rapprochement', *Science & Education* 1, 11–47.

Matthews, M.: 1994, *Science Teaching*, Routledge, New York.

McCord, D.: 1936, *Notes on the Harvard Tercentenary*, Harvard University Press, Cambridge.

Novak, J.D.: 1998, *Learning, Creating, and Using Knowledge: Concept Maps as Facilitative Tools in Schools and Corporations*, Lawrence Erlbaum Associates, London.

Randall, Jr., J.H.: 1962, *Nature and Historical Experience: Essays in Naturalism and in the Theory of History*, Columbia University Press, New York.

Scribner, S.: 1995, 'Vygotsky's Uses of History', in J.V. Wertsch (ed.), *Culture, Communication and Cognition: Vygotskian Perspectives*, Cambridge University Press, Cambridge, pp. 119-145.

Stangos, N. (ed.): 1994, *Concepts of Modern Art: From Fauvism to Postmodernism*, Thames and Hudson, London.

Van Doren, C.: 1991, *A History of Knowledge – The Pivotal Events, People, and Achievements of World History*, Ballantine, New York.

Vygotsky, L.S.: 1978, in M. Cole, V. John-Steiner, S. Scribner & E. Souberman (eds.), *Mind in Society: The Development of Higher Psychological Processes*, Harvard University Press, Cambridge, MA.

Wertsch, J.W.: 1985, *Vygotsky and the Social Formation of Mind*, Harvard University Press, Cambridge, MA.

Wertsch, J.W. (ed.): 1995, *Culture, Communication, and Cognition: Vygotskian Perspectives*, Cambridge University Press, Cambridge.

History, Philosophy and Sociology of Science in Science Education: Results from the Third International Mathematics and Science Study

HSINGCHI A. WANG
US TIMSS, College of Education, Michigan State University, East Lansing, MI 48824, USA; E-mail: wanghs@pilot.msu.edu

WILLIAM H. SCHMIDT
US TIMSS, College of Education, Michigan State University, East Lansing, MI 48824, USA; E-mail: bschmidt@pilot.msu.edu

Abstract. Throughout the history of enhancing the public scientific literacy, researchers have postulated that since every citizen is expected to have informal opinions on the relationships among government, education, and issues of scientific research and development, it is imperative that appreciation of the past complexities of science and society and the nature of scientific knowledge be a part of the education of both scientists and non-scientists. HPSS inclusion has been found to be an effective way to reach the goal of enhancing science literacy for all citizens. Although reports stated that HPSS inclusion is not a new educational practice in other part of the world, nevertheless, no large scale study has ever been attempted to report the HPSS educational conditions around the world. This study utilizes the rich data collected by TIMSS to unveil the current conditions of HPSS in the science education of about forty TIMSS countries. Based on the analysis results, recommendations to science educators of the world are provided.

1. Introduction

The mission of science education has been to prepare individuals who would develop a certain level of scientific understanding after their formal education in school. These scientifically literate individuals would be capable of applying their knowledge and skills acquired in science, whenever personal or socially relevant issues demanded such understanding. For instance, by having an understanding of science contents such as Physiology, Biology, and Chemistry, scientifically literate individuals would be able to use reason to form their opinions and draw their conclusions about such health-related issues as nutrition awareness and medicine usage, rather than being misled or duped by propaganda or positions not supported by evidence. Scientifically literate citizens would know how to evaluate cases when DNA evidence was involved in criminal trials. They would also be able to understand who the qualified scientists are and what they are doing, what processes they anticipate will be involved in their research investigations, and how their findings

F. Bevilacqua et al. (eds.), Science Education and Culture, 83–102.
© 2001 *Kluwer Academic Publishers. Printed in the Netherlands.*

matter to the welfare of society. Some of these scientifically literate individuals might develop a passion for and confidence in science and decide to become scientists. Perhaps some of these scientifically literate individuals who are capable of making reasonable judgements would become policymakers, and they might then decide to provide support for the budget of some critical science research and development projects.

The history of science, philosophy of science, and sociology of science (HPSS) inclusion has been found to be an effective way to reach the goal of enhancing science literacy for all citizens (Anderson and Smith 1986; Brush 1974; Conant 1964; Finley 1983; Klopfer 1969; Klopfer and Watson 1957; Matthews 1994; 1999; Quattropani 1977; Rutherford and Ahlgren 1990; Villani and Arruda 1998; Wandersee 1985; 1990). HPSS were included in the first nationwide content standards document for American K-12 school science, *Benchmarks for Science Literacy* (American Association for the Advancement of Science 1993). Other national standards documents in science education, including the *National Science Education Standards*, also state that students should know the HPSS (National Research Council 1996). Wang and Marsh (2001) report that the recommendation to include the history of science, in recent science education reform reports, is based on a specific rationale: to provide a meaningful context for both scientific information and the operation of the scientific enterprise. The rationale also applied to the inclusion into science education of the philosophy of science and sociology of science.

The three eras in American science education reviewed by Wang and Marsh (2001) – *The Golden Age of Science Education: Post-Sputnik Reactions*, *Science Education for Enlightened Citizenry*, and *Standards-Based Science Education Reform* – have shown that to include the HPSS is not a new proposal to humanize science in American science education. Although Matthews (1994) reports that HPSS inclusion is also not a new educational practice in other parts of the world, nevertheless, no large scale study has ever been attempted to report the HPSS educational conditions around the world.

This study utilizes the rich data collected by TIMSS to unveil the current conditions of HPSS in the science education of about forty TIMSS countries. Specifically, the report of HPSS educational conditions in this study includes: (1) educational officials' reports of HPSS coverage, (2) curriculum guides' HPSS coverage, (3) science textbooks' HPSS coverage, and (4) teachers' report of HPSS practices. Through this report, it addresses the questions concerned by science education worldwide: How much are students expected to learn in HPSS? What are the educational opportunities in delivering HPSS? The results from this study can serve as a framework to further probing serious issues, such as: What have students learned in HPSS worldwide? How does the way HPSS is delivered or learned relate to students' general achievement in the sciences?

2. Third International Mathematics and Science Study

The Third International Mathematics and Science Study (TIMSS) is the most extensive and far-reaching cross-national comparative study of mathematics and science education ever attempted (Beaton et al. 1996; Schmidt and McKnight 1995; Schmidt et al. 1997). It includes comparisons of the official curricula, textbooks, teacher practices, and student achievement of 20 to 50 countries (the number depending on the particular comparison subject groups) Thousands of official documents and textbooks were analyzed. Thousands of teachers, principals, and other experts responded to survey questionnaires. More than half a million children in over 40 countries were tested in mathematics and science. These tests were conducted for nine-year-olds (grades three and four, in the US), thirteen-year-olds (grades seven and eight, in the US), and students in the last year of secondary school (twelfth grade, in the US).

Meaningful measurement of educational systems requires a comprehensive conceptual framework and a corresponding array of measures designed to relate the various parts of the system to each other and with their outcomes. The Conceptual Framework (Figure 1) behind TIMSS was developed to achieve this objective and has been widely presented in various TIMSS publications (Schmidt et al. 1996; Schmidt et al. 1997; Schmidt and McKnight 1995). This model describes delivery of content-related educational experiences that was used to in the design of measures and analyses for the TIMSS. The links in this model were accomplished through the TIMSS. It is our belief that this design can assist us in delineating the relationships we intend to study about the HPSS educational conditions and their relationships toward achievement in the TIMSS countries.

TIMSS data were coded using multiple methods. In this study, we report the HPSS conditions based on the HPSS areas included in the *Content Codes* (Robitaille et al. 1993), which are:

1.4.3.1	Influence of Science, Technology on Society
1.4.3.2	Influence of Society on Science, Technology
1.5	History of Science & Technology
1.7.1	Nature of Scientific Knowledge
1.7.2	The Scientific Enterprise

The data used in this study originated from TIMSS: (1) Grades one through twelve's HPSS curriculum coverage in the educational officials' reports, which is known as the *General Topic Tracing Map (GTTM)*, (2) HPSS coverage in the curriculum guides for the middle school years, (3) HPSS coverage in the science textbooks for the middle school years, and (4) HPSS instructional practices in terms of the proportion of the middle school teachers who have reported that they taught HPSS, and the percentage of their time spent teaching HPSS topics. Following

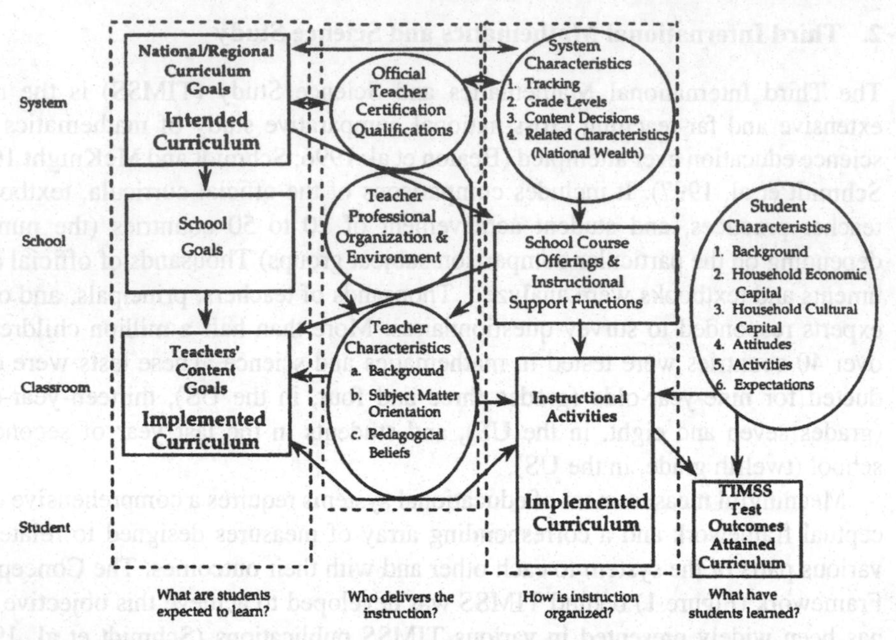

Figure 1. TIMSS conceptual framework: A model of educational experiences.

are further descriptions of the data sources. More detailed information as to how these data were collected, organized, and analyzed is provided in TIMSS technical reports:

1. *General Topic Tracing Map (GTTM).* Respondents to the GTTM were educational officials of each nation utilizing their national content standards or an aggregate of regional standards. They indicated intended coverage of a content area in a given grade by circling the age corresponding to the age most students were at the beginning of that grade. Therefore, the upper grade of Population One in TIMSS should correspond to age 9 (lower grade should be age 8), the upper grade of Population Two in TIMSS is age 13 (lower grade is age 12). Population Three GTTM data may be more difficult to read because, unless otherwise noted in their identifier, it was to be for students not in the specialist group. Thus, the last grade reported in the GTTM data may not be the last grade in school, because it may not be typical in all countries for nonspecialists to take math or science in their last year(s) of school.

2. *Official Curriculum Guides.* TIMSS national document samples of the curriculum guides comprised, as appropriate for each country, (a) the national science curriculum guide or guides (if any) covering each grade, and (b) regional, provincial, state, or cantonal science curriculum guides covering each grade (if needed). The resulting sample for the 48 TIMSS countries included 77 science curriculum guides for the upper grade of Population One, 111 for the upper grade of Population Two, and 62 for the physics specialists in Population

Three, for a total of 250 curriculum guides (Schmidt et al. 1997). The data in this study originated from the coding results of 111 curriculum guides at Population Two.

3. *Science Textbooks*. TIMSS national document samples of the science textbooks included (a) official national science textbooks (if any), and (b) the most widely used commercial textbooks if "officially" provided books were not used. The resulting sample for the 48 TIMSS countries included 75 science textbooks for the upper grade of Population One, 155 for the upper grade of Population Two, and 60 for the physics specialists in Population Three, for a total of 290 sampled science textbooks (Schmidt et al. 1997). Only the Population Two data were applied in this study.
4. *Teachers Responses*. Data collected from teachers were mainly based on their returned responses to TIMSS Teacher Questionnaires. The questionnaires were given to the teachers of the sample student populations. The questionnaires were extensive in terms of areas to be explored and required an average of 60 minutes to fill out. In this study, we only used the data of the Science Topics section from Population Two, and specifically the questions of:

   0037: How long did you spend teaching each of these topic areas to your class this year?
   Topic (t) Science, technology, and society
   Topic (u) History of science and technology
   Topic (v) Nature of science

   The data collected from teachers were organized into: (1) the proportion of teachers who reported that they taught HPSS, and (2) the percentage of time they spent teaching HPSS.

TIMSS science test was designed to test students' general science achievement; HPSS areas were only one small part of the test. Compared to the other two Populations, the test items in Population Two included relatively more HPSS areas. In a follow-up study, students' achievement results used are those of Population Two. Table I displays the sample size and the average age of the eighth graders who participated in TIMSS achievement tests.

3. Intended and Implemented HPSS

3.1. THE HPSS CURRICULUM COVERAGE IN GRADES ONE THROUGH TWELVE

While advocacy is growing for spending more time enhancing students' experiences in HPSS, the actual state of HPSS areas in the school science curricula of 42 TIMSS countries is illustrated in Figures 2 to 6; these data originated from the GTTM survey, to which each nation's educational officials responded. The topic of *History of Science and Technology* was reportedly included, in at least one grade level in 29 out of 42 countries, as the most popular topic of HPSS areas.

Table I. Participating TIMSS countries (Population Two)

COUNTRY	SIZE	AGE of 8th graders
Australia	17843.0	14.2
Austria	8028.0	14.3
Belgium (Fl)	10116.0	14.1
Belgium (Fr)	10116.0	14.3
Bulgaria	8435.0	14.0
Canada	29248.0	14.1
Colombia	36330.0	15.7
Cyprus	726.0	13.7
Czech Republic	10333.0	14.4
Denmark	5205.0	13.9
England	48533.0	14.0
France	57928.0	14.3
Germany	81516.0	14.8
Greece	10426.0	13.6
Hong Kong	6061.0	14.2
Hungary	10261.0	14.3
Iceland	266.0	13.6
Iran	62550.0	14.6
Ireland	3571.0	14.4
Israel	5383.0	14.1
Japan	124961.0	14.4
Korea	44453.0	14.2
Kuwait	1620.0	15.3
Latvia	2547.0	14.3
Lithuania	3721.0	14.3
Netherlands	15381.0	14.3
New Zealand	3493.0	14.0
Norway	4337.0	13.9
Portugal	9902.0	14.5
Romania	22731.0	14.6
Russian Federation	148350.0	14.0
Scotland	5132.0	13.7
Singapore	2930.0	14.5
Slovak Republic	5347.0	14.3
Slovenia	1989.0	14.8
South Africa	40539.0	15.4
Spain	39143.0	14.3
Sweden	8781.0	13.9
Switzerland	6994.0	14.2
Thailand	58024.0	14.3
USA	260650.0	14.2

Influence of Science, Technology on Society was reported as a curriculum topic by 28 countries as the second most popular topic to science classrooms worldwide. Whereas the topic *The Scientific Enterprise* was only reported by 13 countries as a curriculum topic, the least likely HPSS topic being introduced to science classrooms worldwide.

Most countries, as shown in Figures 2 to 6, have exhibited a trend, which is that any HPSS topic could be "introduced" (the half-shaded circle) in any grade level, but rarely has any particular topic become a "focused" (the full-shaded circle) topic in science classrooms for a nation. This trend has few exceptions; The French curriculum official reported that the *Nature of Scientific Knowledge* topic was introduced in grades one through three and started to be a focused topic from grades four through twelve. France, as shown in the Figure 5, was the only country to exhibit such an emphasis on an HPSS topic. The Israeli curriculum official reported the *History of Science and Technology* topic as being a focused science topic from grade one through five and staying in the curriculum for sixth through twelfth grades, but Figure 2 indicates that it became a focused topic in grades eight and eleven. Philippine curriculum officials indicated that in grades seven through ten, every HPSS topic except *The Scientific Enterprise* was a focused science topic. These HPSS topics in the Philippines were introduced at grade three and ended at tenth grade. *The Scientific Enterprise* was introduced at grades seven and eight and focused at grades nine and ten in the Philippines (Figure 4). The Danish curriculum official indicated that for more than five years the topic of *Influence of Society on Science, Technology* had been a focused science topic later on in Denmark students science education (Figure 6).

There are countries that covered some HPSS topics at every grade level. *Influence of Science, Technology on Society* was introduced in every grade level in China and Slovenia (Figure 3). *History of Science and Technology*, as stated above, was reported to be covered in every grade in Israel and introduced to every grade in China (Figure 2). *Nature of Scientific Knowledge* was covered in every grade reported by curriculum officials of Canada, Cyprus, France, Portugal, the United States, and Slovenia (Figure 5). Canada and Slovenia are the only two countries reported to be covering *The Scientific Enterprise* in every single grade's science education (Figure 4).

According to Figures 2 to 6, it would be difficult to infer what is the appropriate grade to introduce HPSS topics to the science classroom; yet as reported by the officials, the most common years of introducing any HPSS topic are grades one,[1] three, [2] and seven. [3] There are some countries' curriculum officials, however, who indicated that HPSS topics have never been covered in any grade of pre-college science education curricula; they are Argentina, Hong Kong, Iran, and Tunisia.

COUNTRY/GRADE	1	2	3	4	5	6	7	8	9	10	11	12
Argentina	○	○	○	○	○	○	○	○	○	○	○	○
Australia	○	○	○	○	○	○	○	○	○	○	○	○
Belgium (Fl)	○	○	○	○	○	○	○	○	○	◑	◑	●
Belgium (Fr)	○	○	○	○	○	○	○	○	○	◑	◑	●
Bulgaria	○	○	○	◑	◑	◑	◑	◑	◑	◑	◑	○
Canada	○	○	◑	◑	◑	◑	◑	◑	◑	●	●	◑
Cyprus	○	○	○	○	○	○	○	◑	◑	◑	◑	●
Czech Republic	○	○	○	○	○	◑	◑	◑	◑	◑	◑	◑
Slovak Republic	○	○	○	◑	◑	◑	◑	◑	◑	◑	◑	◑
Denmark	○	○	○	○	○	○	◑	◑	◑	●	●	●
Dominican Republic	○	○	○	○	○	○	○	○	○	○	○	○
France	○	○	○	○	○	○	○	◑	◑	◑	◑	◑
Germany	○	○	○	○	○	○	○	○	○	○	◑	◑
Greece	○	○	○	○	◑	●	◑	◑	◑	◑	◑	○
Hong Kong	○	○	○	○	○	○	○	○	○	○	○	○
Hungary	○	○	○	○	○	○	○	○	○	○	○	○
Iceland	○	○	○	○	○	○	○	◑	◑	◑	◑	◑
Iran	○	○	○	○	○	○	○	○	○	○	○	○
Ireland	○	○	◑	◑	◑	◑	◑	◑	◑	○	◑	◑
Israel	●	●	●	●	●	◑	◑	●	◑	◑	●	◑
Italy	◑	◑	◑	◑	●	◑	◑	◑	○	○	○	○
Japan	○	○	○	○	○	○	○	○	○	●	●	○
Korea	○	○	○	○	○	○	○	○	○	○	●	◑
Latvia	○	○	○	○	○	○	○	◑	◑	◑	◑	◑
Mexico	○	○	○	○	◑	◑	●	●	●	◑	●	○
Netherlands	○	○	○	○	○	○	○	○	○	○	○	○
New Zealand	○	○	○	○	○	○	○	○	○	○	○	○
Norway	○	○	○	○	○	○	◑	◑	◑	◑	●	◑
Philippines	○	○	◑	◑	◑	◑	●	●	●	●	○	○
Portugal	○	○	○	○	○	○	○	●	◑	◑	◑	◑
Romania	○	○	○	○	○	○	○	○	◑	◑	◑	◑
Russian Federation	○	○	○	○	○	◑	◑	◑	◑	◑	◑	○
Singapore	○	○	○	○	○	○	○	○	○	○	○	○
Spain	○	○	○	○	○	◑	◑	◑	◑	◑	◑	◑
Sweden	○	○	○	◑	◑	◑	◑	◑	◑	○	○	○
Switzerland	○	○	○	○	○	○	○	○	◑	◑	◑	◑
Tunisia	○	○	○	○	○	○	○	○	○	○	○	○
USA	○	○	○	○	○	○	○	○	○	○	○	○
Austria	○	○	○	○	○	◑	◑	◑	○	○	○	○
China	◑	◑	◑	◑	◑	◑	◑	◑	◑	◑	◑	◑
Colombia	○	○	○	○	○	○	○	○	○	○	○	○
Slovenia	○	○	○	○	○	○	○	○	○	○	○	○

○ Not Included in the Curriculum; ◑ Included in the Curriculum; ● Focused in the Curriculum

Figure 2. Presence of the topic *History of Science & Technology* in grades 1–12 for the TIMSS countries.

COUNTRY/GRADE	1	2	3	4	5	6	7	8	9	10	11	12
Argentina	○	○	○	○	○	○	○	○	○	○	○	○
Australia	○	○	○	○	○	◑	◑	◑	◑	◑	◑	◑
Belgium (Fl)	○	○	○	○	○	○	○	○	○	○	○	○
Belgium (Fr)	○	○	○	○	○	○	○	○	○	○	○	○
Bulgaria	○	○	○	○	○	○	○	○	○	○	○	○
Canada	○	○	◑	◑	◑	◑	◑	◑	◑	●	●	◑
Cyprus	○	○	○	○	○	○	○	○	◑	◑	◑	◑
Czech Republic	○	○	○	○	○	○	○	○	○	○	○	○
Slovak Republic	○	○	○	◑	◑	◑	◑	●	◑	◑	◑	◑
Denmark	○	○	○	○	○	◑	◑	●	○	◑	●	●
Dominican Republic	○	○	◑	◑	◑	◑	◑	◑	○	○	○	○
France	○	○	○	○	○	○	○	◑	◑	◑	◑	◑
Germany	○	○	○	○	○	○	○	○	○	◑	◑	◑
Greece	○	○	○	○	○	○	○	○	○	○	○	○
Hong Kong	○	○	○	○	○	○	○	○	○	○	○	○
Hungary	○	○	○	○	○	○	○	○	○	○	○	◑
Iceland	○	○	○	○	○	○	○	○	○	○	○	○
Iran	○	○	○	○	○	○	○	○	○	○	○	○
Ireland	○	○	○	○	○	○	○	○	○	○	◑	◑
Israel	○	○	○	○	○	○	○	○	○	○	○	○
Italy	○	○	◑	◑	●	◑	◑	◑	○	○	○	○
Japan	○	○	○	○	○	○	○	○	◑	○	○	○
Korea	○	○	○	○	○	○	○	○	○	●	◑	◑
Latvia	○	○	○	○	○	○	○	○	○	○	○	○
Mexico	○	○	○	○	○	○	◑	◑	◑	◑	◑	●
Netherlands	○	○	○	○	○	○	○	○	○	◑	◑	◑
New Zealand	○	○	○	○	○	○	◑	◑	◑	◑	◑	○
Norway	○	○	○	○	○	○	◑	◑	◑	◑	●	○
Philippines	○	○	◑	◑	◑	◑	●	●	●	●	○	○
Portugal	○	○	◑	◑	◑	◑	◑	◑	◑	◑	◑	◑
Romania	○	○	○	○	○	○	○	○	○	○	○	○
Russian Federation	○	○	○	○	○	◑	◑	◑	◑	◑	◑	○
Singapore	○	○	◑	◑	◑	◑	◑	◑	◑	◑	◑	○
Spain	○	○	○	○	○	○	◑	◑	◑	◑	◑	◑
Sweden	○	○	○	◑	◑	◑	○	○	○	○	○	○
Switzerland	○	○	○	○	○	◑	◑	◑	◑	◑	◑	◑
Tunisia	○	○	○	○	○	○	○	○	○	○	○	○
USA	○	○	○	○	○	○	○	○	○	○	○	○
Austria	◑	◑	◑	◑	◑	◑	◑	◑	○	○	○	○
China	◑	◑	◑	◑	◑	◑	◑	◑	◑	◑	◑	◑
Colombia	◑	◑	◑	◑	◑	◑	◑	●	◑	○	○	○
Slovenia	◑	◑	◑	◑	◑	◑	◑	◑	◑	◑	◑	◑

○ Not Included in the Curriculum; ◑ Included in the Curriculum; ● Focused in the Curriculum

Figure 3. Presence of the topic *Influence of Science, technology on Society* in grades 1–12 for the TIMSS countries.

COUNTRY/GRADE	1	2	3	4	5	6	7	8	9	10	11	12
Argentina	○	○	○	○	○	○	○	○	○	○	○	○
Australia	○	○	○	○	○	○	○	○	○	○	◑	◑
Belgium (Fl)	○	○	○	○	○	○	○	○	○	○	○	○
Belgium (Fr)	○	○	○	○	○	○	○	○	○	○	○	○
Bulgaria	○	○	○	○	○	○	○	○	○	○	○	○
Canada	◑	◑	◑	◑	◑	◑	◑	◑	◑	◑	◑	◑
Cyprus	○	○	○	○	○	○	○	○	○	○	○	○
Czech Republic	○	○	○	○	○	○	○	○	○	○	○	○
Slovak Republic	○	○	○	○	○	○	○	○	○	○	○	○
Denmark	○	○	○	○	○	○	◑	◑	◑	◑	◑	◑
Dominican Republic	○	○	○	○	○	○	○	○	○	○	○	○
France	○	○	○	○	○	◑	◑	◑	◑	◑	◑	◑
Germany	○	○	○	○	○	○	○	○	○	○	◑	●
Greece	○	○	○	○	○	○	○	○	○	○	○	○
Hong Kong	○	○	○	○	○	○	○	○	○	○	○	○
Hungary	○	○	○	○	○	○	○	○	○	○	○	○
Iceland	○	○	○	○	○	○	○	○	○	○	○	○
Iran	○	○	○	○	○	○	○	○	○	○	○	○
Ireland	○	○	○	○	○	○	○	○	○	○	○	○
Israel	○	○	○	○	○	○	○	○	○	○	○	○
Italy	○	○	○	○	○	◑	◑	◑	○	○	○	○
Japan	○	○	○	○	○	○	○	○	○	○	○	○
Korea	○	○	○	○	○	○	○	○	○	◑	●	◑
Latvia	○	○	○	○	○	○	○	○	○	○	○	○
Mexico	○	○	○	○	○	○	◑	◑	◑	◑	◑	◑
Netherlands	○	○	○	○	○	○	○	○	○	◑	◑	◑
New Zealand	○	○	○	○	○	○	○	○	○	○	○	○
Norway	○	○	○	○	○	○	○	○	○	○	○	○
Philippines	○	○	○	○	○	○	◑	◑	●	●	○	○
Portugal	○	○	○	○	○	○	○	○	○	○	○	○
Romania	○	○	○	○	○	○	○	○	○	○	○	○
Russian Federation	○	○	○	○	○	○	○	○	○	○	○	○
Singapore	○	○	○	○	○	○	○	○	○	○	○	○
Spain	○	○	○	○	○	○	○	○	○	○	○	○
Sweden	○	○	○	○	○	○	○	○	○	○	○	○
Switzerland	○	○	○	○	○	○	○	○	○	◑	●	◑
Tunisia	○	○	○	○	○	○	○	○	○	○	○	○
USA	○	○	○	○	○	○	○	○	○	○	○	○
Austria	◑	◑	◑	◑	◑	◑	◑	◑	○	○	○	○
China	○	○	○	○	○	○	○	○	○	○	○	○
Colombia	○	○	○	○	○	○	○	○	○	○	○	○
Slovenia	◑	◑	◑	◑	◑	◑	◑	◑	◑	◑	◑	◑

Not Included in the Curriculum; ◑ Included in the Curriculum; ● Focused in the Curriculum

Figure 4. Presence of the topic *The Scientific Enterprise* in grades 1–12 for the TIMSS countries.

COUNTRY/GRADE	1	2	3	4	5	6	7	8	9	10	11	12
Argentina	○	○	○	○	○	○	○	○	○	○	○	○
Australia	○	○	○	○	○	○	◑	◑	◑	◑	◑	◑
Belgium (Fl)	○	○	○	○	○	○	◑	◑	◑	◑	◑	◑
Belgium (Fr)	○	○	○	○	○	○	◑	◑	◑	◑	◑	◑
Bulgaria	○	○	○	○	○	○	○	○	◑	●	◑	○
Canada	◑	◑	◑	◑	◑	◑	◑	◑	◑	◑	◑	◑
Cyprus	◑	◑	◑	◑	◑	◑	◑	●	◑	◑	◑	◑
Czech Republic	○	○	○	○	○	○	○	○	○	○	○	○
Slovak Republic	○	○	○	○	○	○	◑	◑	◑	◑	◑	◑
Denmark	○	○	○	○	○	○	◑	◑	◑	◑	◑	◑
Dominican Republic	○	◑	◑	◑	◑	○	○	○	○	○	○	○
France	◑	◑	◑	●	●	●	●	●	●	●	●	●
Germany	○	○	○	○	○	○	○	○	○	○	◑	●
Greece	○	○	○	○	○	○	○	○	○	○	○	○
Hong Kong	○	○	○	○	○	○	○	○	○	○	○	○
Hungary	○	○	○	○	○	○	○	○	○	○	○	○
Iceland	○	○	○	○	○	◑	○	●	◑	◑	◑	◑
Iran	○	○	○	○	○	○	○	○	○	○	○	○
Ireland	○	○	○	○	○	○	◑	◑	◑	○	◑	◑
Israel	○	○	○	○	○	○	○	○	○	○	○	○
Italy	○	○	○	○	○	○	○	○	○	○	○	○
Japan	○	○	○	○	○	○	○	○	○	○	○	○
Korea	○	○	○	○	○	○	○	○	○	◑	●	◑
Latvia	○	○	○	○	○	○	○	○	○	○	○	○
Mexico	○	○	○	○	○	○	◑	◑	◑	●	●	◑
Netherlands	○	○	○	○	○	○	◑	◑	◑	◑	◑	◑
New Zealand	○	○	○	○	○	○	○	○	○	○	○	○
Norway	○	○	○	○	○	○	◑	◑	◑	◑	◑	◑
Philippines	○	○	◑	◑	◑	◑	●	●	●	●	○	○
Portugal	◑	◑	◑	◑	◑	◑	◑	◑	◑	◑	◑	◑
Romania	○	○	○	○	○	○	○	○	○	○	○	○
Russian Federation	○	○	○	○	◑	◑	◑	◑	◑	◑	◑	○
Singapore	○	○	○	○	○	○	○	○	○	○	○	○
Spain	○	○	○	○	○	◑	◑	◑	◑	●	◑	◑
Sweden	○	○	○	○	○	○	○	○	○	○	○	○
Switzerland	○	○	○	○	○	○	○	○	◑	●	◑	◑
Tunisia	○	○	○	○	○	○	○	○	○	○	○	○
USA	◑	◑	◑	◑	◑	◑	◑	◑	◑	◑	◑	◑
Austria	◑	◑	◑	◑	◑	◑	◑	◑	○	○	○	○
China	○	○	○	○	○	○	○	○	○	○	○	○
Colombia	○	○	○	○	○	○	○	●	◑	○	○	○
Slovenia	◑	◑	◑	◑	◑	◑	◑	◑	◑	◑	◑	◑

○ Not Included in the Curriculum; ◑ Included in the Curriculum; ● Focused in the Curriculum

Figure 5. Presence of the topic *Nature of Scientific Knowledge* in grades 1–12 for the TIMSS countries.

COUNTRY/GRADE	1	2	3	4	5	6	7	8	9	10	11	12
Argentina	○	○	○	○	○	○	○	○	○	○	○	○
Australia	○	○	○	○	○	◑	◑	◑	◑	◑	◑	◑
Belgium (Fl)	○	○	○	○	○	○	○	○	○	○	◑	●
Belgium (Fr)	○	○	○	○	○	○	○	○	○	○	◑	●
Bulgaria	○	◑	◑	◑	◑	◑	◑	◑	◑	◑	◑	◑
Canada	○	○	◑	◑	◑	◑	◑	◑	◑	●	●	◑
Cyprus	○	○	○	○	○	○	○	◑	○	◑	◑	●
Czech Republic	○	○	○	○	○	○	○	○	○	○	○	○
Slovak Republic	○	○	○	○	○	○	○	○	○	○	○	○
Denmark	○	○	○	○	○	◑	●	●	○	●	●	●
Dominican Republic	○	○	○	○	○	○	○	○	○	○	○	○
France	○	○	○	○	○	○	○	○	○	○	○	○
Germany	○	○	○	○	○	○	○	○	○	◑	◑	◑
Greece	○	○	○	○	○	○	○	○	○	○	○	○
Hong Kong	○	○	○	○	○	○	○	○	○	○	○	○
Hungary	○	○	○	○	○	○	○	○	○	○	○	◑
Iceland	○	○	○	○	○	○	○	○	○	○	○	○
Iran	○	○	○	○	○	○	○	○	○	○	○	○
Ireland	○	○	○	○	○	○	○	○	○	○	◑	◑
Israel	○	○	○	○	○	○	○	○	○	○	○	○
Italy	○	○	◑	◑	◑	◑	◑	◑	○	○	○	○
Japan	○	○	○	○	○	○	○	○	○	○	○	○
Korea	○	○	○	○	○	○	○	○	◑	◑	●	◑
Latvia	○	○	○	○	○	○	○	○	○	○	○	○
Mexico	○	○	○	○	○	○	◑	◑	◑	◑	◑	●
Netherlands	○	○	○	○	○	○	○	○	○	◑	◑	◑
New Zealand	○	○	○	○	○	○	◑	◑	◑	◑	◑	○
Norway	○	○	○	○	○	○	◑	◑	◑	◑	●	◑
Philippines	○	○	◑	◑	◑	◑	●	●	●	●	○	○
Portugal	○	○	◑	◑	◑	◑	◑	◑	◑	◑	◑	◑
Romania	○	○	○	○	○	○	○	○	○	○	○	○
Russian Federation	○	○	○	○	○	◑	◑	◑	◑	◑	◑	○
Singapore	○	○	◑	◑	◑	◑	◑	◑	◑	◑	◑	○
Spain	○	○	○	○	○	○	◑	◑	◑	◑	◑	◑
Sweden	○	○	○	◑	◑	◑	○	○	○	○	○	○
Switzerland	○	○	○	○	○	◑	◑	◑	◑	◑	◑	◑
Tunisia	○	○	○	○	○	○	○	○	○	○	○	○
USA	○	○	○	○	○	○	○	○	○	○	○	○
Austria	◑	◑	◑	◑	◑	◑	◑	◑	○	○	○	○
China	○	○	○	○	○	○	○	○	○	○	○	○
Colombia	○	○	○	○	○	○	○	○	○	○	○	○
Slovenia	○	○	○	○	○	○	○	○	◑	◑	◑	◑

○ Not Included in the Curriculum; ◑ Included in the Curriculum; ● Focused in the Curriculum

Figure 6. Presence of the topic *Influence of Society on Science, Technology* in grades 1–12 for the TIMSS countries.

3.2. THE HPPS COVERAGE IN CURRICULUM GUIDES AND SCIENCE TEXTBOOKS OF EIGHTH GRADERS

In TIMSS countries, curricular decisions on national goals, instructional content, examinations, and so on were made by groups, agencies, individuals in authority, or some combination of these (Schmidt et al. 1997). Despite the various conditions in the decision making, virtually all educational systems within TIMSS use some form of curriculum guide to structure science education. These guides set forth the system's goals for a nation's science education. "Countries differ widely in the structure and details of their guides and in how they are meant to be used" (Schmidt et al. 1997, p. 38). Furthermore, as described by Schmidt et al. (1997), TIMSS data showed that the decisions concerning textbooks were made jointly in about 42 percent of the systems and subsystems, most often involving a central authority and the school using the text. Individual teachers chose the textbooks for their classes in only about 9 percent of the systems and subsystems. Textbook data have indicated that the lengths of content vary, the sizes of the books vary, the formats vary, and the levels of decoration vary from country to country.

Figure 7 pins down issues regarding the educational intentions in HPSS areas from 36 TIMSS countries, that is, the countries that have both guide and textbook information for eighth grade. Nine out of 34 countries have included every HPSS topic in their guides. Only two out of these nine countries also covered every HPSS topic in their science textbooks.[4]

Issue of alignment between the textbooks and guides. Since 1986, the year the first national mathematics educational standards document was published in the United States, there has been an increase in the production of every subject's educational standards for American education. This exercise has further brought the American educational community to a discussion of the alignment between the intended educational objectives (standards or curriculum guides) and implemented education (instruction and instructional resources). Textbooks are perceived as "potential" implemented education because textbooks have persistently had a great influence on what is taught and how it is delivered in science. American science teachers were known to rely heavily on textbooks when they delivered science instruction (Harms and Yager 1981; Weiss 1978). Despite the new wave of science education reform that advocated that science instruction shift away from textbook-based instruction to kit-based instruction (Wang and Marsh 2001), TIMSS data indicate that teachers throughout the world base about 50 percent of their weekly teaching time on textbooks. Thus, there exists the need to examine the appropriateness of science textbooks in terms of their alignment with the guides.

The symbol "+"in Figure 7 represents countries that were found to have covered an HPSS topic in their textbooks; the symbol "○" represents countries with an HPSS topic in their guides; and the symbol "⊕" represents countries with an HPSS topic covered in both their guides and their textbooks. The topic *History of Science and Technology* was found again to be the most consistent topic covered by both

	Influence of S, T on Society	Influence of Society on S, T	History of S & T	Nature of Scientific K.	Sci. Enterpr.
Australia	○	○	⊕	⊕	○
Austria	⊕			○	
Canada	⊕	⊕	⊕	⊕	○
Colombia	○		○		
Czech Republic	+				
Denmark	○	○	○	○	○
Dominican Republic					
France	⊕	○	○		
Germany	○	○	+	○	
Greece					
Hong Kong				○	
Hungary	○	○	⊕	○	
Iceland	⊕	+	+	⊕	⊕
Ireland	○	○	⊕	⊕	
Israel			+	○	
Japan					
Korea	○		⊕		
Latvia	○	○	⊕	○	○
Mexico	○		⊕	○	
Netherlands	○	○		⊕	○
New Zealand	⊕	⊕	⊕	⊕	⊕
Norway	○	○	⊕	○	⊕
Philippines	+		+		
Portugal	⊕	⊕	⊕	⊕	○
Romania			○		○
Russian Federation	○	○	⊕	○	
Scotland	+		+	○	○
Singapore	○	○		○	
Slovak Republic	○	○		○	○
Slovenia	⊕	○	⊕	○	○
Spain	⊕		⊕	○	⊕
Sweden	○	○	⊕		
Switzerland	○	○	⊕	○	
USA	⊕	⊕	⊕	⊕	⊕

+ Covered in Textbook; ○ Covered in Guide; ⊕ Covered in Both Guide & Textbook

Figure 7. Coverage of HPSS topics in 8th grade as found in curriculum guides and textbooks.

guides and textbooks (16 countries showed this consistency). *Influence of Science, Technology on Society* is another topic that has relatively more consistency between guides and the textbooks (9 out of 34 countries). New Zealand and USA are the only two countries showing perfect alignment in all five HPSS topics between the guides and the textbooks. Additionally, four out of 34 countries showed at least 60 percent alignment between the guides and textbooks in HPSS areas; they are Canada, Iceland, Portugal, and Slovenia. Denmark had every HPSS topic in its guide, yet nothing was found in its textbooks. Similarly, Singapore and the Slovak

Republic were found to have HPSS topics in their curriculum guides, yet, on examining their science textbooks, no HPSS was found. Conversely, Czech Republic, Dominican Republic, Greece, Japan, and the Philippines were found to have no coverage in their guides. Czech Republic's science textbooks included one HPSS topic – *Influence of Science, Technology on Society*, while the Philippines science textbooks included both *Influence of Science, Technology on Society* and *History of Science and Technology*, despite no HPSS topic being covered in both countries' guides.

Figure 8 are representations of the percentage of books devoted to each HPSS topic from 42 TIMSS countries that had their science textbooks coded by TIMSS researchers.

Overall, Canada, and USA were found to have over 15 percent of their textbooks' space devoted to HPSS areas. he grand average coverage of the five HPSS topics in science textbooks of these 42 countries was 4 percent. Twenty-seven of the 42 countries were found to have less than 4 percent of their science textbooks related to any HPSS topic. On the topic of *Influence of Science, Technology on Society*, Canada's textbooks were found to have 7 percent; Cyprus's textbooks had an average of 6 percent; and USA's textbooks had an average of 5 percent. Compared to this topic, *Influence of Society on Science, Technology* was found to have almost *no* coverage in science textbooks worldwide. Furthermore, science textbooks in Iceland, Italy, and Slovenia covered more than 5 percent of the content topic *History of Science and Technology*. The HPSS topic *Nature of Scientific Knowledge* took up 5 percent or more textbook space in the countries of Canada, Portugal, and USA.

3.3. INSTRUCTIONAL CONDITIONS OF HPSS AREAS

There were 36 TIMSS countries who completed the teacher instructional questionnaire. Two out of the 36 (Israel and Kuwait) only had eighth grade teachers responded to the questionnaire. The remaining 34 countries gave responses from teachers of both seventh and eighth graders. Table II shows a ranking of the percentage of teachers per country, who reported that they taught at least one or more HPSS topics, and it also shows the average percentage of instructional time devoted to HPSS areas for both seventh and eighth graders.

As indicated in Table II, 84 percent of American eighth grade teachers reported that they taught HPSS areas in their science classrooms; this made the US the number one country in terms of HPSS coverage with around 20 percent more teachers teaching these topics than the second country the Russian Federation (65%). Percentage of yearly instructional time devoted to HPSS topics in the US was five percent for eighth grade, which along with Canada ranked as the top country. Overall, 32 percent of seventh grade teachers and 36 percent of eighth grade teachers worldwide reported that they have spent time in teaching HPSS content. Nearly 60 percent of the countries have more than one-third of their teachers reported to be

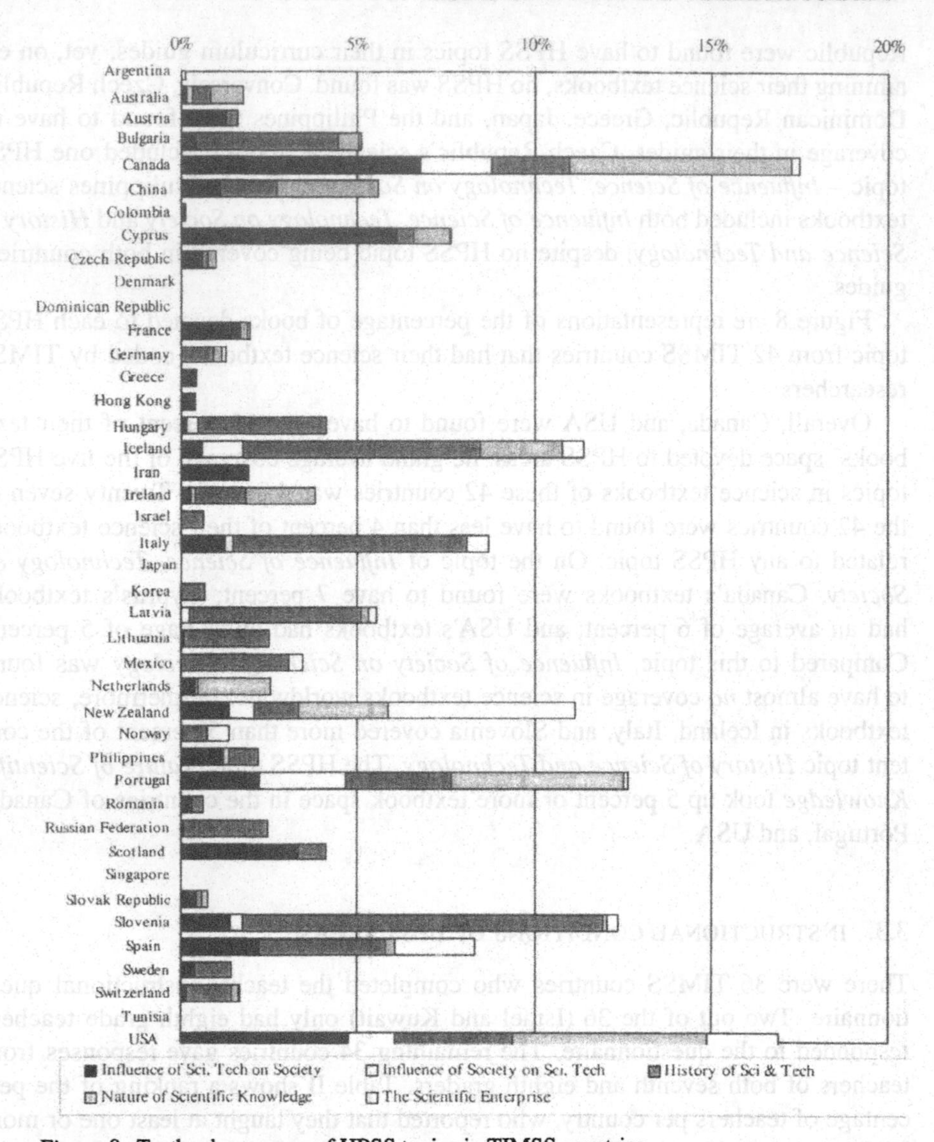

Figure 8. Textbook coverage of HPSS topics in TIMSS countries.

giving HPSS instruction to the eighth graders of the country. However, the average time worldwide for HPSS instruction per school year is 3 percent for eighth grade, and 2 percent for seventh grade. According to Table II, those high performing countries in TIMSS on the science test, the Czech Republic, Japan, Netherlands, and Singapore are not among the top tier in terms of percentage of teachers or percentage of instructional time devoted to HPSS content.

Table II. **The percentage of teachers and the percentage of instrutional time associated with HPSS topics in TIMSS**

% of 7th Graders' Teacher		% of 8th Graders' Teacher		Inst. Time for 7th Graders'		Inst. Time for 8th Graders'	
USA	82%	USA	84%	USA	6%	USA	5%
Russian Fed.	61%	New Zealand	65%	Thailand	5%	Canada	5%
Australia	54%	Canada	64%	Canada	5%	Hungary	4%
Canada	53%	Russian Fed.	61%	New Zealand	4%	Iceland	4%
New Zealand	49%	Australia	51%	Hungary	4%	Slovenia	4%
Slovak Republic	46%	Colombia	50%	Hong Kong	4%	Cyprus	4%
Thailand	46%	Cyprus	50%	Australia	4%	New Zealand	4%
Slovenia	45%	Slovenia	49%	Russian Fed.	3%	Germany	4%
Colombia	44%	Slovak Republic	44%	Slovenia	3%	Slovak Republic	3%
Korea	40%	Sweden	41%	France	3%	Sweden	3%
Singapore	40%	Iceland	41%	Cyprus	3%	Russian Fed.	3%
Spain	35%	Germany	40%	Slovak Republic	3%	Portugal	3%
Cyprus	34%	Lithuania	40%	Singapore	3%	Australia	3%
Hungary	33%	Norway	40%	Germany	3%	Lithuania	3%
Hong Kong	32%	Spain	39%	Colombia	2%	France	3%
Average	32%	Portugal	39%	Korea	2%	*Average*	3%
Ireland	30%	Czech Republic	36%	Sweden	2%	Romania	2%
Germany	30%	*Average*	36%	Czech Republic	2%	Norway	2%
Sweden	29%	Hungary	35%	Netherlands	2%	Czech Republic	2%
Lithuania	29%	Singapore	35%	Romania	2%	Thailand	2%
Czech Republic	28%	Israel	32%	Norway	2%	Austria	2%
Romania	27%	France	28%	Spain	2%	Colombia	2%
Portugal	24%	Korea	27%	Ireland	2%	Netherlands	2%
Norway	22%	Hong Kong	27%	Belgium (Fl)	2%	Israel	2%
Netherlands	21%	Thailand	27%	*Average*	2%	Greece	2%
France	21%	Ireland	26%	Lithuania	1%	Spain	2%
Japan	21%	Netherlands	25%	Austria	1%	Korea	2%
Switzerland	18%	Iran	25%	Belgium (Fr)	1%	Latvia	2%
Belgium (Fr)	16%	Romania	24%	Portugal	1%	Hong Kong	2%
Iran	16%	Japan	23%	Iceland	1%	Belgium (Fr)	1%
Belgium (Fl)	14%	Greece	23%	Japan	1%	Belgium (Fl)	1%
Iceland	13%	Latvia	21%	Switzerland	1%	Ireland	1%
Austria	10%	Switzerland	20%	Iran	1%	Singapore	1%
Greece	9%	Austria	16%	Latvia	1%	Japan	1%
Latvia	8%	Belgium (Fr)	16%	Greece	1%	Iran	1%
		Belgium (Fl)	14%			Switzerland	1%
		Kuwait	14%			Kuwait	1%
AVERAGE	33%		37%		3%		3%

4. Discussion and Conclusions

Throughout the history of enhancing the public scientific literacy, Conant (1951), one of the important advocators, has postulated that since every citizen is expected to have informal opinions on the relationships among government, education, and issues of scientific research and development, it is imperative that some appreciation of the past complexities of science and society be a part of the education of both scientists and non-scientists. Because of the increasingly scientific nature of our society and the individual needs of its members, every person must be scientifically literate in order to function effectively. Furthermore, the generation of scientific knowledge is a dynamic process with social, historical, psychological, and other contextual rather than purely abstract and formal determinants. Science is an enterprise in which dynamic change and alteration are the rules rather than the exceptions. The dynamic characteristic of science can help individuals to cultivate scientific habits of perception and to be capable of practicing rational thinking and logical reasoning.

In science education, the critical role of HPSS has been continuously identified as a powerful way to enhance the public's scientific literacy. The findings from TIMSS reflect one crucial message – little of it is done worldwide other than in the US and a handful of other countries, and the top achieving countries are not among them. In addition, according to the attempt of TIMSS researchers to understand the relationships among the *intended*, *implemented*, and *attained* curriculum, students' performance in the HPSS area *does* have a significant effect on general school science performance. However, there are more lessons needed to be learned from the TIMSS results.

1. *To teach HPSS alone will not result in greater performance; the science content knowledge also has an impact*. According to the results presented above, we found that countries such as Japan, Korea, and the Czech Republic were not among the top tier of countries in any aspect of the *intended* or *implemented* curriculum for HPSS. However, their performance on both the general science test and for HPSS items in particular were more satisfactory than those other countries that had significant indications of HPSS coverage in textbooks, curriculum guides, or on the part of teachers. This seems to imply that without the substance of science, classrooms with a heavier focus on HPSS might confuse students more and may just be another social studies class in a science disguise.

2. *Teacher preparation and training may be the key to reach the goal of enhancing scientific literacy through HPSS*. For advocates of HPSS, good news was found in the results: despite the fact that very little instructional time was allocated, science teachers worldwide did practice the inclusion of HPSS to some extent. One critical factor that may need to be addressed to make this instruction more effective may be the quality of teachers, that is the teachers may not have adequate training in HPSS worldwide. With inadequate training and insufficient background knowledge, teachers' misconception in HPSS may do more harm than good.

Lastly, the information reported in this paper is not to have every country conform and adopt identical strategies to improve the scientific literacy of the citizens worldwide. Some people would suggest that what a country should do is to figure out who is the top achieving country, say Singapore. What you then do is to find out what Singapore does educationally and copy it – bring it to your country and put it into the educational system and everything will be fixed. This is a fairly naïve approach. One cannot simply take what is done in one cultural context and lift it out of that context and place it into another one and somehow expect it to work. That is a misguided implication; which is *not* the way we learn from an international study. What we can learn is that there is not only one way to provide quality science education. HPSS inclusion to achieve scientific literacy can be approached in so many diverse ways. What appeared sensible and successful in one nation (to emphasize more history of science yet downplay the sociology of science) may not be an effective way for another nation (where the sociology of science or not teaching HPSS directly may evidently assist students to learn more effectively). HPSS approach has increasingly received satisfactory learning outcomes, findings presented in this study can serve as a direction or topic of dialogue about HPSS in science education for every country worldwide.

Notes

[1] At first grade, three countries started to introduce *History of Science & Technology*; seven countries started to introduced *Nature of Scientific Knowledge*; and three countries reported introducing the topic of *The Scientific Enterprise.*

[2] At grade three, six countries reported starting to introduce *Influence of Science, Technology on Society*; five countries started to introduce *Influence of Society on Science, Technology*; three countries introduced the topic of *History of Science & Technology*.

[3] At grade seven, nine countries started to introduce *the topic of Nature of Scientific Knowledge*; and three countries reported introducing the topic of *The Scientific Enterprise*.

[4] Notice that the information about the guides is different from what the educational officials reported in the GTTM survey, as shown in Figures 2 to 6. Despite the differences in the collection of the data of GTTM and guides (see previous paragraph for detailed information), a discrepancy exists between these two sources of *intended* curriculum in HPSS areas. In contrast to the ten countries found in the guides, only five countries' educational officials reported that every HPSS topic was covered at the eighth grade level. Only two of these five countries are like those ten countries in having that information in their guides: Canada and Denmark. This result signaled a critical implication for future research methods of this sort, a choice needs to be made between "self-reporting" and "content analysis of written documents".

References

American Association for the Advancement of Science: 1993, *Benchmarks for Science Literacy*, Oxford University Press, New York.

National Research Council: 1996, *National Science Education Standards*, National Academy Press, Washington, DC.

Anderson, C. W. & Smith, E.: 1986, 'Teaching Science', in V. Koehler (ed.), *The Educator's Handbook: A Research Perspective*, Longman, New York.

Beaton, A. E., Martin, M. O., Mullis, I. V. S., Gonzalez, E. J., Smith, T. A. & Kelly, D. L.: 1996, *Mathematics Achievement in the Middle School Years: IEA's Third International Mathematics and Science Study (TIMSS)*, Boston College, Chestnut Hill.

Brush, S. G.: 1974, 'Should the History of Science be Rated "X"?' *Science* **18**, 1164–1172.

Conant, J. B.: 1964, 'Introduction', in J. B. Conant and L. K. Nash (eds), *Harvard Case Histories in Experimental Science* (1), vii–xvi.

Finley, F.: 1983, 'Science Processes', *Journal of Research in Science Teaching* **20**, 47–54.

Harms, H. & Yager, R. E.: 1981, *What Research Says to the Science Teacher*, 3, NSTA, Washington DC.

Klopfer, L. E.: 1969, 'The Teaching of Science and the History of Science', *Journal of Research in Science Teaching* **6**, 87–95.

Klopfer, L. E. & Watson, F. G.: 1957, 'Historical Materials and High School Science Teaching', *The Science Teacher* **24**, 264–265, 292–293.

Matthews, M. R.: 1994, *Science Teaching: The Role of History and Philosophy of Science*, Routledge, NY.

Matthews, M. R.: 1999, *Time for Science Education: How Teaching the History and Philosophy of Pendulum Motion Can Contribute to Science Literacy*, Plenum Publishing Company, New York.

Quattropani, D. J.: 1977, *An Evaluation of the Effect of Harvard Project Physics on Student Understanding of the Relationships Among Science, Technology, and Society,* Ph.D. Dissertation. University of Connecticut.

Robitaille, D. F., Schmidt, W. H., Raizen, S., McKnight, C., Britton, E. & Nicol, C.: 1993, *TIMSS Monograph No. 1: Curriculum Frameworks for Mathematics and Science*, Pacific Educational Press, Vancouver.

Rutherford, F. J. & Ahlgren, A.: 1990, *Science for All Americans: Scientific Literacy*, Oxford University Press, New York.

Schmidt, W. H. & McKnight, C. C.: 1995, 'Surveying Educational Opportunity in Mathematics and Science: An International Perspective', *Educational Evaluation and Policy Analysis* **17**(3), 337–353.

Schmidt, W. H., Jorde, D., Cogan, L. S., Barrier, E., Gonzalo, I., Moser, U., Shimizu, K., Sawada, T., Valverde, G., McKnight, C., Prawat, R., Wiley, D. E., Raizen, S., Britton, E. D. & Wolfe, R. G.: 1996, *Characterizing Pedagogical Flow: An Investigation of Mathematics and Science Teaching in Six Countries*, Kluwer, Dordrecht/Boston/London.

Schmidt, W. H., Raizen, S. A., Britton, E. D., Bianchi, L. J. & Wolfe, R. G.: 1997, *Many Visions, Many Aims, Volume II: A Cross-National Investigation of Curricular Intentions in School Science*, Kluwer, Dordrecht/Boston/London.

Villani, A. & Arruda, S.: 1998, 'Special Theory of Relativity, Conceptual Change and History of Science' *Science and Education* **7**(1), 85—100.

Wandersee, J. H.: 1985, 'Can the History of Science Help Science Educators Anticipate Students', Misconceptions?' *Journal of Research in Science Teaching* **23**(7), 581–587.

Wandersee, J. H.: 1990, 'On the Value and Use of the History of Science in Teaching Today's Science: Constructing Historical Vignettes', in D. E. Herget (ed.), *More History and Philosophy of Science in Science Teaching*, Florida State University, Tallahassee, pp. 278–283.

Wang, H. A. & Marsh, D. D.: 2001, 'Science Instruction with a Humanistic Twist: Teachers' Perception and Practice in Using the History of Science in Their Classrooms', *Science & Education* **10**(3), forthcoming.

Weiss, I. R.: 1978, *Report of the 1977 National Survey of Science, Mathematics, and Social Studies Education*, U.S. Government Printing Office, Washington, DC.

PART TWO

Foundational Issues in Science Education

The history and philosophy of science bears upon three domains of science education: pedagogical, curricular and foundational or theoretical. Papers Part Two of the anthology contribute to the third domain of theoretical issues in science education.

Science teachers and science educators are faced by numerous theoretical issues: constructivism, particularly its epistemological claims, is an obvious area; multicultural science education is another; so to are feminist critiques of science, and consequent proposals for the reshaping of science education; the relations between religion and science are another perennial issue faced almost daily by teachers and students, as recent events in Kansas, and numerous episodes associated with Islamic science, testify. Intelligent discussion, let alone resolution, of these theoretical issues is impossible without some knowledge of the history and philosophy of science.

This is not a novel opinion: it has long been argued that science teachers should have some competence in the history and philosophy of the subject they teach. As long ago as 1918, the British *Thomson Report* said 'some knowledge of the history and philosophy of science should form part of the intellectual equipment of every science teacher in a secondary school' (Thomson 1918, p. 3). A 1981 review of the place of philosophy of science in British science-teacher education said:

> This more philosophical background which is being advocated for teachers would, it is believed, enable them to handle their science teaching in a more informed and versatile manner and to be in a more effective position to help their pupils build up the coherent picture of science – appropriate to age and ability – which is so often lacking. (Manuel 1981, p. 771)

The Science Council of Canada, after advocating increased attention to historical and philosophical matters in the science curriculum, said: 'Although Council does not expect children or adolescents to be trained in the philosophy of science, it does expect science educators to be trained in this area' (SCC 1984, p. 37).

In the USA, the Association for the Education of Teachers in Science (AETS) has recognised the foregoing concerns, and a position paper recommended that:

> The beginning science teacher educator should possess levels of understanding of the philosophy, sociology, and history of science exceeding that specified in the [US] reform documents. (Lederman et al. 1997, p. 236)

Despite all the advocacy, preservice teacher-education programmes poorly prepare science teachers for intelligent, or informed, discussion of foundational and theoretical questions in science education. The history and philosophy of science is rarely taught in teacher education programmes; and where it is taught, it is frequently covered in a perfunctory manner. In England, a number of preservice courses deal with the Nature of Science in one day of classes. And as teacher education becomes 'more practical' and school based, it is likely that the current scant treatment of foundational matters will be even further reduced.

James Donnelly discusses one major theoretical question: how can compulsory school science can be justified?. Given the backdrop of the 'Science Wars', this is a pressing problem. In answering the problem, Donnelly goes back to examine the basic ontological claims that science makes about the world - what kind of a world is it that science is describing? Donnelly agrees with Anthony O'Hear's analysis that 'it is not in science that the observer of the world explores the meanings and potential meanings things in the world can have for him or establishes just how he should relate to the world or his fellows' (O'Hear 1989, p. 231). Such a view of science requires considerable reanalysis of arguments for science in the curriculum.

Bo Dahlin resurrects phenomenological interpretations of science that have their contemporary roots in Husserl's *The Crisis of European Sciences and Transcendental Phenomenology* (1954) and Merleau-Ponty's *The Phenomenology of Perception* (1962). This tradition has had little impact on Anglo-American theorising about science education, although it is often recognised that the idealisation and formalism that Galileo and Newton introduced into Western science has problems when it is translated unthinkingly into the classroom. In the 1930s the English educator and philosopher, F.W. Westaway, warned that 'once a science lesson arrives at the stage of symbols, it may cease to be Science altogether' (Westaway 1937, p. 4); and he decried the fact that:

> It is astonishing how few, even of the older pupils of a school, are able to give an intelligent physical interpretation of a formula they have established. (Westaway 1937, p. 4)

Half-a-century later the well known US physics educator, Arnold Arons, echoed these sentiments:

> As physics teaching now stands, there is a serious imbalance in which there is an overabundance of numerical problems using formulae in canned and inflexible examples and a very great lack of phenomenological thinking and reasoning. (Arons 1988, p. 18)

Martin Wagenschein, the German science educator, was concerned with the same problem:

> My deepest motive force is the cleavage between an original feeling-at-home within nature and a strong fascination by physics and mathematics, and the ensuing irritation by the growing alienation between man and nature effected by science, starting early at school. My pedagogical aim is to overcome - still better to avoid - the cleavage by an educationally centered humanistic physics teaching. (Wagenschein 1962, p. 9, W. Jung trans.)

The English critic, F.R. Leavis, is just one of many who have decried the unfortunate cultural effect of idealised and formalised science instruction in schools. He believes that its routine substitution of the universal and abstract is poor training for social life and human interactions (Leavis 1972, pp. 207ff). Dahlin's paper reexamines the epistemological claims of phenomenology and details some of their pedagogical consequences for science teaching.

Edgar Jenkins addresses one of the most widespread, and loudest, theoretical debates in contemporary science education: the philosophical and pedagogical claims of constructivism. The title of his paper nicely captures the two extreme positions in the debate: on the one hand the science educator John Staver's view that constructivism is a powerful model for the guidance of science pedagogy and curriculum, and on the other hand the philosopher John Devitt who thinks that constructivism destroys the intellectual immune system that saves us from silliness.

Sibel Erduran outlines key aspects of the emerging field of philosophy of chemistry and indicates how they impinge on chemical pedagogy and learning. In contrast to physics and, more recently, biology, chemistry has not received much philosophical attention. In the past decade this situation has began to change, with a number of books, and at least one journal, being devoted to the subject. It is timely that this flowering be brought to the attention of educators and teachers, as they have long been faced by students who ask questions - such as: 'Miss, if no one has ever seen an atom, how come we are drawing pictures of them?' - that require philosophical competence.

Fritz Kubli takes seriously the 'story telling' aspect of science instruction and asks how investigations into the theory of narrative might enhance science teaching.

Douglas Allchin deals with the important subject of values in science. He maintains, contrary to widespread popular belief, that science is not value free. It has its own constellation of epistemic values that guide its truth-seeking purposes, and its conduct is subject to cultural and social values. Subjects like animal experimentation and cloning are the most obvious areas where external values impinge on the conduct of science. But science also informs and impacts on social values. It is important for teachers and students to consider the reciprocal interaction of values and science, and to have the experience of puzzling over some of the questions that arise. This is one way of overcoming the divorce between science and humanities that F.R. Levis and C.P. Snow lamented in the 1950s.

Alexander Levine discusses an important part of Thomas Kuhn's theory of science: the idea that historical development in science mirrors or recapitulates the process of individual cognitive development. Kuhn in many places states that Piaget's account of children's developmental stages gave him a key to unlock the meaning of ancient, specifically Aristotelian, physics. The problem that Levine identifies is that Piaget equally claimed that the history of science enabled him to interpret the developmental sequence. There appears to be a vicious circle here. It is one that deserves the careful attention of science educators, the moreso given the enormous influence that Kuhn has exercised on educators' thinking about science learning and the processes of conceptual change.

Robert Nola hopes to rescue Kuhn from the embrace of Edinburgh-inspired sociologists of science; and other champions of the view that the cognitive claims of science are not so much a mirror of reality, but a mirror of the society in which science is conducted. It is well know that Kuhn did distance himself from the more enthusiastic Kuhnians. Nola elaborates how he maintained to the end an engagement with the problem of understanding the rationality of theory choice in science, and in the historic process of historical development even through revolutionary change. These problems are standard methodological issues in the philosophy of science and cannot be reduced to 'mob psychology' or other sociological factors.

Arons, A.B.: 1988, 'Historical and Philosophical Perspectives Attainable in Introductory Physics Courses', *Educational Philosophy and Theory* **20**(2), 13-23.

Leavis, F.R.: 1972, *Nor Shall My Sword*, Chatto & Windus, London.

Lederman, N.G., Kuerbis, P.J., Loving, C.C., Ramey-Gassert, L., Roychoudhury, A. & Spector, B.S.: 1997, 'Professional Knowledge Standards for Science Teacher Educators', *Journal of Science Teacher Education* **8**(4), 233-240.

Manuel, D.E.: 1981 'Reflections on the role of History and Philosophy of Science in School Science Education', *School Science Review* **62**(221), 769-771.

O'Hear, A.: *An Introduction to the Philosophy of Science*, Clarendon Press, Oxford.

(SCC) Science Council of Canada: 1984, *Science for Every Student: Educating Canadians for Tomorrow's World*, Report 36, SCC, Ottawa.

Thomson, J.J. (ed.): 1918, *Natural Science in Education*, HMSO, London. (Known as the *Thomson Report*.)

Wagenschein, M.: 1962, *Die Padagogische Dimension der Physik*, Westermann, Braunschweig.

Westaway, F.W.: 1937, *Scientific Method: Its Philosophical Basis and its Modes of Application*, 5th Edition, Hillman-Curl, New York, (orig. 1919).

Instrumentality, Hermeneutics and the Place of Science in the School Curriculum

JAMES DONNELLY
CSSME, University of Leeds, Leeds LS2 9JT, U.K.

Abstract. This article examines some key characteristics of science, under the headings of: the elemination of the personal; demarcation from ethics; and the denial of reflexivity. It relates these characteristics to an instrumental criterion of knowing, which, it is argued, is pervasive in science. The relationship between this complex whole and the interpretative emphasis within both science and the humanities is then examined. The article suggests that these characteristics may underlie the difficulties which science experiences in the curriculum, and that contemporary curricular innovations can be construed as an attempt to address these difficulties by introducing a more thoroughgoing personal, interpretative and humane dimension to the science curriculum. It suggests that there are severe limits to this project.

1. Introduction

The nature of science is not a neglected topic in writing about the science curriculum. There have been several international conferences around the theme of the relationship between history and philosophy of science and school science (Herget 1989; Hills 1992; Finley et al. 1995). Books and articles are plentiful (e.g., Jenkins 1996; McComas 1998). The present article offers an account of some aspects of the 'nature of science' which is unashamedly essentialist in tone. This essentialism is to be taken strongly but narrowly. It is directed at central ontic and epistemic characteristics of scientific accounts of the world. Though I link the ontic and the epistemic, I should perhaps add that my intention is in part to seek to repair a deficiency in the attention given to the former. Thus, of example, in the National Curriculum for England and Wales, the supposed epistemic characteristics of science have figured extensively, if inchoately' leading to an emphasis on the methods in deriving from these supposed epistemic characteristics. Explicit treatment of its ontic character has been entirely disregarded. By contrast, my concern is, in the first instance, with the kinds of being which science recognizes in the world.

Many years ago E.A. Burtt famously argued that natural science makes strong metaphysical assumptions about the world (Burtt 1932). The focus of this article is on characteristics of broadly this kind. The first three are broadly ontic in orientation, with an increasing degree of contingency. Indeed the third may yet prove to be empirically refuted, at least at the higher (emergent) levels. I will call them,

F. Bevilacqua et al. (eds.), Science Education and Culture, 109–127.
© 2001 *Kluwer Academic Publishers. Printed in the Netherlands.*

respectively: elimination of the personal, demarcation from ethics, and absence of reflexivity. The examination of these three characteristics is complementary to a more epistemologically-orientated discussion which draws on the Habermasian distinction between hermeneutics and instrumentality. This discussion focuses on the extent to which science is premised on an instrumental criterion of knowing, and any such criterion is symbiotic with the ontic characteristics which have been identified.

Although I will at times comment on the social sciences, or more indeterminate fields such as economics, the primary target of the analysis is the natural sciences as they are commonly understood within the school curriculum. My aim is to raise questions about the possibilities and problems of science in the school curriculum, rather than to debate questions about the boundaries of the term 'science'.[1] The ultimate argument of the article is that the characteristics I will explore have important implications for the place of science in the school curriculum. They have conditioned, if somewhat implicitly, key aspects of its recent history in many parts of the world, and the underlying concerns speak to fundamental aspects of school science.

2. Elimination of the Personal

The term 'personal' is preferred here to 'mental', although many of its characteristics would in the Cartesian tradition commonly be called 'mental' phenomena. The point is simple. Natural science offers no place for such characteristics as judgement, purpose or personality in its account of the universe, other than as a possible target of reductionist endeavour. Reductionism is not a fashionable philosophical position (Kim 1998, 4–5) and the previous statement is not to be taken as a claim that the reductionist programme can be fulfilled, or that it is coherent. Nor do I claim that scientists personally need to, or do, subscribe to some kind of reductive position. In any case what people subscribe to intellectually, in this context, is less significant than what they enact in their practice. Categories such as judgement and purpose, which I am taking to be characteristic, if not exhaustively, of the personal, have no explanatory power in natural science.[2]

This assumption, or perhaps rather, this self-denying ordinance, links with the trajectory of science and its claim to objective (understood as mind-invariant) knowledge about the world, and thus to universalism. Science offers a perspective on the world which is independent of any particular human perspective: what has been called by Rorty a 'God's eye view' or by Nagel the 'view from nowhere' (Rorty 1980; Nagel 1986). This claim, that the ontic absence of mind in the (scientific) world is linked to the programmatic independence of science from any particular human viewpoint, is perhaps a large one. It becomes more convincing when one examines the writings of some contemporary philosophers, and their attempts to understand how mind, or what has been called the 'space of reasons', can engage the material world (McDowell 1996: 5–13). From my own perspective it links to the instrumentalism which I will discuss later.

The corollary of the ontological point is simple, and hardly controversial or novel: that within scientific explanation as it is normally understood there is a place only for the material, broadly understood.[3] Science acknowledges only a world of mechanism, again broadly understand. There may be a personal dimension to the practice of science, but it is ultimately extraneous to the understandings to which the scientific project aspires. Whether, contingently, human judgement and creativity enter into the production of well-established scientific knowledge is a question of marginal relevance, despite the weight placed on these matters in much talk about the educational claims of science. Whether existing scientific theories are sufficiently underdetermined by the world for possible alternatives to have any serious claim on our attention and meet other scientific criteria is, I suggest, equally marginal. Any such alternatives must surely share the ontic characteristics which I am identifying.

All of this may seem obvious. But as an account, even programmatic, of how science construes the world, and human life in the world, it raises important questions for the place of science in education (O'Hear 1989: 223–32). What is the significance of such an attenuated view of the universe for children, and how its relationship to ontically richer, commonsensical or ascientific views to be handled? Little appears to be written about this in educational settings. There is a body of work which addresses issues related to this point, under such headings as 'cultural border crossings', but it tends to treat the issue as to do with pedagogic or social relations, rather than the place of science in the curriculum or its meaning for children (e.g., Aikenhead 1996).

3. Demarcation from Ethics

It is generally acknowledged, except by those with an extreme scientistic turn of mind, that, though science might tell us how the world is, and how it might be materially altered, it does not speak of the ethical judgements we might make in intervening in the world: of how, so to speak, it ought to be. The argument is linked to those in the previous section. Human beings are fundamentally at least as much ethical as rational beings. There is no place for ethical categories in the conceptualization of the world offered by science. This is not to say that science does not stand in a relationship with ethical issues. There are many discussions of that relationship, including some in the context of science education (Frazer & Kornhauser 1986; Barbour 1992; Poole 1995; Reiss & Straughan 1996; Fullick & Ratcliffe 1996). But the relationship is of a very specific kind. It might be summed up in the statement that ethical categories are external to the ontic categories of science (Williams 1985: 136). The relationship between the two domains, of science and ethics, is commonly discussed under two aspects, and I will suggest this externality applies under both.

The first aspect pertains to the argument that science can provide a value-neutral knowledge which yet can be put to use in making value-laden judgements. Thus, in assessing the policy we should adopt in relation to, say, the exploitation of energy resources or the creation or disposal of chemical products, scientific practices may provide information and conceptualizations without which no judgement can be made. (This is not to claim that such knowledge is necessarily winnable in uncontested form.) Many would argue that, without scientific knowledge, the types of problem just identified would not have been brought into an intelligible focus. Others, less convincingly, claim that scientific knowledge of itself engendered the problems. But science remains firmly outside the sphere of the ethical judgements themselves – they depend on the sort of world we judge desirable.

Implicit in this position is a fairly robust attitude to arguments such as those about the relative impact on science of 'social' and 'cognitive' values, to be found in the recent issue of *Science and Education* which was devoted to examining the relationship between science and 'values', though most of what was written was about the technical implications of science (*Science and Education* 1999). Throughout the exchanges a fundamental point seems largely to be accepted by all participants: that the potentialities of the material world are not to be altered by any number of social values, though of course such values may well influence which possibilities are realized.[4] It is on this point that I ground my claim of the mutual externality of scientific and ethical categories.

Perhaps there are some who claim that science can offer a model of what a desirable world might look like. It might be argued that concepts such as sustainability, genetic diversity, or whatever, are both scientifically meaningful and possible bases for ethical judgements, about our conduct in the world. But supporters of this perspective are perhaps a minority, and in any case it is not convincing as a claim that scientific categories have ethical content. It is more a manifestation of the use of scientific categories to inform, sharpen and perhaps help in the realization of, ethical judgements. That is to say, even outcomes such as biodiversity or sustainability are not usually represented as intrinsically good, but rather as means to some other good, usually long-term human happiness and flourishing. Programmatic attempts have been made to insert ethics in a more thoroughgoing way into the practice of science, so as to inform its outcomes. But this is a far from common position, particularly if it is claimed that the process will have epistemic and ontic consequences (Maxwell 1984; Irzik 1998).

The second aspect under which the relationship between science and ethics is commonly discussed concerns the ethical status of the objects of scientific endeavour, and the experimental practices directed towards them. I suggest that in scientific practice, *qua scientific*, there is in fact no attribution of ethical status to the objects of attention. On the face of it my claim is clearly false. Most reputable scientists have a concern with ethics at some point in their work. Many devote considerable attention to ethical matters, especially in biological disciplines. But I claim that these concerns do not themselves enter scientific practice, or the reasoning associated with it. They set boundaries to the practices which may be used to gain knowledge. Where ethics impinges on the research practices of science, by requiring the scientist to attribute ethical status to objects of study, this attribution limits what can be known, or perhaps, more circumspectly, redirects the process, with possible impacts on the knowledge which can be achieved.[5] At the risk of overstatement, it can be claimed that it is common, though not universal, to find scientists seeking to push the ethical limits on research possessing a strong connection with ethical questions (such as that involving animals, or human embryos, or genetic manipulation) as far as possible.[6] The attribution of ethical status may be sentimentally conditioned in everyday discussions, and this carries over into science. Furry, macro-scale organisms have a different status from others. Nevertheless, such attributions have an important role in defining the boundaries of scientific practice. To avoid misunderstanding, it is perhaps worth restating that this is not to claim that scientists are necessarily without ethical concerns. The claim is rather that, as scientists, they can acknowledge no intrinsic ethical status in the beings (*qua* scientific objects) they study, or allow any ethical status they do attribute to play a part in the reasoning processes, *qua* scientific, in which they engage.

4. The Absence of Reflexivity

This point is perhaps more contingent than the previous two. It is included because it appears to capture a further dimension of the ontic attenuation within the view of the world taken by natural science. Anthony Giddens has distinguished the natural sciences from the social sciences in terms of the absence, in the former, of what he has called a 'double hermeneutic' (Giddens 1983: 284). Giddens uses this term to refer to the tendency for the conceptualizations of social 'scientists' to be taken up by their subjects into their (thereby transformed) self-understanding. (A good example is the way in which role-talk has entered everyday life from sociological theorising.) Indeed Giddens sees this process as part of the entire modernist project: ultimately self-consciously so, such that we are in a constant process of reflective self-transformation (Giddens 1991: chapter 3). This might be thought to be the modernist version of Heidegger's famous claim that human beings have no essence, but are particularly characterized by the fact that their being is always an issue of self-interpretation for them (Heidegger 1962: H15). Ian Hacking has made a similar point in relation to what he calls 'interactive' forms of knowledge (Hacking 1999: 31–2).[7]

But does this notion of reflexivity have any place in our understanding of the material world? I suggest that it does not. Except amongst those who, for example, place a very strong interpretation on Lovelock's Gaia hypothesis, or on the perspective that to measure is to intervene, the physical universe is held to be independent of our knowledge of it. As I have hinted, this potentially is an empirically refutable claim. There might be some who would claim that the world is responding over time to the interventions made in it. Such phenomena as AIDS, ME, BSE, Gulf War syndrome and so on might be judged by those making this claim to illustrate this process in action. But such claims, even if accepted as referring to real phenomena, would surely be assimilated to some more or less complex mechanism (as in Lovelock's own 'Daisy World'). At some fundamental level the world is not merely independent of our understandings of it. It is *a fortiori* unresponsive except in mechanistic ways, that is to say, ways which are themselves continuous with the material worldview of science. This entire claim might be understood as another variation on the theme that there is no space for the personal in natural science. It supports, though it is perhaps not essential to, the epistemic effectiveness of the instrumentalism to be discussed later.

Each of the three claims above is more or less ontic in its reference. Each speaks of how science construes world. Together, they delimit the entities which science acknowledges, their associated qualities and powers and their ethical status in relation to instrumental practices. It is difficult to know what a natural science would look like which allowed the three categories (the personal, the ethical and the reflexive) to have a non-reducible scientific place in the world. Evidently such limits do not, of themselves, exclude the possibility that other perspectives on the world might exist, and that the space of the personal, the reflexive and the ethical can co-exist with the space of science. Indeed the writings of phenomenologists such as Husserl and Heidegger might be thought of as centrally concerned with this possibility. Even Richard Dawkins seems, somewhat uneasily, to be attempting to occupy something like this position (Dawkins 1998). But science has long had a reductionist tendency. Even if the reductionist programme has fallen into philosophical disfavour, it retains some power at least at a crude ideological level, or perhaps not so crude in the philosophy of mind (Churchland 1989). At the very least, science is thoroughgoingly physicalist in its orientation. While there have been many attempts to integrate science with more humanistic, even mystical, conceptions of the world, such attempts remain peripheral to mainstream science, and certainly to the science of the school curriculum (Lorimer 1998).

Though there is some explicit acknowledgement in the school science curriculum of generic characteristics of science, this acknowledgement is to a remarkable degree epistemological rather than ontological in orientation. The argument of this section of the paper has been that science indeed has generic characteristics, but that these characteristics are more obviously ontic than epistemic in character. They concern the kinds of entities, powers and processes which are acknowledged as the legitimate domain of science. Natural science abrogates key aspects the lived world of human experience, not contingently, but in its programmatic ontic assumptions. However, though I claim that science is characterized primarily by its concepts, and not by its methods, I suggest that this ontic construal of the world has an important epistemic dimension. The methods of science disclose a world under a particular materialistic aspects, and its key criterion of knowledge, from which these methods derive, is similarly characterized.[8] These matters will be the focus of the second, epistemologically-orientated part of the article.

5. The Instrumental and the Hermeneutic

In his book *The Mangle of Practice* Andrew Pickering distinguished what he called 'representational' from 'performative' aspects of science (Pickering 1995). It is the former, he claims, which has received the greatest attention in science studies. Only in recent times has laboratory work in its full material sense received an appropriate measure of attention, in a range of studies (e.g., Radder 1988; Gooding, Pinch & Schaffer 1989; Gooding 1990). The representational is concerned with symbolic and discursive practices. Such practices confer enormous advantages in terms of generalizability, stability, portability, linkage, flexibility and so on. They sustain human interpretations of the world. They are the vehicles of argumentation, discourse and dialogue. In sum, they stand at the centre of most human intellectual activity. But interpretative practices in the humanities resist being driven to any universally accepted agreement. By contrast, one of the key characteristics of science is that, despite assuredly involving interpretation, its practice commonly leads to consensus.

Why should science be different? The view that it embodies some kind of privileged 'method' guaranteeing access to truths has long since been discarded, except perhaps in the stonier reaches of the school curriculum. Part of the answer, it has been implied earlier, lies in what science refuses and, those areas of specifically human experience and practice about which it is therefore silent. The bifurcation in play here can be traced back to the Descartes and beyond (MacIntyre 1987: chapter 7).

But another crucial influence lies in the prominence within science of the performative – which can be identified, as I shall claim, with the instrumental, because of its underlying agenda of prediction and control (Pickering 1995: 7). The perspective can be stated succinctly. In science, to understand the world, to have knowledge about it (to have interpreted it aright) is to be able to predict and control its behaviour at a material level. It reverses the thrust of Bacon's famous claim *Nam et ipsa scientia potestas est* (Bacon 1857–70: 241), and in doing so speaks of what it is to know, scientifically. Instrumentalism in its usual philosophical sense contrasts with realism. Here I am using the term more broadly, related to Deweyan pragmatism and grounded in action in the world. It is consistent with, though it does not imply, realism concerning scientific entities and theories. Similarly, it does not require, though it is consistent with, the narrower sense of philosophical instrumentalism, that scientific entities and theories are mere calculative instruments, the position now most famously associated with van Fraassen's constructive empiricism.

A few further points perhaps should be made in connection with my argument. There are of course areas of natural science where the notion of control is inapplicable, at least at present: most notably astronomy, but also areas of Earth science. I suggest that the logic of prediction and that of control are essentially similar. Each is about predicting future from present states. The difference lies in the practicability of material intervention to promote change, and thus, so to speak, counterfactual prediction. But this is essentially a contingent (sometimes historically contingent) distinction, not an essential one. Moreover I am not claiming that the reverse argument applies, that is, that all domains involving prediction and control are part of science in its strong disciplinary sense. Second, this emphasis on instrumentality may be thought to attenuate the notion of scientific 'understanding' and other criteria for the quality of scientific knowledge, such as simplicity, consistency with other theories, range and so on. No doubt these and many other influences and criteria, including some which might be thought of broadly as social, have an impact on scientific judgements.[9] But in the last resort I concur with Goldman's argument that empirical (read 'instrumental') effectiveness is the fundamental criterion at work within science, including for what it means to 'understand'.[10] (I will return to the interpretative aspect of 'empirical effectiveness' below) (Goldman 1999: 245).

I suggested earlier that despite arguments over the relative impact of so-called cognitive and social values there seems to be general acknowledgement that no set of values can alter the material potentialities of the world. This point may be taken a little further. Despite the limitations of the word 'cognitive', in consequence of its de-emphasizing of the element of materially and socially enacted practice, certain 'cognitive' values are symbiotic with, perhaps even constitutive of, instrumental effectiveness. This, in turn, is arguably constitutive of materiality in its narrow scientific sense. As Heidegger argued in *The Question Concerning Technology*:

> Modern science's way of representing pursues and entraps nature as a calculable coherence of forces. Modern physics is not experimental physics because it applies apparatus to the questioning of nature. The reverse is true. Because physics, indeed already as pure theory, sets nature up to exhibit itself as a coherence of forces calculable in advance, it orders its experiments precisely for the purpose of asking whether and how nature reports itself when set up in this way. (Heidegger 1978: 302–3)

To Heidegger, our enthusiasm for identifying the form of being disclosed by science with all being is one of the great threats to humankind.

It is because natural science *enacts* a broadly instrumental perspective on the world that it is able to avoid the endless cycle that characterizes more unequivocally interpretive practices. The corollary of this is that what cannot be manipulated (either predictively as with astronomy or materially in the laboratory) cannot be known, scientifically. In many fields with a substantial empirical component, such as history, the interpretive character remains dominant, and closure is rarely achieved in part because (*contra* science) human motivations, intentions and actions are implicated non-reductively in the substance of what is studied. These domains, and this point is central to the overall argument of this paper, are not amenable to instrumental forms of practice.

Allowing beings to be the subject of instrumental action, or not, has ethical implications, most obviously in respect of human beings. Human beings are attributed a subjective status and an authority over themselves and their bodies which is inimicable to their being treated merely as manipulable objects, and this limits their treatment as objects of science. Other living creatures, and sometimes the Earth itself, are occasionally attributed this quality in modified form. It may be recalled that a person who seeks to use others as mere tools, means or objects is often described, with a derogatory connotation, as 'manipulative'. The ways of knowing associated with science are strictly limited in their applicability to human beings and human relationships, including within classrooms (Donnelly 1999). There is an intimate link between science's epistemic instrumentality and the ontic characteristics identified in the previous section.

Jurgen Habermas used the term 'instrumental sciences' to signal a great division from what he called the hermeneutic sciences (using 'science' in its broader Continental sense). Hermeneutics may be taken here to refer to a self-consciously interpretive practice, or perhaps even the 'theory' of such a practice. Within instrumental science 'the meaning of the validity of statements is determined with reference to possible technical control of the connection of empirical variables' (Habermas 1972: 94–139). Habermas ultimately contrasts this with that hermeneutic interest which aims at 'the maintenance of the intersubjectivity of mutual understanding within whose horizon reality can first appear as something' (ibid. 176). Note Habermas's Heideggerian placement of 'reality' (and implicitly instrumentality) in a derivative status. This formulation of the relationship is an important strand within phenomenological writing, including that of Husserl (Moran 2000: 144).

To a degree concomitant with their specific subject matter the hermeneutic 'sciences' display, in relation to each of the three characteristics which I identified in the earlier section, a counteremphasis to the instrumental sciences. They are strongly personal in orientation (though this is not to say subjective), indeed they depend for their special characteristics on the different interpretative standpoint of another person. Indeed the hermeneutic relationship often appears to be intrinsically three-way, involving both the object of study and others engaged with it. In natural science, as I have suggested, the 'other' (as the focus of attention) is necessarily a material object or system. The hermeneutic sciences acknowledge the thoroughgoing equivalence in ethical status of the other, one which goes well beyond even that attenuated status which requires a 'duty of care', as when non-human animals are objects of scientific practice. Where possible, these 'sciences' acknowledge reflexivity in the hermeneutic endeavour: that the process may well change the character of its 'object', to the extent that that 'object' can understand the interpretations made, or even when it cannot, as in our deepening understanding of a work of art. These characteristics figure in most humane curricular areas. Two important aspects might be highlighted. Firstly the subject matter is commonly the lives, writings or speech of others. Secondly, teaching is predicated on establishing an interpretive relationship on the part of pupils with the object of study, one which is commonly explored in dialogue with the teacher. Under both of these aspects, in however embryonic and primitive a sense, the ethical, the personal and the reflexive are fundamental and inescapable.

Curricular areas (the boundaries of which are of course, to some degree, arbitrary) exhibit a wide and heterogeneous engagement with the two ideal-typical modes of knowing I have sketched. But it can hardly be doubted that natural science sits at one extreme, in its programmatic elimination of the personal within its objects of knowledge and even its mode of knowing. Disciplines such as history and English literature sit at the other. Even technology, which might appear the archetypal means-ends field, is quite distinct from science in this respect, for two reasons. First, technology does not, at least not necessarily, aspire to the universality, the, as it were, perspectival-independence of science. Second, technology usually engages directly with human wants and needs, and the meaning and significance of those needs and wants involves an inescapable interpretive engagement with the humanity of those who bear them.

I am not seeking to deny that natural science possesses a hermeneutic dimension (Eger 1993a, b; Ihde 1998). The practice of science involves extended, symbolically mediated interaction. It is difficult to imagine any form of scientific activity, even the most severely instrumental, which would not involve a very substantial symbolic component. As Radder suggests: 'experimental action is discursively mediated right from the start' (Radder 1988, 170). The point might be put differently: what it is to be instrumentally effective, even at a material level, is a matter of interpretation. But instrumentality has a way of asserting itself, as in Alan Sokal's invitation to any sceptic about realism to step out of his (21st floor) window.[11] Jon Ogborn argument for realism is: 'You can think what you like but you cannot do what you like'. Yet the attenuation that goes with this argument must be acknowledged. What else are we to understand by Ogborn's statement but an instrumental criterion of (recognizing) reality. The elision of the boundary between knowing and doing is clear enough (Ogborn 1995).

Despite the ineluctability of the hermeneutic, instrumentality is the ultimate court of appeal in science. It might be formulated somewhat metaphorically as a limiting case which is never reached in its pure form. Even the death which might be the outcome of leaving Sokal's flat by the window has been shown in recent years to be not quite the limiting case it once was, but to be, in Radder's terms, discursively mediated. Even so, without instrumentality science would be unable to escape from the endless (though of course not necessarily fruitless) hermeneutic movement of the humanities. 'Thinking with empirical content' would, to appropriate McDowell's vivid image, 'degenerate into a picture of a frictionless spinning in a void' (McDowell 1996, 50).

In sum, I wish to argue that these symbiotic ontic and epistemic characteristics are the necessary, though perhaps not sufficient, conditions under which natural science achieves its special consensual and technically effective character. It might be said that all of this is simply an elaborate way of saying that empirical science is empirical. I accept the point in part, but I think my formulation sharpens and disaggregates the issue. It also speaks directly to the curriculum.

6. Science in the Curriculum

Natural science, except as interpreted by affiliates to a reductive physicalism, does not deny the personal or the ethical. It merely stands aside from it. Occasionally, consciousness of the problems of such an ethically and personally bereft view of the world surface. We were blandly told in one version of the National Curriculum for England and Wales that 'while science is an important way of thinking about experience, it is not the only way' (DES/WO 1991: 22). But this acceptance of what might be called an ontic parallelism is only bland if it is assumed that pupils take an equally disinterested attitude towards these various construals of the world. The possibility must at least be entertained that they do not. Many pupils' central interest, as they mature, is likely to be increasingly in the humane, the personal and the aesthetically and ethically significant.[12] The statement cited above has vanished from the National Curriculum for England and Wales, in the cavalier way of these documents with intellectual issues. One finds in the current version that pupils' 'spiritual development' is to be promoted within the science curriculum by 'exploring questions such as when does life start and where does life come from?' (DfEE/QCA 1999: 8). What these grand-sounding statements mean in terms of the science curriculum remains to be seen.

There has been a wide-ranging impetus towards curriculum development within science education since at least the 1950s, supported in recent years by arguments which suggest that science as a field of study is failing to attract pupils. It is not always easy to distinguish between shifts in pedagogy and shifts in curricular aims within the changes that have been envisaged. Their character has been superficially diverse. I suggest that they can be classified under three broad, though not always distinct, orientations. There has been an effort to increase the emphasis on inquiry by pupils themselves. There has been the Science Technology and Society (STS) 'movement'. Finally, there has been sustained exploration of so-called constructivist pedagogy, in which attention is given to the transformation of what are represented as pupils' spontaneous or everyday ideas, and to pupils' self-conscious achievement of understanding. I will argue in this third section that these apparently divergent themes reflect a common impetus within the recent history of the science curriculum: to oppose that closure of the hermeneutic dimension, which, I have suggested, is the distinguishing epistemic characteristic of science. More specifically, the impetus is to undermine the perception of science as divorced from ethical and personal concerns.

According to Black and Atkin, an emphasis on inquiry-orientated laboratory activity is a commonplace in curriculum development around the world (Black & Atkin 1996, 32). The rationale for this emphasis is largely a variation on two themes. It is claimed that inquiry-orientated pedagogy offers children insight into the methods by which science achieves reliable knowledge. Beyond expressing some scepticism, I will make no further comment on this claim. It is also often suggested that such methods pass ownership of learning, or of the science to be learnt, to children themselves. The thought appears to be that undertaking independent inquiries enables pupils to engage at a personal level with science. The process might thus be termed a 'personalization' or perhaps 'individuation' of science in the curriculum, in which pupils' interpretive activity has a legitimate and central place.

A range of issues could be raised here, some of which have been explored in commentaries on the British 'Nuffield' projects. At the most concrete level, it is doubtful, at least as matters have progressed within English science education, whether it is possible to identify areas of experimental activity in secondary schools which are in some realistic sense open, and in which may be enacted, more or less authentically, the practices of inquiry (Donnelly et al. 1996). Allowing the possibility that such activity might be identified on any scale,[13] such a pedagogic strategy might, perhaps, show that the process of scientific investigation is exciting, complex and difficult, and sometimes ineffective. It might be possible to enable pupils to demonstrate a form of creativity. But, even if all of these desirable outcomes were to be achieved, the epistemic imperative, and ontic programme, of science would remain. The points which I have raised would remain untouched. Furthermore, the trajectory of scientific inquiry is in the reverse direction from individuation. The investigatory process serves to demonstrate how science removes knowledge from the domain of the individual, making it universal and depersonalized. It is difficult to see how such a process could also serve the purpose of individuating science in its relationship with pupils.[14] It might be argued against this position that all intellectual endeavour, including that of the humanities, seeks intersubjectivity, and even universality. Some part of this response I accept, though I suggest that the emphasis on universality is close to being specific to science. But the intersubjectivity of the humanities is quite different from that of the sciences. Deeply hermeneutic practices remain at the heart of the process, as does the individual human response. The intersubjectivity which characterizes the humanities does not achieve, or even seek, that closure which characterizes the natural sciences: the humanities are chronically dialogic. I suggest that they have this character because they have no instrumental moment. Scientific knowledge is very different. It becomes, to use Bruno Latour's metaphor, a black box which is difficult, if not impossible, to re-open. Let us also recall that the possibility of such instrumentality rests on the ontic refusal of the personal and the ethical in the self-chosen subject matter of the natural sciences. These two characteristics of natural science are not merely contingently related.

It is not an accident that the primary function of laboratory work within school science, beyond its contingent motivational effects, has historically been, and usually still is, quite other than the project of 'inquiry-based' science education suggests. Most of the science to which pupils in school are exposed is at present well- established. Laboratory work demonstrates, or is intended to demonstrate, the instrumental power and material grounding of science. To put the point as emphatically as may be, the ubiquity of the laboratory, and the great extent to which it conditions science teachers' work, in England at least, might be seen precisely as a celebration of the distinction between natural science and the hermeneutic disciplines (Donnelly 1998). It is a celebration of closure, of the means of closure and of the end of interpretation.

The second form of curriculum development, the STS curriculum is, at one extreme, little more than a motivational tool for more traditional curricular aims, based on social and technological issues in which science is implicated. At the other, it embodies an effort to transform the central aim of science education into that of promoting critical understanding of the creation, use and implications of scientific knowledge. This too is a form of personalization, in which individual judgement, and the absence of closure, are central. It is significant that the STS movement draws substantially on historical dimensions of science, and thus appeals to a discipline, history, with a radically hermeneutic aspect[15] (Solomon & Aikenhead 1994; Donnelly 1999).

The attempt to 'hermeneuticize', or perhaps de-instrumentalize, school science is indeed most convincing in the context of the more radical forms of the STS movement. But it is convincing precisely because the focus of the pupils' attention is not scientific knowledge in its usual sense. The issues of ethical and political judgement which STS curricula invite pupils to address are of irreducibly human significance within the world. There is no possibility of a instrumental resolution. STS curricula might introduce the pupil to the view that the success achieved by science, through bracketing the complexity of the world, is dangerous. It might demonstrate that the process of scientific investigation is, as Lacey claims, apt to interact destructively with the social circumstances surrounding it (Lacey 1999). Yet in all of this there is some sleight of hand. What such curricula invite is the replacement of education in science with curricula in what might be loosely called the political sociology of science. Such a shift has a range of implications. To sloganize: it is not an education in science, but an education about science. This may be a desirable change in emphasis, though I venture to doubt it. But let us not pretend that it is other than a major shift in the foundations and character of the science curriculum. It would involve major shifts in the expertise which teachers must display.

The sleight of hand operates also in a different direction. On what basis does science sustain its place in the curriculum? (Let it not be forgotten that, in England and Wales, late secondary age children commonly spend more time studying science than any other school subject.) I suggest that the political support for its place derives from the claims that science makes as an incomparable human intellectual achievement, from the authoritative knowledge of the material world which it provides, and from its supposed economic benefits. Whether the claims of science in the curriculum would be sustained by the argument that it provides a vehicle for political education, or a vehicle for education in ethical judgement, or a preparation for critical encounters with expertise is, at best, not proven. None of these domains is uniquely linked to science, and there are other areas of the curriculum which might prove better educational vehicles for their exploration. If, on the other hand, it is suggested that the application of these qualities in specifically scientific settings merits the dedication of a major proportion of our educational resources, one can perhaps only observe that the case has yet to be made on the necessary scale, and so as to address the population at large, who in the end provide the resources for science education.

Of constructivism I have little to say, in part because it is difficult to identify any clear theme within the wide-ranging issues, arguments and controversy surrounding this term, but also because it can often appear as no more than a glorified pedagogic technique. Nevertheless, I suggest that its advocates commonly display an emphasis on the individual pupil's relationship with scientific knowledge which shows at least some overlap with the themes of individuation, judgement and interpretation that have been my concern here.

In each of these three approaches to curricular change there is a strand which simply seeks to find more effective methods of teaching scientific knowledge. But, in each, with the possible exception of 'constructivism', there is also a strand which seeks to transform the deeper aims of the science curriculum. I have suggested that these innovations, particularly in their radical manifestations, but even as mere instruments for the promotion of better scientific learning, share a common impetus. In Habermasian terms, that impetus can be construed as breaking down the barriers between instrumental and hermeneutic disciplines, so as to radicalize, strengthen and individuate the interpretative moment within science education. It might also be seen as putting at a more individual level the relationship between pupils and scientific knowledge. To use more traditional terms, it might be judged as seeking to find a place within science for the highest aims of 'liberal' or 'humane' education. I have made little use of these terms, but I understand them to refer to an education which focuses on the development of pupils' capacity to bring to bear independent powers of ethical, intellectual and other judgement, and to do so facing other human beings exercising similar powers. Central to this point is the requirement that the subject matter under study should be authentically open to such judgement and, what is the corollary of this point, intrinsically resistant to closure. Can this be the case with the natural sciences to be taught in schools?

7. Conclusion

The construal of the world sketched in the first part of this paper is, as I claim, a necessary, though perhaps not sufficient, condition of the effectiveness and power of science. The characteristics identified are, fundamentally, ontic, though their ontic character is mediated by an epistemic practice which reveals the world under an instrumental aspect. The personal, the ethical and the reflexive cannot be addressed through a practice dominated by instrumentality, in its engagement with the world, and by closure, in its intended outcomes. This difficulty has implications for the character and limitations of the educational aims to which the science curriculum might aspire. As a result, despite its claim to be perhaps the greatest of human intellectual and material achievements, natural science can often appear to occupy a marginal place within those higher aims of education which extend beyond engagement with and mastery of an authoritative, pre-given body of knowledge.

Contemporary curricular developments in science tacitly acknowledge this difficulty. But they are severely disabled by a lack of intellectual self-awareness, a piecemeal and superficial approach, and a failure to understand the challenges offered by the ontic and epistemic character of science. It is, then, ironic that it is precisely these challenges which, as I claim, underly much of the motivation for change. It seems to me inescapable that the first and most obvious educational purpose of science in education is to teach pupils about the most authoritative knowledge we have of the world, under its material aspect. The coherence and legitimacy of this purpose must be acknowledged, both intellectually and politically, rather than side-stepped or denied. It is only against this background, and the difficulties which it brings, that a project to reposition science education within the wider aims of a humane and liberal education can realistically be countenanced. Indeed it might be argued that the starting point for such a project ought to be explicitly to confront pupils, when they are mature enough, not with an ersatz political sociology or epistemology of science, but with the perils and challenges of a materialistic account of the world.

Notes

[1] I am aware that in some settings, e.g., the AAAS *Benchmarks for Science Literacy* (AAAS, 1993) an effort is made to include the human and social sciences within the notion of science, or at least science literacy. This seems to me to be an unhelpful approach, because it appears to ignore arguments over whether or to what degree these latter areas can adopt an empirical approach, because it ignores the issues raised by the emphasis within the social sciences on what has been called the double hermeneutic and interactive forms of knowledge (see below) and because, at least in the AAAS document, it results in statements of supposed 'scientific knowledge' which are little more than platitudes. (For example: 'Sometimes social decisions have unexpected consequences ...'. AAAS, 1993: 165).

[2] On empirical psychology, see note 5 below.

[3] I should perhaps stress that I am only here referring to the programmatic assumptions of science, and not committing myself to a dualistic theory of mind and matter.

[4] Lacey has provided a subtle book-length development of his argument, though I would still wish to claim that the final point just made stands (Lacey 1999). To anticipate a later section, it might be claimed that the emphasis within science on prediction and control is a kind of 'value', and one which preferentially articulates with projects which similarly stress prediction and control (such as those of the 'military-industrial complex'), and which are taken to be less socially desirable than other projects (such as supporting 'organic' farming). This argument, if sustainable, undermines the independence which I am claiming. But the argument assumes too great a similarity between the

place of prediction and control in the context of scientific practice and other settings. Science values prediction and control (as a measure of understanding) as nearly as possible for its own sake, as an indicator of knowledge. That is as true of 'industrial' or 'military' science as 'pure' science. There would be no purpose in creating science for industrial purposes which proved to be incapable of instrumentally enabling the necessary technical activity. To return to the point made in the main text: no number of 'values' can alter the material *potentialities* of the world. The valuing of prediction and control of the material world in other domains seems to me to be largely subservient to some other value, such as military or economic superiority. Further, I suggest that all material practices to some degree value prediction and control. What makes a difference is the wider ethical environment in which these 'values' are set. Prediction and control have a place in organic farming, the conservation of biodiversity and so on.

[5] I am again setting aside the question of whether scientific understandings are sufficiently underdetermined by the world for these influences to have substantive impact on the conceptualizations arrived at. Should this be the case, then ethical categories might be construed as not external to those of science, and influential on scientific knowledge. But it is a weak and, as it were, contingent relationship, still some distance from the claim that scientific categories have intrinsic ethical content. It leaves untouched the claim that the material/performative possibilities of the world are independent of our ethical judgements.

[6] My argument also raises more specific questions about the human sciences. A recognition of the irreducible and ineluctable individuality of human subjects is a necessary precondition for socially acceptable research in this area, certainly in the non- clinical area, and probably also in the clinical. An instrumentally-orientated empirical psychology which construed human beings as merely the targets for manipulative intervention (while perhaps technically possible) would be highly questionable ethically. The argument here derives in part from later points relating to the instrumentalism which I take to be a central aspect of science. Only under very specific circumstances, such as clinical medical practice, or highly safeguarded voluntaristic settings, can ethically significant beings be subject to a thoroughgoing instrumental practice.

[7] This analysis stands in a somewhat problematic relation to a field such as economics. Economic behaviour might be construed naturalistically, but the case that the categories of economics are arbitrary human creations is stronger than it is in the natural science. Certainly the phenomena reported under those categories are likely to be reflexively influenced by economists' reporting of them, as they become known to economic actors. This is another reason for the restriction of my argument to the natural science.

[8] When using this argument in seminars I have been surprised to find it interpreted as a direct advocacy of a science curriculum based entirely on what educationalists call 'content' (by which term I understand them to mean scientific knowledge). As a corollary, I am thought to be claiming that attempts to introduce pupils to the methods, limitations and implications of science are mistaken. This is not my direct intention, but I must add two qualifications. First, I hope that the arguments I am putting forward about science will be judged on their merits, rather than on their supposed implications for whatever is currently fashionable or unfashionable within science curriculum development. Second, much of what passes for education in the methods, limitations and implications of science appears platitudinous or contentious, and to have little to do with the real needs of citizens confronted with contentious issues with a scientific dimension. As I indicate later, such innovations often appear more like an attempt to find a more convincing, or at least fashionable, purpose for a science curriculum apparently in difficulties.

[9] Paul Dirac once famously, and quotably, argued that '(i)t is more important to have beauty in one's equations than to have them fit experiment'. The quotation might be judged an example of the hubris of the theoretician, though when read in context it appears to suggest that Dirac was making a more subtle point, about the historical development of scientific ideas (Dirac 1963: 47).

[10] To anticipate, hermeneutics has commonly distinguished between scientific causal *explanation* and that *understanding* which is more properly the domain of the humanities. Whether it is called 'understanding' or 'explanation', the key distinction is contained in the reference to causality, with the key possibility, and criterion of acceptability, of using these systems of causality for predictive or manipulative purpose (Moran 2000: 276).

[11] Sokal was the perpetrator of the now infamous *Social Text* hoax.
http://www.physics.nyu.edu/faculty/sokal/lingua_franca_v4/lingua_franca_v4.html

[12] In certain circles of UK science education it is a conventional wisdom that pupils in primary schools are more enthusiastic about science than those in secondary schools. This is usually attributed to the quality of the science teaching they experience. So far as can be judged the hypothesis that the shift is due to the growing maturity of pupils has never been seriously considered (Ford & Baxter 1998, 29).

[13] I am not referring to such 'investigations' as 'the effect of acid concentration on the reaction between marble chips and dilute hydrochloric acid', the staple of 'investigatory' work in English schools.

[14] In other words, the difficulties which the introduction of investigatory work into the National Curriculum for England and Wales has experienced is no mere contingency of bungled implementation, and wilful ignorance of the realities of science teachers' work, but reflects fundamental philosophical difficulties (Donnelly et al. 1996).

[15] A careful analysis of the raiding of some of the science studies literature by science educationists has been provided by Turner and Sullenger (1999).

References

Aikenhead, G.S.: 1996, 'Science Education: Border Crossing into the Sub-culture of Science', *Studies in Science Education* **27**, 1–52.

American Association for the Advancement of Science: 1993, *Benchmarks for Science Literacy*, Oxford University Press, New York.

Bacon, F.: 1857–70, 'Meditationes Sacrae. De Hæresibus', in James Spedding, Robert Leslie Ellis and Douglas Denon Heath (eds), *The Works of Francis Bacon*, Vol. VIII (part II), Longman & Co, London.

Barbour, I.: 1992, *Ethics in an Age of Technology*, SCM Press, London.

Black, P. & Atkin, J.M.: 1996, *Changing the Subject. Innovations in science, mathematics and technology education*, Routledge/OECD, London.

Burtt, E.A.: 1932, *The Metaphysical Foundations of Modern Physical Science*, Routledge & Kegan Paul, London.

Churchland, P.M.: 1989, *A Neurocomputational Perspective on the Nature of Mind and the Structure of Science*, MIT Press, Cambridge, MA.

Dawkins, R.: 1998, *Unweaving the Rainbow. Science Delusion and the Appetite for Wonder*, Allen Lane, London.

DES/WO: 1991, *Science in the National Curriculum (1991)*, HMSO, London.

DfEE/QCA: 1999, *Science. The National Curriculum for England*, HMSO, London.

Dirac, P.A.M.: 1963, 'The Physicist's Picture of Nature', *Scientific American* **208**(5), 45–53.

Donnelly, J.F.: 1998, 'The Place of the Laboratory in Science Teachers' Work', *International Journal of Science Education* **20**, 585–596.

Donnelly, J.F.: 1999, 'Interpreting Differences: The Dducational Aims of Teachers of Science and History, and Their Implications', *Journal of Curriculum Studies* **31**, 17–41.

Donnelly, J.F.: 1999, 'Schooling Heidegger: On Being in the Classroom', *Teaching and Teacher Education* **15**, 933–949.

Donnelly, J.F., Buchan, A., Jenkins, E., Laws, P. & Welford, G.: 1996, *Investigations by Order. Policy, Curriculum and Science Teachers' Work under the Education Reform Act*, Studies in Education, Driffield.

Eger, M.: 1993a, 'Hermeneutics as an Approach to Science, part I', *Science & Education* 2, 1–30.

Eger, M.: 1993b, 'Hermeneutics as an Approach to Science, part II', *Science & Education* **2**, 303–328.

Finley, F. et al.: 1995, *Proceedings, Third International History, Philosophy, and Science Teaching Conference*, University of Minnesota, Minneapolis.

Ford, S. & Baxter, M: 1998. 'Comments from the 11-16 Committee', *Education in Science* **179**, 29.

Frazer, M.J. & Kornhauser, A.: 1986, *Ethics and Social Responsibility in Science Education*, Pergamon/ICSU, Oxford.

Fullick, P. & Ratcliffe, M.: 1996, *Teaching Ethical Aspects of Science*, Bassett Press, Southampton.

Giddens, A.: 1983, *The Constitution of Society*, Polity, Cambridge Press.

Giddens, A.: 1991, *Modernity and Self-Identity. Self and Society in the Late Modern Age*, Stanford University Press, Stanford.

Goldman, A.I.: ????, *Knowledge in a social world*, Clarendon Press, Oxford.

Gooding, D., Pinch, T. & Schaffer, S. (eds): 1989, *The Uses of Experiment*, Cambridge University Press, Cambridge.

Gooding, D.: 1990, *Experiment and the Making of Meaning*, Kluwer Academic Publishers, Dordrecht.

Habermas, J.: 1972, *Knowledge and Human Interests*, Heinemann, London.

Hacking, I.: 1999, *The Social Construction of What?*, Harvard University Press, Cambridge, MA.

Head, J.: 1985, *The Personal Response to Science*, Cambridge University Press, Cambridge.

Heidegger, M.: 1962, *Being and Time*, Blackwell, London.

Heidegger, M.: 1978, 'The Question Concerning Technology', in *Basic Writings*, Routledge & Kegan Paul, London, pp. 283–318.

Herget, D.E.: 1989, *The History and Philosophy of Science and Science Teaching, Proceedings of the First International Conference*, Science Education and Department of Philosophy, Florida State University, Tallahassee, Florida.

Hills, S.: 1992, *The History and Philosophy of Science in Science Education, Proceedings of the Second International Conference on the History and Philosophy of Science and Science Teaching*, Faculty of Education, Queens University at Kingston, Ontario.

Hodson, D.: 1998, *Teaching and learning science : towards a personalized approach*, Open University Press, Buckingham.

Ihde, D.: 1998, *Expanding Hermeneutics. Visualism in Science*, Northwestern University Press, Evanston, IL.

Irzik, G.: 1998, 'Philosophy of Science and Radical Intellectual Islam in Turkey', in W.A. Cobern (ed.), *Socio-cultural Perspectives on Science Education. An International Dialogue*, Kluwer Academic Publishers, Dordrecht.

, E.W.: 1996, 'The 'Nature of Science' as a Curriculum Component', *Journal of Curriculum Studies* **28**, 2, 137–50.

Kim, J.: 1998, 'The Mind-body Problem after Fifty Years', in A. O'Hear (ed.), *Current Issues in Philosophy of Mind*, Cambridge University Press, Cambridge, pp. 3–23.

Lacey, H.: 1999, 'Scientific Understanding and the Control of Nature', *Science & Education* **8**(1), 13–35.

Lacey, H.: 1999, *Is Science Value Free? Values and Scientific Understanding*, Routledge, London.

Lorimer, D.: 1998, *The Spirit of Science. From Experiment to Experience*, Floris Books, Edinburgh.

MacIntyre, A.: 1987, *After Virtue. A Study in Moral Theory*, Duckworth, London.

Maxwell, N.: 1984, *From Knowledge to Wisdom. A Revolution in the Aims and Methods of Science*, Blackwell, Oxford.

McComas, W.F.: 1998, *The Nature of Science in Science Education, Rationales and Strategies*, Kluwer Academic Publishers, Dordrecht.

McDowell, J.: 1996, *Mind and World*, Harvard University Press, Cambridge, MA.

Merton, R.K.: 1973, 'The Normative Structure of Science', in *The Sociology of Science. Theoretical and Empirical Investigations*, University of Chicago Press, Chicago, pp. 267–278.

Moran, D.: 2000. 'Husserl and the Crisis of European Science', in M.W.F. Stone & J. Wolff (eds), *The Proper Ambition of Science*, Routledge, London, pp., 122–150.

Nagel, T.: 1986, *The View from Nowhere*, Oxford University Press, Oxford.

O'Hear, A.: 1989, *Introduction to the Philosophy of Science*, Oxford University Press, Oxford.

Ogborn, J.: 1995, 'Recovering Reality', *Studies in Science Education* **25**, 25–38.

Pickering, A.: 1995, *The Mangle of Practice. Time, Agency, and Science*, University of Chicago Press, Chicago.

Poole, M.: 1995, *Beliefs and Values in Science Education*, Open University Press, Buckingham, UK.

Radder, H.: 1988, *The Material Realization of Science*, van Gorcum, Maastricht.

Reiss, M., & Straughan, R.: 1996, *Improving Nature? The Science and Ethics of Genetic Engineering*, Cambridge University Press, Cambridge.

Rorty, R.: 1980, *Philosophy and the Mirror of Nature*, Blackwell, Oxford.

Science & Education: 1999, Special Issue on Values in Science and in Science Education **8**(1).

Solomon, J. & Aikenhead, G.: 1994, *STS Education. International Perspectives on Reform*, Teachers College Press, New York.

Turner, S. & Sullenger, K.: 1999, 'Kuhn in the Classroom, Lakatos in the Lab: Science Educators Confront the Nature of Science Debate', *Science, Technology and Human Values* **24**, 5–30.

Williams, B.: 1985, *Ethics and the Limits of Philosophy*, Fontana/Collins, London.

The Primacy of Cognition – or of Perception? A Phenomenological Critique of the Theoretical Bases of Science Education

BO DAHLIN
Department of Educational Sciences, Karlstad University, Sweden

Abstract. This paper is a phenomenological critique of a particular trend in educational research and practice, which is identified as "cognitivism". The basic feature of this trend is a one-sided and exclusive focus on conceptual cognition and concept formation, with a simultaneous neglect of sense experience. It is argued that this kind of thinking is the result of the reception by education of epistemological theories, which have an objective alien to that of education, which is the all-round development of human personality. The discussion draws mainly upon the philosophies of Dewey, Husserl and Merleau-Ponty. It is argued that present, mainstream theories of science education need to be complemented with phenomenological perspectives. This would make the transition from immediate lifeworld experience to the idealizations of scientific theories less difficult for students. It would also contribute towards less alienation between man and nature.

1. Introduction

It seems that one of Dewey's main interests was to find a way out of the dilemma of the apparently necessary choice between inhuman rationality and human irrationality. In order to do so, the assumption of a primary dualism between subject and object had to go (Biesta 1994). In the first chapter of *Experience and Nature* (1929/1997), Dewey discusses this dualism in terms of the commonly accepted distinction between the two categories that constitute the title of the work. According to dualism, experience and nature are external to each other. Experience is regarded as "too casual and sporadic to carry with it any important implications for the nature of Nature", whereas nature "is said to be complete apart from experience" (ibid., p. 1). Dewey attempted to (re)establish a non-dualistic perspective on the relation between human experience and nature, or the world in general. His purpose could also be rendered in terms of the alienation of human beings from nature. If all significant human concerns are conceived as infinitely distant from what is considered to be objective, real, and true, it becomes very difficult for sensitive people to feel at home in the "real" world. Such a concern also lies behind this paper. It is a critique, from this point of view, of the theoretical bases of mainstream research and practice in science education.

F. Bevilacqua et al. (eds.), Science Education and Culture, 129–151.
© 2001 *Kluwer Academic Publishers. Printed in the Netherlands.*

Dewey admits that he could not find any direct arguments to disprove the dualistic position. One can only hope, he says, in the course of one's discussion to produce a change in the previously attached significations to fundamental terms like "subject" and "object", or "experience" and "nature" (ibid., pp. 1–2). A long tradition of discourse and practice have produced and solidified the significations attached to such terms. These signifying links cannot be cut off in an instance. However, through the creation and establishment of an alternative discourse they may gradually change. This paper is also an attempt in this direction.

What I try to show is the need for phenomenological and "aesthetic" (see below) perspectives in the research and practice of science education. Without the implementation of such theoretical visions, the major trend in present science teaching – based (I will argue) on the primacy of conceptual cognition – will not only contribute to students' alienation from nature, but from science as well. The main sources of my discussion are the non-dualistic philosophies of Dewey, Husserl and Merleau-Ponty.[1]

Dewey substituted the classic epistemological dualism between subject and object with a triadic relation between subject matter, objective and inquirer. The same subject matter can be inquired into with different purposes: scientific, aesthetic, religious, and political. Different objectives lead to different knowledge. Dewey would certainly hold that an investigation of natural phenomena from the point of view of art would be as educationally legitimate as a scientific inquiry into them. It would just have a different objective. However, I contend that we can keep the same objective as science, and still put more emphasis on the aesthetic dimension of knowledge formation. By aesthetic I mean a point of view which cultivates a careful and exact attention to all the qualities inherent in sense experience.[2] The objective of such an approach to natural phenomena would be not merely to appreciate their beauty, but also to *understand* them. Nature "speaks" through the gestures it makes in its forms, colours, sounds, smells, and tastes. From ancient times, human inquiry has tried to understand this "language" of nature. Galileo also wanted to understand it, but for various reasons he assumed that the only language nature was capable of speaking was that of mathematics. His approach was, as we know, very effective and successful. Therefore, after the so-called scientific revolution of the 17th century, philosophers began to regard non-quantifiable sense experience as irrelevant for true, i.e., scientific knowledge. Philosophers and natural scientists started to listen to the voice of nature through very thick walls as it were, walls which let through only the thin and abstract sound of numbers and formulas. Husserl (1970) called this cultural and historical process the "mathematization of nature". A phenomenological approach to nature, following Husserl's imperative to "return to the things themselves", calls upon us to tear down these thick walls and start to listen to *all* that nature has to say. It is as if nature has a hundred languages, but we have become deaf to ninety-nine of them. In order to (re)discover these languages, we have to intentionally and attentively explore all aspects of sense experience.[3]

Is the science education going on in our schools today a part of this deafening process? To the extent that these educational activities are informed by epistemologies and theories of learning which do not pay due attention to the aesthetic dimension of knowledge formation, I believe that it is. The main part of this paper is a critique of those aspects of such theories, which I believe contribute to an "anaesthetic" and alienated view of nature.

2. Dewey, Merleau-Ponty, and the Rejection of Dualism

In his philosophical thinking, Dewey was radically committed to "lived experience". His was a historical, contextual, and qualitative theory of experience, very different from the sense data atomism of the British empiricists (Boisvert 1998). Dewey identified and opposed a major negative trend in Western philosophical thinking, which he called intellectualism, in which experience was misunderstood as a form of knowledge or knowing. Thus, by intellectualism he meant

> . . . the theory that all experiencing is a mode of knowing, and that all subject-matter, all nature, is, in principle, to be reduced and transformed till it is defined in terms identical with the characteristics presented by refined objects of science as such. (1997, p. 21)

For Dewey, experience is always embodied and immediate, enjoyed or suffered, whereas knowledge is the mediated product of inquiry, such as the "refined objects of science". His critique of intellectualism was not intended as a denigration of science or reason. It was the inherent reductionism that was his target. Intellectualism reduced the manifold forms of experience to a mere knowing, as well as the rich complexity of nature to what a single type of inquiry, viz., science, can say about it.

Boisvert (1998) summarizes Dewey's philosophical critique as directed against three major "dogmas" of Western thought, which he labels "the Plotinian temptation", "the Galilean purification" and "the asomatic attitude" (p. 5ff). The Plotinian temptation has been present ever since Plotinus and the neo-platonics. It is the tendency to reduce everything to a single, underlying "unity". This can be seen for instance in the Cartesian search for a single, irrefutable idea, or in the simple sense data of Locke. "Oneness" has always been at least a regulative ideal of philosophical thinking. Dewey opposed this "temptation" because it could not represent the irreducible multiplicity, which characterizes the world and human life.

Galileo's law of free falling bodies exemplifies the Galilean purification, which ignores such factors as the friction of the air and other accidental circumstances. It was Galileo's great discovery that the muddled context of everyday experience could be substituted for an idealized situation in which all accidental and contingent factors were rendered invisible. This made it possible to construct the mathematical principle regulating a free falling body. The mathematical law is certainly useful. However, there is a serious but very common mistake to take it as more real than the concrete phenomena it refers to. According to Dewey, the success of science has made philosophers look for Galilean purifications in their own field and to consider them as more real than life itself.

The asomatic attitude is the mind-body dualism, the assumption that cognition takes place only in and by the mind. Knowledge, according to Dewey, is derived from *embodied* intelligence, not from an asomatic reason having a life and existence independent of the senses and affections of the lived body.

Dewey's (1997) critique was concerned with philosophy. However, as Boisvert (1998) points out, since the 17th century the Galilean purification and the asomatic attitude have had their effect on educational thought and practice as well. This is not surprising, since education was long a subdiscipline of philosophy. Philosophical theories of the nature, origin and grounds of knowledge have had a continuous influence on educational theory, research and practice. If the dualistic and intellectualistic traits of Western philosophy is taken as the basis of educational thought, the result is what Dewey called the "spectator theory of knowledge". According to this view, learning is best accomplished when the student is detachedly and objectively watching or listening, with the other senses, the body, and the feelings, as little involved as possible. Dewey, however, was concerned that intellect, senses, feelings, manual skills and moral development were all integrated in the learning process. Instruction should never focus on the intellect alone, at least with younger students. Dewey's efforts to reform education therefore went hand in hand with his philosophical critique.

The spectator theory of knowledge is no longer prominent in educational thought. The word of the times is "constructivism", with its emphasis on students' active knowledge construction.[4] However, most theories of constructivism remain within the dualistic framework which Dewey opposed, in particular those theories that focus on individual psychological processes. Reality, or nature, is seen as external to and independent of both knowledge and experience. For example, von Glasersfeld (1990) writes:

> [Constructivism] treats both our knowledge of the environment and of the items to which our linguistic expressions refer as subjective constructs of the cognizing agent. This is frequently and quite erroneously interpreted as a denial of a mind-independent ontological reality, but even the most radical form of constructivism does not deny that kind of independent reality. (p. 37)

Furthermore, such theories often have an intellectualistic flavour in that they tend to collapse all difference between experience and knowledge (cf. Suchting 1992). I would also maintain that traits of the Galilean purification and the asomatic attitude still exist, particularly in science education.

In order to explicate further these claims, Dewey's thinking has to be complemented with Husserl's phenomenological critique of science and Merleau-Ponty's phenomenology of perception. There are two basic connections between Dewey and phenomenology: one is the anti-dualism, the other is the critique of intellectualism. As for anti-dualism, both Husserl and Merleau-Ponty view subject and object, experience and world, as internally related. The world and human consciousness mutually constitute each other; one is unthinkable without the other. They are two sides of one whole, connected by the intentionality of consciousness. According to Merleau-Ponty, the phenomenological concept of intentionality entails that this unity is *lived* as "ready-made or already there", even before our reflective knowledge of it (1992, p. xvii). Consciousness is seen as "meant for a world which it neither embraces nor possesses, but towards which it is perpetually directed" (ibid.). Dewey, on his side, although not using the concept of intentionality, expresses basically the same idea when saying that experience is a "double-barrelled" word, because "it recognizes in its primary integrity no division between act and material, subject and object, but contains them both in an unanalyzed totality" (1997, pp. 10–11). In comparison, "thought" and "thing" are "singe-barrelled" since they refer to products discriminated by reflection out of immediate experience.

Merleau-Ponty's critique of intellectualism has a different tone from that of Dewey's. Nevertheless, it addresses basically the same trait of philosophical and psychological theories. For Merleau-Ponty, the weakness of intellectualism (as well as its opposite stance, empiricism) is the inability to understand and account for human perceptual consciousness. Intellectualism renders consciousness itself "too rich for any phenomenon to appeal compellingly to it" (1992, p. 28). That is, what we already know intellectually is allowed to play too great a role in explaining perceptual awareness. For instance, intellectualism does not see that we are necessarily ignorant of what we are trying to learn, otherwise we would not look for it. This means that intellectualism cannot "grasp consciousness *in the act of learning*" (ibid., italics in original),[5] since according to it, all experience is preformed by the concepts already inherent in consciousness. The failure arises out of an external, dualistic and non-dialectical view of the relation between the subject and object of perception. We have to realize, says Merleau-Ponty, that there is no reason "hidden *behind* nature, but that reason is *rooted* in nature" and that the role of intellectual judgement in perception is "not the concept gravitating towards nature, but nature *rising to the concept*" (ibid., pp. 41, 42; italics mine).

For Dewey on his part, intellectualism entailed as we have seen a reduction of experience to the "refined objects of science as such". Thereby, experience is divorced from nature: "When intellectual experience and its material are taken to be primary, the cord that binds experience and nature is cut" (1997, p. 23). If this point of view is generalized, other aspects of our everyday lifeworld experience is considered less real (or even unreal):

> [T]he discoveries and methods of physical science, the concepts of mass, space, motion, have been adopted wholesale in isolation by philosophers in such a way as to make dubious and even incredible the reality of the affections, purposes and enjoyments of concrete experience. The objects of mathematics, symbols of relations having no explicit reference to actual existence [...]

have been employed in philosophy to determine the priority of essences to existence ... (ibid., p. 33)

At this point Dewey's thinking is in total agreement with Husserl's (1970) later philosophy. When mathematically formulated physical laws are seen as *explanations* of observed phenomena they tend to be taken as more real than the phenomena themselves. The result, according to Husserl, is

> ... [a] surreptitious substitution of the mathematically substructed world of idealities for the only real world, the one that is actually given through perception, that is ever experienced and experienceable – our everyday life-world. (1970, pp. 48–49)

Harvey (1989) calls this "the Ontological Reversal": abstract models for a supposedly hidden reality behind concretely experienced phenomena takes on a higher ontological status than these experiences themselves. According to Husserl, the problem with this reversal is that science thereby divests itself of the very basis for its verifications. Scientific theories can only be tested and verified in and by immediate sense experience. But if this experience is by definition not (or less) real, what evidence can it provide? This is an aspect of the intellectualistic fallacy and the Galilean purification, which Dewey does not seem to discuss, presumably because he had a purely pragmatic view of scientific knowledge, meaning that scientific theories cannot be regarded in a realist way, or in accordance with the correspondence theory of truth.

The Galilean purification and the ontological reversal both have to do with the idealizations carried out in science, that is, the abstraction from the accidents and contingencies of everyday experience. This is now recognized as one of the major stumbling blocks in students' acquisition of scientific understanding (Matthews 1994, pp. 211ff). In science teaching and learning – consciously or subconsciously – idealizations in science are easily turned into Galilean purifications and ontological reversals, creating a rupture between students' intuitive lifeworld experience and scientific knowledge, which is hard to bridge. I will return to this issue in the conclusion of the paper.

If, as Dewey maintains, all our intellectual pursuits have a specific purpose or goal, then the objective of traditional epistemology would be to give a logically coherent, formal theory of (scientific) knowledge and knowing. But is such an objective altogether in harmony with *educational* concerns? Perhaps we make a serious mistake if we base all our educational theories on such philosophical grounds. The next section will illustrate this suggestion.

3. Kant, Piaget, and Intellectualism

Kant is sometimes called Newton's philosopher, because he set out to formulate the epistemological foundations for Newtonian science. His epistemology was concerned with the foundations of *scientific* knowledge. Yet, in his writings he most often used examples of everyday knowledge, such as the experience of seeing a dog and knowing that "this is a dog". As Böhme (1980) has pointed out, Kant's epistemology fails to distinguish between our everyday lifeworld experience on the one hand, and the purely theoretical knowledge of science on the other:

> It is a general weakness of the Kantian theory of knowledge, that it does not distinguish between lifeworld experience and scientific experience. Since its ultimate aim is the grounding of scientific experience, especially that of physics, yet continually works with examples from everyday experience, and since Kant always speaks about "experience in general", the impression is engendered that this distinction does not exist at all. (p. 71)[6]

The lack of distinction between these two realms of knowing and experience contributed to the reception of Kant's epistemology – in particular that of *Kritik der reinen Vernunft (KdrV)* – as dealing with all kinds of knowledge, not only science. Thereby he also contributed to the intellectualism described in the previous section, making *all* experience dependent on theoretical or conceptual knowledge.

From an educational point of view, it is interesting to observe that a similar confusion seems to have arisen about the works of Piaget, who, by the way, owes some of his central concepts to Kant. (For instance, Kant in *KdrV* introduced the notion of schema, and Piaget's distinction between figurative and operative knowledge is prefigured in the same work.) As Herzog (1991) has pointed out, many educationalists seem not to have realized that the object of Piaget's research was never cognitive or psychological development in general, but the psychological genesis of *scientific* knowledge:[7]

> What Piaget has presented as a psychologist is namely far from a representation of "the" cognitive development. From the very beginning Piaget's analysis of human development stands in the light of his question from the theory of science, which he took over in a more or less unmodified form from the natural sciences. (ibid., p. 290)

Thus, for Piaget, the kind of knowledge constituting modern science was the more or less taken-for-granted *telos* of the individual's intellectual development. It was from this particular point of view that he described the development of intelligence and knowledge. However, when Piaget's theory was taken up in education, it was taken as a theory of the development of all kinds of knowledge, not just science. Sometimes, for instance in teacher education, it even appears to have been received as a theory of the general psychological development of children. In this way, I believe, Piaget's thinking has come to contribute to a particular variety of intellectualism within educational thought, and in practice as well. I call this *cognitivism*. Intellectualism is a philosophic, in particular an epistemological stance, cognitivism is an educational one. In education we are primarily concerned with learning and development, and when this concern is solely focused on the formation and development of *concepts*, I call it cognitivism. Intellectualism, on the other hand, is the ontological overinterpretation of the role of abstract, conceptual elements in our knowledge, understanding and experience of the world. The two are closely linked, but not identical.

This is not to say that theories of the development of scientific cognition are irrelevant to educational research and practice. They can certainly be of help in fostering students' understanding of science. Therefore, my critique is not directed towards such cognitive theories as such, but towards cognitiv-*ism*. Cognitivism means letting conceptual, theoretical cognition constitute the central theme of all research or practice dealing with teaching, learning and the development of knowledge. The acquisition of concepts then becomes the primary and most important aim of all schooling. As Säljö (1995) has noted, concept formation is almost turned into a "pedagogical drug".

In an interview with Martin Wagenschein (1981), a well known German science educator (see for instance Wagenschein 1965), the interviewer recounts how his little daughter once refused to eat a piece of bread. The helpful father cut the bread in two, meaning to help her eat by making the pieces smaller. But the girl cried out that now she had to eat "even more" bread. The father "realized" that this was an expression of the child's lack of the concept of quantity invariance, as described by Piaget. However, Wagenschein, being a wise educational phenomenologist, answered by pointing out that this was not the only possible interpretation of the girl's protest. Perhaps she meant that now there was more to eat in the sense that it would take a longer time, or that several more mouthfuls would be needed. What the child really meant could only reveal itself in further dialogue, not by jumping to conclusions from a pre-established theory.

The father's reaction is a typical example of cognitivism: the child's behaviour is seen as an expression of a *lack of concepts*, a conceptual deficiency. This is what her behaviour must look like, if seen from the point of view of our scientific knowledge of nature. Naturally, the *quantity* of the bread is the same, whether it is cut up in pieces or in one whole. However, from a qualitative, aesthetic point of view, two pieces of bread are *not* equal to one piece, even if they are that same piece cut in two. Aesthetically, two pieces are *two* pieces, not reducible to *one* piece.[8]

By viewing mental development exclusively from a Piagetian, or cognitivistic perspective, the stages of development preceding that of formal operations easily become something the child merely has to pass through and outgrow, in order to reach "real" thinking, i.e., the ability to carry out abstract logical and mathematical operations. But the rootedness of thought in immediate sense experience takes time to overcome. Thus, Piaget (1950) reports how children up to the age of 11 may still consider a piece of clay rolled into a thin thread as lighter and less voluminous than the same amount formed into a ball. This illustrates how thinking in the sensorimotor and (to some extent) the concrete operational stages is strongly linked to immediate sense perception.[9] This kind of thinking is rooted in aesthetic perception because it starts from the immediately perceived gestures that things make. A lump or ball of clay makes a gesture of heaviness, whereas the same clay rolled out into a thin string makes a gesture of "lightness".

In present day theories of education, particularly of science education, Piaget's theory has been assimilated within the more general framework of constructivism. Just as Piaget took up central concepts from Kant, so constructivism can be seen as a form of neo-Kantianism (Boyd 1991). For instance, Devitt (1991) holds that constructivism is based on two Kantian ideas: "first, that we make the known world by imposing concept, and, second, that the independent world is (at most) a mere "thing-in-itself" forever beyond our ken' (p. ix). It seems that constructivism in general has a strong cognitivistic bias, due to its narrow focus on conceptual change (cf. Tarsitani 1996). However, some of the forms it has taken are not incompatible with a more aesthetic perspective on learning and knowledge formation, see for instance Abercrombie (1960).

4. Schema – Machines of Experience Production?

In his *KdrV*, Kant seems to look upon sense impressions as the raw materials out of which understanding produce, first of all, *experience*. The following quote is illuminating:

> For how is it possible that the faculty of knowledge should be awakened into exercise otherwise than by means of objects which affect our senses, and partly of themselves produce representations, partly rouse our powers of understanding into activity, to compare, to connect, or to separate these, and so to convert the raw material of our sense impressions into a knowledge of objects, which is called experience. (1993, p. 30 [BI])

Perhaps this statement was the starting point for Horkheimer and Adorno's critique of Kant's epistemology. Horkheimer and Adorno (1944) describe Kant's theory of knowledge as entailing that our conceptual apparatus predetermines our senses even before perception takes place. They contend that this theory is an unconscious reflection of the material conditions of production in bourgeois society, where nature is treated as raw material for factory production. The bourgeoisie "look a priori upon the world as the stuff out of which they fabricate it" (ibid., p. 103). In the same vein, Horkheimer himself later (1967) compared Kant's "pure understanding" to a machinery of boxes and levers, imprinting its forms on the raw material of sense impressions. Thus, according to Horkheimer and Adorno, Kant's philosophy of knowledge is modelled on industrial production. Just as a factory brings in raw material to work on in order to produce consumable goods, so our "mental apparatus" takes in the "raw stuff" of sense impressions, which is worked on by our "schemata" in order to produce – in the first instance – "experience". Experience may then be further worked on, by the schemata of higher order concepts, to produce "knowledge".

This critique of Kant is surely one-sided and unfair, not paying due attention to all the complexities of his thinking.[10] I take it up not in order to associate Kant's (or neo-Kantian) epistemology with "bad capitalism", and then throw one out with the other. The reason is rather that I believe Horkheimer and Adorno very accurately identified a basic metaphor for the way many people tend to picture the relation between intelligence or understanding (*Verstand*) on the one hand, and sensation or sense experience (*Sinnlichkeit*) on the other. What we receive through our senses is looked upon as "data", which are "treated" in various ways by our conceptual system(s). This treatment is subconsciously modelled on the image of how raw materials are processed in our factories, to produce commodities. The result of these processes is knowledge or representations *(Vorstellungen)* of the "outer world". Indeed, in our present information society, knowledge *is* becoming more and more like a commodity, being "packed" and "sold" like any other marketable goods (cf. Frohmann 1992). But if these epistemological notions are subconscious reflections of outer socio-material conditions, do we not need to stop and ask whether they are as valid and true as we take them to be?

Intellectualism (which Horkheimer and Adorno on their part call the "intellectuality of perception") and cognitivism both imply such an external and mechanical relation between the senses and the understanding. The consequence is a tendency to neglect the significance of more aesthetic modes of experience, as illustrated in the previous section. As for the theory and philosophy of science, a relatively recent example of this stance can be found in Bogen and Woodward (1992). With reference to Churchland, they recommend that human beings learn to employ "the plasticity of perception" in order to replace

> ... the present old-fashioned [sic!] framework in which, for example, we "observe the western sky redden" with a more scientifically up-to-date framework in which we "observe the wavelength distribution of incoming solar radiation shift towards the large wavelengths". (p. 610)

From such and similar arguments, Bogen and Woodward come to the conclusion that the distinction between what can be perceived by our senses and what cannot be so perceived "corresponds to nothing of fundamental epistemological interest" (ibid., p. 610).[11] Thus, sense experience as an epistemological factor is more or less abolished. The strong trust in electronic data registrations and a parallel distrust in human sense experience reveal the mechanistic stance of such a perspective:

> ... many advances in reliability come, not by improving perception at all (and still less by loading it with better theory), but rather by replacing perception entirely with mechanical detection and recording devices, or by redesigning the detection process so that perception plays a less central roll. (ibid., p. 608)

The message seems to be that we can profitably reduce the role of sense perception in scientific research. It is a rather good example of how the asomatic attitude and the Galilean purification have entered into the epistemology of science. The suggestion that we use the presumed "plasticity" of perception to produce sense experiences that are more in accord with scientific concepts illustrates also the metaphor of industrial production described above. What will happen if such ideas are turned into starting points for educational theories of learning? I do not believe that they or similar conceptions are generally accepted, neither among teachers nor among educational thinkers (at least not yet). However, they would be logical consequences of intellectualism in epistemology and of an exclusive and one-sided focus on concepts and concept formation in education.

5. The Phenomenological Alternative

In the phenomenological analyses of perception and knowing, above all in those carried out by Merleau-Ponty (1992), the conception of the relation between our conceptual systems and our sense experience is very different from that described in the previous section. Merleau-Ponty's writings are extensive and complex, and I do not claim to expound the whole and true intent of his work. There is, however, one paragraph in one of his books (1964), which, to my mind, captures the problem and suggests its solution in a particularly interesting way. It is when he defines the meaning of "the primacy of perception", which is

> ... that the experience of perception is our presence at the moment when things, truths, values are constituted for us; that perception is a nascent *logos*; that it teaches us, outside all dogmatism, the true conditions of objectivity itself, that it summons us to the tasks of knowledge and action. (p. 25)

Many comments could be made on these few lines of pregnant thought. Concerning the question of the relation between understanding and perception, we may note first of all the expression that "perception is a nascent *logos*". "Logos" is meaning, order, structure, and knowledge. Perception is thus potential knowledge, or knowledge in the process of being born.[12] It has not yet come into daylight, i.e., into the clarity of conceptual understanding. Thus, perception is not yet fully developed knowledge, but it is nevertheless "pregnant" with meaning.[13] The kind of perception that Merleau-Ponty describes here could be called aesthetic, in the sense that I have defined it above. Such perception is holistic as well as synesthetic. It does not restrict itself to one sensory modality at a time. In aesthetic perception, we "see" what a thing sounds like if we strike it or what it feels like if we touch it, etc. These are examples of the inherent structures of this deep level of perceptual awareness.

However, in everyday life, another kind of perception dominates our experience. Merleau-Ponty calls it "empirical" or "second-order" perception. This is loaded with habitually established meanings and conceals from us the basic, aesthetic level of perceptual experience. It plays, Merleau-Ponty says, "on the surface of being" (1992, p. 43). It is regulated by the pragmatic needs of everyday life, where we simply identify the general meaning of objects while our practical intention is directed elsewhere. In contrast to this play "on the surface of being", there is the other, more basic kind of perception:

> *But when I contemplate an object with the sole intention of watching it exist and unfold its riches before my eyes*, then it ceases to be an allusion to a

> general type, and I become aware that each perception re-enacts on its own account the birth of intelligence and has some element of creative genius about it: in order that I may recognize the tree as a tree, it is necessary that, *beneath* this familiar meaning, the momentary arrangement of the visible scene should begin all over again, *as on the very first day of the vegetable kingdom*, to outline the individual idea of this tree. (ibid., pp. 43–44; italics mine)

Contemplating something in order to watch its riches unfold – this could be called the intentional cultivation of aesthetic perception. It involves attentive listening to all the qualities inherent in sense experience.

Dewey makes a similar distinction between two kinds of perceptual experience, calling the habitual and more superficial one "recognition". Recognition is "perception arrested before it has a chance to develop freely" (1981, p. 570). However, sometimes we may be struck by something that catches our attention and interest. Then genuine perception can replace mere recognition, and there can be "an act of reconstructive doing, and consciousness becomes fresh and alive" (ibid.). Such perception is creative, because "to perceive, a beholder must *create* his own experience" (ibid., p. 571; italics in original).

Similarly, for Merleau-Ponty, this basic level of perceptual experience takes on the pristine character of the first day of creation: it is seeing as if for the first time, without any of the formerly acquired, habitual meanings interposing themselves as a veil between consciousness and its object. If we possessed this quality of awareness more often, we would probably not feel alienated from the world and from nature to the extent that we generally do. However, our *presence* at the moment when truths or values are constituted for us is rare. Yet, such presence is surely a basic characteristic of genuine learning and radical insight. At such moments, a new understanding of things can be consciously realized, whether new to the whole of mankind, or only to one particular individual. Learning may happen without such presence, but can it be as deep, convincing and satisfying?

The distinction between aesthetic and "second order" perception has a certain parallel in information-processing constructivist terminology, where one distinguishes between a top-down and a bottom-up processing of sensory "input" (Rumelhart and Ortony 1977). A top-down processing imposes a pre-established schematic structure on the incoming data. It is similar to the second order perception, which does not stop to contemplate the sense-qualities of the object, but quickly labels it according to the practical needs of the situation. A bottom-up processing, on the other hand,

> ...occurs when aspects of the input directly suggest or activate schemata which correspond to them and when these schemata themselves activate or suggest dominating schemata of which they are constituents. (ibid., p. 128)

The similarity between this and aesthetic perception lies in opening oneself to the object and letting the sensory input itself suggest which schemata to use in order to describe it to oneself, or to others. Presumably, this reduces the role of habitual and purely pragmatic descriptive categories.[14] However, apart from the theoretical difficulties connected with schema theories of this kind, which have already been noted by others (see for instance Bickhard 1995), the *existential* significance of the radical difference between aesthetic and second order perception is lost in such theoretical discourses. Educational thought and practice has to do with human existence. Therefore, they must be informed by theoretical visions in which existential dimensions are not neglected, but, on the contrary, made explicit.

It is said about Kaspar Hauser that in the beginning of his association with people – just after he was discovered – he could distinguish between apple- and pear-trees merely by listening to the sound that the wind made in their leaves. His senses were extremely alert and sensitive. However, he gradually lost this capacity as he learnt to speak, write, and assimilated other knowledge. His own writings are a good source for investigating the relations between the pre-social, "silent" and purely sense-perceptual world, and the socialized world of second order perceptions that we generally inhabit as grown ups (cf. Mollenhauer 1985). The object of science is nature. What if science teaching deliberately tried to break our common play "on the surface of being" when looking at natural phenomena? Could we not (re)discover some of the perceptual abilities of a Kaspar Hauser, and at the same time learn what science (in its present form) really is about?

6. Phenomenology in Educational Theory and Practice

It has been the argument of both scientists and educational thinkers that our conventional forms of schooling are rather poor from the point of view of sense experience and active, conscious use of our senses (to name but a few: Caraher 1982; Dale 1990; Egan & Nadaner 1988; Jardine 1990; Martin 1974; Murphy 1985; Sperry 1983; Wagenschein 1965). It seems that most teaching and instruction today takes place on the grounds of what Martin (1974) calls "the spectating experience", in which conceptual understanding forms the framework within which the perceived thing is fitted. He contrasts this to the "participative" mode of experience, in which "ideas vivify the thing because *the thing initiates and controls every idea*" (ibid., p. 93; italics mine). The spectating experience rests upon the implicit assumption that our relation to things can only be of an external kind. Our ideas and concepts then function like the Kantian categories, bringing order to the world from without. Order and meaning are "imposed" on phenomena by the thinking of human beings. The world in itself has no order, no meaning. This epistemological conception is presumably at the bottom of a large portion of teaching and learning today, both in science and in other subjects. It makes for an aesthetically poor knowledge formation, because the qualities of sense experience are either disregarded, or only attended to as a passive material, to be structured and put in order by intellectual concepts.

An aesthetically rich knowledge formation, on the other hand, may be said to arise when we "let the thing think" in us. "Only then will the depth dimension of our world come to presence explicitly in our experience" (ibid., p. 92). This kind of attentive learning, which Martin explains with reference to the phenomenology of Heidegger, has its roots in aesthetic perception. "The thing thinks" in the sense that *logos*, the meaning which thinking grasps in the thing, is not imposed from without, but born out of the sense experience that the thing evokes in us. This mode of "thinking Being" is not something extra, without educational or even philosophical significance; "it is the *ground* of all other modes, of all experiences" (ibid., p. 98; italics in original).

Johann W. Goethe was one of the few (and largely forgotten) people in the history of science, who tried to establish and uphold such a participative mode of experience even in the scientific study of nature (cf. Bortoft 1996). Goethe's approach to research actually has some basic features in common with phenomenology (Heinemann 1934). His idea of "anschauendes Denken" – perhaps translatable as "thoughtful observation" – implies that there is a sensitive (and highly cultivated) surrender to sense experience, and *at the same time* a sharp and clear conceptual interpretation of this experience. That is, thinking and seeing/perceiving/experiencing go together all the way, they are never separated.

Bortoft (1996) explores Goethe's theory of knowledge and science in the context of present day phenomenology and hermeneutics. He comments on Kant's saying that reason "must adopt as its guide ... that which it has itself put into nature" when researching the natural world:[15]

> Thus nature is compelled to provide answers to the questions *we* set, which means to be frameworked in *our* conceptual scheme. (p. 240, italics in original)

The Goethean approach to nature is the opposite, because

> ... the organizing idea in cognition comes from the phenomenon itself, instead of from the self-assertive thinking of the investigating scientist. It is not imposed *on* nature but received *from* nature. (ibid., p. 240; italics in original)

This organising idea is the "intrinsic necessity" of the phenomenon and it "comes to expression in the activity of thinking when this consists in trying to think the phenomenon concretely" (ibid., p. 240). To think concretely means letting perception itself be thinking, and thinking perception, as Goethe did (ibid., p. 240). Thus, Goethe's science is a kind of hermeneutic phenomenology of nature, where phenomena are understood in terms of themselves, not in terms of imposed "schemata". It perfectly illustrates what Martin calls the participative mode of experience, in which "the thing initiates and controls every idea".

The implications of a Goethean, hermeneutic-phenomenological approach to the study of nature for teaching and learning in science have been explicated by Rumpf (1991, 1993; see also Buck and Kranich 1995). With reference to a Swiss physics teacher, Peter Stettler, Rumpf quotes the collective summary of an experiment with colour formation, informed by Goethe's theory of colour and performed in a class of 8th grade students:

> Then we threw a thin white thread in the water. When we carefully watch this thread from above, it appears white to us. If we let our heads down a bit, the thread rises from the black bottom. At the same time it assumes the colour blue at its upper edge and at the lower edge yellow to red. The lower one lets one's head down, the broader and clearer become the stripes of colour. It is wondrous that these colours only appear at the edges. The colours change beautifully and continuously into each other. If one puts one's eyes at the level of the water surface the whiteness of the thread is pressed together, that is, it is completely covered with colours. It goes so far that the yellow and the blue touch each other. Good observers discover a green shimmer therein. These colours remind strongly of a rainbow. (Stettler, as quoted in Rumpf 1991, p. 322)

This is a good example of aesthetic perception. One most probable consequence of such cultivated attention to the qualities inherent in sense experience when learning about natural phenomena is the awareness of the difference between descriptions of immediate experience on the one hand, and theoretical interpretations on the other. Careful attention to what is actually given to the senses, and the description of this, presumably increases the awareness of the point where a verbal account of events transcends what is thus given, and rises to the level of interpretation. Abercrombie (1960, p. 85) gives some telling examples of how students do not distinguish clearly between these two levels. Yet the distinction is of fundamental importance to science.[16] This may be taken as an illustration of how habitual second order perception neglects the aesthetic potentials of sense experience. As Abercrombie puts it,

> ... a conclusion about "meaning" had limited the perception of the observers, causing them to ignore information which did not fit the ordained pattern, the chosen schema. (ibid., p. 88)

One area in science where aesthetic perception plays a fundamental role is classification. Scientific classification actually depends upon both aesthetic perception and judgement, as Abercrombie maintains:

> Judgement of the suitability of a system of classification is presumably based on the perception, not necessarily conscious, of a pattern of correlated features, and seems to involve the same kind of processes of aesthetic judgement. (ibid., p. 118)

Classification is not a superficial aspect of science. It can lead to new discoveries. When Newlands arranged the material elements known at his time according to atomic weight instead of their initial letters, because it seemed more appropriate to him, he prepared the ground for the discovery of further elements (ibid., p. 118). Thus, to constitute a new classification system is not a trivial thing, and it demands a certain aesthetic sensibility. At the same time, once constituted and established, such systems become the basis of our habitual second order perceptions, that is, our everyday "anaesthetic" experience.

7. Conclusion

The importance of attending to the aesthetic dimensions of science in science teaching has been argued for by Flannery (1992). However, her focus is on the personal and informal side of scientific research. Her concept of aesthetics is also of a more conventional kind, having to do with beauty and harmony more than with sense qualities in general. This is fine as far as it goes, but the phenomenological approach, which I have shortly sketched above, goes further. It brings aesthetic perception (in its original sense) into the formal and objective aspect of both learning and research in science. More precisely, it establishes an internal, dialectic relation between the personal, subjective aspect, and the formal, objective one.

Dewey maintained that all inquiry has a definite purpose or objective. What then is the objective of a phenomenological-aesthetic approach to science education? It is, as indicated in the introduction, to alleviate students' alienation from both nature and science. Intellectualism and cognitivism, with their accompanying asomatic attitudes and Galilean purifications – as well as the "ontological reversal" connected with the latter – all contribute to the establishment of a dualistic, external and unmediated relation between our subjective experience on the one hand, and objective nature on the other. Certainly, people still occasionally have wonderful experiences of oneness with nature, especially in emotionally enhancing natural surroundings. However, basically, such experiences are considered as "merely subjective" because science has no place for them. At the same time, popular science books about the "unity of nature" – from atoms, molecules and genes to stars and galaxies – are regarded as objective and true, although the unity they describe is but an object of thought, not a lived experience.[17] Science education is neglecting its general educational responsibility if it does not consider these questions when trying to guide children into a scientific understanding of nature.

As for the alienation from science itself, it has to do with its idealizations. As already noted, this has proved a major stumbling block for students, many of whom can neither understand nor appreciate a knowledge system which does not deal with concrete reality as experienced and lived, but with abstract idealizations (which are often, on top of that, mathematical). Matthews (1994) argues that a science teacher well informed by the history and philosophy of science "can assist students to grasp just how science captures, and does not capture, the real, subjective, lived world" (p. 213). I completely agree. What I want to add with this paper is that in theories and research on science education we also need a deeper understanding of the role of aesthetic perception in knowledge formation. In teaching practice, we need an increased emphasis on such perception when we let students observe and make experiments, especially in the lower grades. By a careful, explicit thematizing of all the aspects of sense experience, the potential shock of mathematical idealizations can be prevented or assuaged. Thereby, the relation between the idealized model and the immediate experience, and how the former grows out of the latter, can be made clear in each particular case.[18] Students can then more readily come to understand what science is doing. The mistakes of the Galilean purification and the ontological reversal need not be made.

Cognitivistically flavoured texts on science education do not seem to grasp the necessity of this theoretical revision. Instead, one may sense a certain hidden agenda, viz., to *replace* children's original, spontaneously formed "schemata" with those which science has established as more correct and "true". For instance, Gunstone et al. (1988) write:

> Research on the effectiveness of a teaching program must [...] not be satisfied in testing for acquisition of new knowledge, but must also *ensure that other beliefs have been discarded*, a much more difficult measurement task. (p. 522; italics mine)

In a phenomenological-aesthetic approach to science teaching and learning, there is no notion of schemata having to be discarded, or even "developed". Experience speaks, and inquiry tries to interpret the voice of experience from different angles, with different interests and purposes. The voice of science is one of many such voices, and children should certainly learn to understand it, even appreciate it. But interpretations other than what is scientifically established as "correct" must also be allowed to exist and to speak, even within science teaching, because it is recognized that one does not know beforehand which interpretations are conducive to the flourishing of a good, fully developed human life. For, as Aristotle said, a human being who thinks well and *perceives* well, lives well (Oksenberg Rorty 1980). However, our computerized information society runs the risk of producing an even greater bias towards purely conceptual cognition, supporting both Galilean purifications and asomatic attitudes. Our sense-perceptual capacities tend to be neglected, and wither from lack of exercise. This, in the end, would mean a lopsided development of the possibilities of human life and experience.

What I have tried to show in this paper is that the formal theories of knowledge, which have hitherto been taken as starting points for systematic educational thinking, seem to have a content and an objective which is less in agreement with the general educational endeavour: to contribute to an all-rounded human development of the growing generation. I have focused on the notion of an external, dualistic relation between the knowing subject and the object known as an illustration of this thesis. A central concept emerging from this notion is that of "sense data" as the "raw material" which is being "processed" by our mental conceptual system(s). But, as Hamlyn (1961) has pointed out,

> ... the notion of a sense-datum was introduced in the first place to fulfil certain logical or epistemological requirements, and these requirements have always seemed more fundamental to sense-datum theorists than the requirement that the notion should be given content by reference to *the facts of experience*. (p. 174; italics mine)

In phenomenology, on the other hand, it is precisely these "facts of experience" which are put into focus, elucidated, and interpreted in order that we may better understand them, and ourselves. This is particularly the case with the non-foundational phenomenology of Merleau-Ponty, who, unlike Husserl, did not have the ambition to formulate any "absolute certainties".[19] The objective of phenomenological reflection in science education is to elucidate and clarify our *experience* of knowledge and learning about nature – through thinking, feeling, perception, imagination, or whatever. Such inquiry takes as back to our immediate lifeworld, the ultimate ground out of which all genuine, human learning must grow.

Acknowledgements

The research behind this paper was sponsored by the Swedish Council for Research in the Human and Social Sciences. I am grateful for valuable comments by the Editor, Michael Matthews, and by the reviewers Jim Garrison, Peter Machamer and Peter Slezak of an earlier version of the paper.

Notes

[1] There are in fact many similarities between Dewey and phenomenology, in partcular with Merleau-Ponty, but it is not part of the aim of my paper to specifically point these out.

[2] This is (arguably) the original sense of aesthetics, referring to reflections on sense experience in general and not just dealing with art and beauty. The term aesthetics goes back to the 18th century German philosopher Alexander Baumgarten, who inaugurated the discipline of aesthetics as a complement to that of logic (Baumgarten 1954). If logic is the description and evaluation of the processes of thinking, aesthetics was to be the description and evaluation of the processes of sense perception. These are not the exact words used by Baumgarten, but it is the meaning that can be gathered from his text (for a similar interpretation, see Schweizer 1973). More recently, Böhme (1995) has argued for a similar reinterpretation of aesthetics: "The new aesthetics is first of all that, which its name says, namely a general theory of perception" (p. 47; my translation).

[3] To say that the main interest of science is to understand the "language" of nature may be an overgeneralisation. The actual interest behind particular scientific investigations probably has to be studied from case to case. For instance, the immediate motive behind Newton's optics appears to have been the desire to make more effective telescopes (Böhme 1980). Nevertheless, in a general sense, Galileo's claim to have discovered the true language of nature, viz., mathematics, seems to be approved of by most scientists.

[4] Naturally, there are various forms of constructivism, see for instance Steffe and Gale (1995). In the following, I disregard social constructivism and refer only to those types dealing with mental processes within the individual.

[5] In order to make sense, the learning Merleau-Ponty refers to must imply a qualitative change of experience, awareness, or perspective. This kind of learning has recently been dealt with extensively by Marton and Booth (1997).

[6] All quotes from German sources have been translated by me.

[7] Of course, Piaget dealt also with other aspects of children's psychological development, such as play and imagination, affections, and moral judgment. However, these can be seen as branches of his main and overriding interest, 'genetic epistemology'. In 1976, Piaget himself wrote that

... my efforts directed toward the psychogenesis of knowledge were for me only a link between two dominant preoccupations: the search for the mechanism of biological adaptation and *the analysis of that higher form of adaptation which is scientific thought, the epistemological interpretation of which has always been my central aim.* (quoted in Gruber and Vonéche 1982, p. xi; italics mine)

In this respect it is also interesting to read Piagets' novel *Recherche*, written 1918, which is a more or less autobiographical account of how he was led into his main research interest (for a summary, see ibid., pp. 42ff).

[8] Merleau-Ponty gives the following example to illustrate the two different points of view:

For the understanding a square is always a square, whether it stands on a side or an angle. For perception it is in the second case hardly recognizable. [. . .] There is a significance of the percept which has no equivalent in the universe of understanding, a perceptual domain which is not yet the objective world, a perceptual being which is not yet determinate being. (1992, pp. 46–47)

In a similar vein, if I draw a rectangle lying down, and another standing upright, from an abstract, geometrical point of view, there is no difference. Both are rectangles, and the formula for the calculation of their areas is the same. But from a concrete, aesthetic perspective, the two figures are entirely different, because they make different *gestures*.

[9] As for concrete operations, they are more independent of the senses, but not completely so. According to Piaget (1950), children in this phase still need to see the things they are reasoning about. A purely verbal representation of phenomena is often not enough for them to be able to reason logically.

[10] For how Kant's *KdrV* permits several readings, see Neujahr (1995). In an early essay from 1884, Dewey also deals with this problem in Kant's philosophy, but in a more nuanced way (see Dewey 1981, pp. 13–23). His conclusion is that even though the distinction between the senses and the understanding for Kant was purely analytic, he formally retained the error of looking on their relation as external, i.e., non-dialectical.

[11] Rudolf Steiner, the founder of Waldorf education, once said that if the materialistic world view inherent in science will continue to dominate our culture for some generations more,

. . . so will really the red of the rose disappear. The human being will really see the small grey atoms out there swirling, the atomic swirl, not because he must see them because they are there, but because he has prepared himself to see them. (quoted in Suchantke 1998, p. 101)

It seems that Churchland and his adherents are doing their best to make Steiner's prophecy come true. The psychological consequences of this for our human being-in-the-world can be imagined by reading Boss's (1978) description of some modern neuroses. He tells about one client, who was suffering from depression, and who could see no beauty in the flowers of a blossoming cherry tree, but only a swarm of molecules.

[12] In a similar vein, Dewey says that it is "in the concrete thing *as experienced* that all the grounds and clues to its own intellectual or logical reification are contained" (1981, p. 245; italics in orignal).

[13] The metaphor of pregnancy is used by Merleau-Ponty himself (1968).

[14] Naturally, from the point of view of the 'finished product', there is no difference between aesthetic and second order perception. Both must be described in pre-established and familiar linguistic categories. The distinction refers to *the way they have been produced*, the process that constitutes them.

[15] The similarity between the spectating mode of experience and Kant's vision of the scientific approach to nature can be illustrated by the following quote from the preface to the second edition of *KdrV*, where he talks about the 'natural philosophers':

They learned that reason only perceives that which it produces after its own design: that it must not be content to follow, as it were, in the leading-strings of nature, but must proceed in advance with principles of judgment according to unvarying laws, and compel nature to reply to its questions.[...] Reason must approach nature with the view, indeed, of receiving information from it, not, however, in the character of a pupil, who listens to all that his master chooses to tell him, but in that of a judge, who compels the witnesses to reply to those questions which he himself thinks fit to propose. (1993, pp. 13, 14 [BXII, BXIII])

[16] It may be objected that description is already theory laden, but this is only relatively true. It refers primarily to observations within hypothetico-deductive experimental designs, since such experiments often build on (explicit or implicit) theoretical frameworks. But if *all* observational descriptions already imply full blown theories, how can one explain disagreements between researchers about how certain observed facts should be interpreted?

[17] See Thomas (1997) for a more extensive discussion of this argument.

[18] Husserl (1970) seems not to doubt the possibility to reconstruct a continuity between original lifeworld experience and the idealized models of science. Brady (1998, pp. 88f) has indicated how such a reconstruction could be carried out in the field of Newtonian mechanics. However, with the ever more increasing mathematization of physics and chemistry, e.g. the introduction of non-Euclidian geometry, this is no longer selfevident (Rang 1997). Still, a careful attention to perceptual experience could reveal exactly where the break between lived experience and ideal mathematical modelling occurs.

[19] As Madison (1990) puts it:

... the rediscovery of the *Lebenswelt* underlying the objectifying thought of science furnishes us with the means of overcoming modern dualism. This is precisely the lesson Merleau-Ponty drew from Husserl. However, Husserl's (never fulfilled) aim was to go on to show how the *Lebenswelt* is itself the product of a constituting Ego, and this is something Merleau-Ponty refused to accept. (p. 60)

References

Abercrombie, M. L. J.: 1960, *The Anatomy of Judgement. An Investigation into the Processes of Perception and Reasoning*, Hutchinson, London.

Baumgarten, A. G.: 1954, *Reflections on Poetry*, University of California Press, Berkeley & Los Angeles.

Bickhard, M. H.: 1995, 'World Mirroring Versus World Making: There's Gotta Be a Better Way', in L. P. Steffe & J. Gale (eds.), *Constructivism in Education*, Lawrence Erlbaum, Hillsdale, NJ, pp. 229–268.

Biesta, G. J. J.: 1994, 'Education as Practical Intersubjectivity: Towards a Critical-Pragmatic Understanding of Education', *Educational Theory* **44**, 299–317.

Bogen, J. & Woodward, J.: 1992, 'Observations, Theories and the Evolution of the Human Spirit', *Philosophy of Science* **59**, 590–611.

Boisvert, R. D.: 1998, *John Dewey: Rethinking Our Time*, SUNY Press, Albany, NY.

Bortoft, H.: 1996, *The Wholeness of Nature. Goethe's Way toward a Science of Conscious Participation in Nature*, Lindisfarne Press, Hudson, NY.

Boss, M.: 1978, 'Der neue Wandel der Neurosen-Erkenntnisse der Psychoterapie', *Universitas* **33**, 1023–1029.

Boyd, S.: 1991, 'Confirmation, Semantics, and the Interpretation of Scientific Theories', in S. Boyd (ed.), *The Philosophy of Science*, MIT Press, Cambridge, MA, pp. 3–35.

Böhme, G.: 1980, *Alternativen der Wissenschaft* [Alternatives of Science]. Suhrkamp, Frankfurt am Main.

Böhme, G.: 1995, *Atmosphäre. Essays zur neuen Ästhetik* [Atmospheres. Essays towards a New Aesthetics]. Suhrkamp, Frankfurt am Main.

Brady, R. H.: 1998, 'The Idea in Nature: Rereading Goethe's Organics', in D. Seamon & A. Zajonc (eds.), *Goethe's Way of Science. A Phenomenology of Nature*. SUNY, Albany, NY, pp. 83–114.

Buck, P. & Kranich, E.-M. (eds.): 1995, *Auf der Suche nach dem erlebbaren Zusammenhang. Übersehene Dimensionen der Natur und ihre Bedeutung für die Schule* [Searching for the Experienceable Connection. Neglected Dimensions of Nature and their Consequences for Schools]. Beltz Verlag, Weinheim and Basel.

Caraher, B. G.: 1982, 'Construing the Knowledge Situation: Stephen Pepper and a Deweyan Approach to Literary Experience and Inquiry', *Journal of Mind and Behavior* **3**, 385–401.

Dale, E. D.: 1990, *Kunnskapens Tre og Kunstens Skjønnhet. Om den Estetiske Oppdragelse i det Moderne Samfunn* [The Tree of Knowledge and the Beauty of Art. On the Aesthetic Education in Modern Society]. Gyldendal, Oslo.

Devitt, M.: 1991, *Realism and Truth*, Basil Blackwell, Oxford.

Dewey, J.: 1981, *The Philosophy of John Dewey*. The University of Chicago Press, London and Chicago.

Dewey, J.: 1997, *Experience and Nature*, Open Court, Chicago and La Salle, IL (2nd edn., originally published 1929).

Egan, K. & Nadaner, D. (eds.): 1988, *Imagination & Education*, Open University Press, Milton Keynes, UK.

Frohmann, B.: 1992, 'The Power of Images: A Discourse Analysis of the Cognitive Viewpoint', *Journal of Documentation* **48**(4), 365–386.

Flannery, M.: 1992, 'Using Science's Aesthetic Dimension in Teaching Science', *Journal of Aesthetic Education* **26**(1), 1–15.

Glasersfeld, E. v.: 1990, 'Environment and Communication', in L. Steffe & T. Wood (eds.), *Transforming Children's Mathematic Education: International Perspectives*, Lawrence Erlbaum, Hillsdale, NJ, pp. 3–38.

Gruber, H. E. & Vonèche, J. J. (eds.): 1982, *The Essential Piaget. An Interpretive Reference and Guide*, Routledge & Kegan Paul, London.

Gunstone, R. F., White, R. T. & Fensham, P. J.: 1988, 'Development in Style and Purpose of Research on the Learning of Science', *Journal of Research in Science Teaching* **25**, 513–529.

Hamlyn, D. W.: 1961, *Sensation and Perception. A History of the Philosophy of Perception*. Routledge & Kegan Paul, London.

Harvey, C. W.: 1989, *Husserl's Phenomenology and the Foundations of Natural Science*, Ohio University Press, Athens, OH.

Heinemann, F.: 1934, 'Goethe's Phenomenological Method', *Philosophy* **9**, 67–81.

Herzog, W.: 1991, 'Piaget im Lichte der Phänomenologie: Eine pädagogische Erkundung', in M. Herzog & C. F. Graumann (eds.), *Sinn und Erfahrung. Phänomenologische Methoden in den Humanwissenschaften*, Roland Asanger Verlag, Heidelberg, pp. 288–312.

Horkheimer, M.: 1967, 'Kants Philosophie und die Aufklärung' in M. Horkheimer (ed.), *Zur Kritik der instrumentellen Vernunft (Teil II)* [Towards the Critique of Instrumental Reason], Fischer, Frankfurt am Main.

Horkheimer, M. & Adorno, T.: 1944, *Dialektik der Aufklärung* [The Dialectics of Enlightenment]. Querido, Amsterdam.

Husserl, E.: 1970, *The Crisis of the European Sciences and Transcendental Phenomenology*. Northwestern University Press, Evanston, IL.

Jardine, D.: 1990, 'On the Humility of Mathematical Language', *Educational Theory* **40**, 181–191.

Kant, I.: 1993, *Critique of Pure Reason. A Revised and Expanded Translation Based on Meiklejohn*. Everyman's Library, London and Vermont (originally published in 1781).

Madison, G. B.: 1990, *The Hermeneutics of Postmodernity. Figures and Themes*. Indiana University Press, Bloomington and Indianapolis.

Martin, F. D.: 1974, 'Heidegger's Being of Things and Aesthetic Education', *Journal of Aesthetic Education* 8(1), 87–105.

Marton, F. & Booth, S.: 1997, *Learning and Awareness*. Lawrence Erlbaum, Mahwah, NJ.

Matthews, M. R.: 1994, *Science Teaching. The Role of History and Philosophy of Science*. Routledge, New York and London.

Merleau-Ponty, M.: 1964, *The Primacy of Perception*, Northwestern University Press, Evanston, IL.

Merleau-Ponty, M.: 1968, *The Visible and the Invisible*, Northwestern University Press, Evanston, IL.

Merleau-Ponty, M.: 1992, *The Phenomenology of Perception*. Routledge, London (first English publication in 1962).

Mollenhauer, K.: 1985, *Vergessene Zusammenhänge. Über Kultur und Erziehung* [Forgotten Connections. About Culture and Education]. Juventa, München.

Murphy, J.: 1985, 'Consideration of Computer Mediated Education: A Critique', *Phenomenology & Pedagogy* 3, 167–176.

Neujahr, P. J.: 1995, *Kant's Idealism*. Mercer University Press, Macon, GA.

Oksenberg Rorty, A: 1980, 'The Place of Contemplation in Aristotle's Nicomachean Ethics', in A. Oksenberg Rorty (ed.), *Essays on Aristotle's Ethics*, University of California Press, Berkeley, pp. 377–394.

Rang, B.: 1997, 'Der systematische Ansatz von Husserls Phänomenologie der Natur', in G. Böhme & G. Shiemann (eds.), *Phänomenologi der Natur*. Suhrkamp, Frankfurt am Main, pp. 85–119.

Rumelhart, D. E. & Ortony, A.: 1977, 'The Representation of Knowledge in Memory', in R. C. Anderson & W. E. Montague (eds.), *Schooling and the Acquisition of Knowledge*. Lawrence Erlbaum, Hillsdale, NJ, pp. 99–136.

Rumpf, H.: 1991, 'Die Fruchtbarkeit der phänomenologischen Aufmerksamkeit für Erziehungsforschung und Erziehungspraxis', in M. Herzog and C. F. Graumann (eds.), *Sinn und Erfahrung. Phänomenologische Methoden in den Humanwissenschaften*. Roland Asander Verlag, Heidelberg, pp. 313–335.

Rumpf, H.: 1993, 'Anfängliche Aufmerksamkeiten. Beispiele und Begründungen', in H.-G. Herrlitz and C. Rittelmeyer (eds.), *Exakte Phantasie. Pädagogische Erkundungen bildender Wirkungen in Kunst und Kultur*. Juventa, Weinheim and München, pp. 123–145.

Säljö, R.: 1995, 'Begreppsbildning som Pedagogisk Drog' [Concept Formation as a Pedagogical Drug], *Utbildning och Demokrati* 4, 5–22.

Schweizer, H. R.: 1973, *Ästhetik als Philosophie der Sinnlichen Erkenntnis. Eine Interpretation der "Aesthetica" A. G. Baumgartens mit teilweiser Widergabe des lateinischen Textes und Deutscher Übersetzung*, Schwabe & Co, Basel and Stuttgart. (Aesthetics as the Philosophy of Sensual Knowledge. An Interpretation of A. G. Baumgarten's "Aesthetica" with Partial Reproductions of the Latin Text and German Translation.)

Sperry, R.: 1983, *Science and Moral Priority. On Merging Mind, Brain and Human Values*. Basil Blackwell, Oxford.

Steffe, L. P. & Gale, J. (eds.): 1995, *Constructivism in Education*. Lawrence Erlbaum, Hillsdale, NJ.

Suchantke, A.: 1998, 'Die Wahrnehmung der Wahrnehmung. Zur Aktualität des Goetheanismus', in R. Dorka, R. Gehlig, W. Schad & A. Scheffler (eds.), *'Zum Erstaunen bin ich da', Forschungswege in Goetheanismus und Anthroposophie*. Verlag am Goetheanum, Dornach, pp. 84–103.

Suchting, W. A.: 1992, 'Constructivism Deconstructed', *Science & Education* 1, 223–254.

Tarsitani, C.: 1996, 'Metaphors in Knowledge and Metaphors of Knowledge: Notes on the Constructivist View of Learning', *Interchange* 27, 23–40.

Thomas, P.: 1997, 'Leiblichkeit und eigene Natur. Naturphilosophische Aspekte der Leibphänomenologie', in G. Böhme & G. Shiemann (eds.), *Phänomenologi der Natur*. Suhrkamp, Frankfurt am Main, pp. 291–302.

Wagenschein, M.: 1965, *Ursprüngliches Verstehen und exaktes Denken* [Original Understanding and Exact Thinking]. Ernst Klett, Stuttgart.

Wagenschein, M.: 1981, 'Ein Interview zu seinem Lebenswerk mit P. Buck und W. Köhnlein', *Zeitschrift Chimica Didactica* **7**, 161–175.

Constructivism in School Science Education: Powerful Model or the Most Dangerous Intellectual Tendency?

E. W. JENKINS
School of Education, University of Leeds, Leeds LS2 9JT, UK (e.w.jenkins@education.leeds.ac.uk)

Abstract. This paper explores and challenges a number of the assumptions and claims commonly associated with a constructivist approach to school science education, e.g., that constructivist ideas about learning require a progressive pedagogy or that 'active learning' demands engaging students with practical activities. It suggests that constructivist ideas have a particular appeal within primary education because they help to justify classroom practices and activities that primary school teachers, for a variety of other reasons, regard as important. It is suggested that the recent dominant emphasis upon constructivism in science education has narrowed both the professional and the research agenda relating to school science teaching. The paper argues for greater clarity and precision when referring to constructivist ideas in science education and for a better understanding of the role that learning theories should play in influencing the ways in which science is taught in schools.

Key words: Constructivism, teaching, learning, science

> The constructivist view of teaching and learning has proved to be a powerful model for describing how conceptual change in learners might be promoted.
>
> Keogh and Naylor 1997, p. 12

> [Constructivism is a candidate for] *the* most dangerous contemporary intellectual tendency . . . [because] it attacks the immune system that saves us from silliness.
>
> Devitt 1991, p. ix

1. Introduction

The two quotations above clearly imply different estimations of constructivism as a contemporary intellectual phenomenon and it is possible to make at least some attempt at reconciliation only by acknowledging that the authors are writing from different standpoints, namely those of pedagogy and learning theory on the one hand and of philosophy on the other. The philosophical dimensions of constructiv-

F. Bevilacqua et al. (eds.), Science Education and Culture, 153–164.
© 2001 *Kluwer Academic Publishers. Printed in the Netherlands.*

ism have generated a substantial and still burgeoning literature and several scholars have written about constructivism in science education from a philosophical point of view, with Matthews providing a convenient and up-to-date introduction to the field (Matthews 1998). This paper is concerned with what might be called the implications of constructivism for science education and, more particularly, with the claims made for it as 'a theory of teaching and learning'. Some of these claims can hardly be described as modest, e.g., 'Children learning science through the constructivist approach are noticeably different from children learning by a more passive method' (Wadsworth 1997, p. 24) and '*Learning* science and *doing* science proceed in the same way' (Harlen 1996, p. 5). It is, of course, acknowledged that, in some form or other, constructivist perspectives have influenced contemporary intellectual debate in fields as diverse as literature, the arts and the social and natural sciences. It is also acknowledged that a constructivist stance in education is often bound up with political, ethical or moral claims, especially when constructivist ideas are intimately linked with such issues as 'the emancipation of student learning', 'socially empowering groups or individuals', 'having respect for' students or their ideas, or, as the above quotation from Wadsworth indicates, the promotion of a 'child centred/progressive' pedagogy. These claims are not examined in detail in this paper, although some reference to them will be made.

There is little doubt that constructivist ideas in some form have come to dominate much of educational discourse, if not necessarily practice. Phillips has commented that 'Across the broad fields of educational theory and research, constructivism has become something akin to a secular religion'. Noting that 'whatever else it may be', constructivism is a '"powerful folk tale" about the origins of human knowledge', he adds that 'Like all religions, [it] has many sects[1] – each of which harbors some distrust of its rivals' (Phillips 1995, p. 5). Fensham has identified 'the constructivist view of learning' as the 'most conspicuous psychological influence on curriculum thinking in science since 1980' (Fensham 1992, p. 801), and the American Association for the Advancement of Science has described the widespread acceptance of constructivism as a 'paradigm change' in science education (Tobin 1993). Wadsworth, writing of primary teacher education in the UK, claims that constructivism is 'generally accepted by teacher educators as the most effective way of teaching the ideas of science' (Wadsworth 1997, p. 23). Duit and Treagust, taking a wider perspective, assert that constructivism 'has the ascendancy among learning theories in the 1990s', acknowledging that in the United States the earlier emphasis had been upon behaviourist approaches to learning (Duit and Treagust 1997, p. 3).

2. Responding to Constructivist Claims

A not inconsiderable difficulty in responding to many of the claims made on behalf of constructivism within science education is the difficulty of knowing what interpretation to give to 'constructivism' or 'constructivist ideas' when used in this context. There are many variants of 'constructivism' and the educational literature on constructivism has been described as 'enormous and growing rapidly' (Phillips 1995, p. 5). In addition, debates that are ostensibly about teaching or learning readily become confused with others that are essentially epistemological or philosophical disputes about the nature of science, of scientific knowledge, or about the existence of an external 'reality' which it is the business of science to describe. Phillips, in a valuable attempt to impose a degree of conceptual order among the various constructivist 'sects', draws a distinction between those who focus their attention on the 'cognitive contents of the minds of individual learners' and others who emphasise the growth of the 'public' subject-matter domains, adding a third category of 'brave groups' who 'tackle both – thus doubling the amount of quicksand to be negotiated' (Phillips 1995, pp. 5–6). To further complicate matters, constructivists who focus their attention upon, for example, how individuals learn, may differ quite profoundly about the mechanisms they suggest are involved.

> Piaget and Vygotsky ... gave quite different accounts of this matter; one stressed the biological/psychological mechanisms to be found in an individual learner, whereas the other focussed on the social factors that influenced learning. (Phillips, *op. cit.*, p. 7)

If there is common ground among constructivists of different persuasion it presumably lies in a commitment to the idea that the development of understanding requires active engagement on the part of the learner. Put another way, knowledge cannot be 'given' or handed over and received in the same way as a parent might give a child a book, a toy or a tool. When characterised in this way, constructivism has a long ancestry and accommodates considerable flexibility, with even someone like John Locke admitted to the constructivist camp by allowing that the mind can 'put together those ideas it has, and make new complex ones' (Locke 1947, p. 65). What this characterization means for teaching and learning, however, is by no means as straightforward as many of those arguing for constructivist approaches to teaching and learning commonly assume, imply or assert.

The notion of the mind actively constructing knowledge does not, for example, lead in any logical way to a rejection of the world as an external reality. Nor does it require the problematic idea that science education is about 'making sense' of the world rather than about establishing a valid scientific understanding of natural phenomena. More than enough has now been written to expose the subtlety and complexity of scientific ideas and their frequent divorce from common-sense understanding and experience (e.g., Wolpert 1992). To establish heliocentricity or to prefer uniform motion in a straight line to rest in understanding the Newtonian universe involves more than 'making sense' of the world, unless this phrase is asked to bear a greater and more qualified meaning than is usually the case. From this perspective, progressivist claims such as 'children are natural scientists' and 'everyone engages in scientific activity during the course of their everyday activities' are not only beguiling but, from the point of view of science education, misleading.

Care is also need in coupling so-called constructivist learning and teaching, as in the first quotation at the head of this article. A theory of teaching (however that may be defined) is necessarily more complex than a theory of learning, not least because it must accommodate what is known about a range of matters not embraced by studies of how students learn. In addition, while the large volume of empirical data about students' understandings of a range of scientific phenomena (see, for example, Pfundt and Duit 1994) is of interest, comparatively little is known about how teachers can most effectively respond to it (Claxton 1986). For example, are *eliciting* and *reorganising* students' ideas to be seen as distinct steps or, as some writers suggest (e.g., Harlen 1996) better regarded as part of a continuous process? The pedagogical consequences of having elicited students' ideas are also far from clear. Likewise, are ideas, once elicited, meant to assist a teacher to plan what he or she must now do in response or is their principal purpose to help students clarify their own thinking?

In more general terms, the question being asked here is what, in practical terms, is a science teacher to do if, as constructivism would seem to dictate, he or she must take students' 'existing ideas' into account in planning science teaching activities? Certainly, students' understandings of natural phenomena are to be valued and treated with respect, and, in many cases, they can be used as a starting point for a range of activities ranging from class discussion to experimental work in the laboratory. If the students' understandings of natural phenomena are wrong, science teachers would argue that they are to be corrected.[2] Constructivism, however, offers little in the way of guidance about how this may best be done, despite the fact that a range of so-called constructivist curriculum materials have been produced (Driver and Oldham 1985). Science evolved very late in human history and it seems more than optimistic to assume that young students can construct scientific explanations simply by observing phenomena and generating and testing hypotheses. Even if this were possible, the question would remain of whether engaging students in the necessary practical activities is the most efficient way of promoting their learning.

It is also important to ask what it is that 'constructivist teachers' wish their students to 'construct' during the course of their science lessons. If it is assumed that one purpose, perhaps the principal purpose, of school science education is to help students learn some of the ways in which the scientific community understands the natural world, then what is to be 'constructed' by students is not simply the understanding that might flow from their interactions with a range of natural phenomena. The significant omission, as Driver et al. have acknowledged, is students' 'interactions with *symbolic* realities, the cultural tools of science' (Driver et al. 1994, p. 7). Rectifying this omission has a number of implications for what might be called naïve constructivist approaches to teaching science. It requires an acknowledgement of the importance of expert scientific knowledge on the part of the teacher, and it shifts the debate away from learning as an individual construction towards learning as a social activity, i.e., towards so-called social constructivism. What remains a matter for debate, however, is the way in which the expert scientific knowledge of the teacher can be most effectively deployed to help students learn something of the ways in which the world is understood by the scientific community. Solomon's comments are of interest here. In a perceptive article, entitled 'The Rise and Fall of Constructivism', she suggests that constructivism 'obscures other perspectives', and offers an 'alternative picture of pupil learning'.

> . . . a young student sits outside a circle of disputing scholars picking up fragments of conversation and trying to piece them together. Once we were all that child, the family was the circle, and we turned over the phrases that we heard until they built up into an idea. We tried out the sense of it, and occasionally we were amusingly wrong. If we were lucky, no one laughed. Then it was explained once more in helpful ways and with good games to go with the learning of it. When we tried it again and the half-formed idea seemed to be accepted by others, it became stronger. Kindly adults encouraged us to use it in new ways: our understanding and pride in using it grew. The idea gradually became ours, and, by the same token, we became part of the privileged and knowing circle who used it. (Solomon 1994, pp. 17–18)

It is perhaps significant that much of the research that has been done within the 'misconceptions' tradition has been concerned with concepts such as force, energy, power, gravity or mass, in which everyday words and notions are given highly specialised and often mathematical meanings within a scientific context. It is not difficult, indeed it is to be expected, that young people's own experiences will have led them to have ideas about at least some of these concepts. It is more difficult, however, to understand how young students might have developed an out-of school understanding of concepts such as ion, electromagnetic radiation, oxidation, free energy or chemical equilibrium. With these, and many other scientific concepts far removed from everyday experience, 'eliciting' students' ideas becomes more difficult, if not impossible. This, of course, does not mean that it is no longer possible or desirable for a teacher to engage in a conversation with students, to explore analogies intended to promote intellectual growth, to probe their understanding, or to challenge their assumptions, arguments or conclusions by whatever strategies he or she judges to be effective. Such strategies, however, depend upon constructivist ideas only to the extent that they acknowledge that learning requires the active engagement of the learner.

Equally, a constructivist view of learning does not demand a pedagogy that might be described as 'progressive', any more than 'active learning' necessarily entails engaging students in practical activities. If, as constructivism requires, learning presupposes the active engagement of the mind of the learner, then the notion of 'passive learning' lacks meaning. As any teacher knows, it is possible to engage the minds of learners by a wide variety of teaching strategies, some of which might be described as formal and didactic, rather than informal and exploratory. Indeed, selecting a strategy that is more, rather than less, likely to interest students and promote their learning is central to a teacher's professional competence.

3. The Prominence of Constructivism in Science Education

It is pertinent to ask why constructivist assumptions and claims have come to figure so prominently in much of the professional and academic literature of science education, especially that relating to the education of children of primary school age. Writing of primary science education in the UK, Harlen has claimed that the SPACE Project[3] was 'largely responsible for bringing "constructivism" into common discourse in primary science teaching, although, like many ideas in education, it had roots in earlier work, particularly that of Piaget' (Harlen 1997). In the present context, the validity of this claim is less important than the issue of why constructivism has become part of the common discourse to which Harlen refers. It has been suggested elsewhere (Jenkins and Swinnerton 1998, p. 223) that in the case of primary science education, constructivist ideas may simply be being raided to sustain classroom practices and activities (such as group work and projects) that many primary school teachers, for a variety of other reasons, regard as important. These activities and practices (conveniently, if unhelpfully) are sometimes labelled 'progressive', and, in England and Wales they derive most recently from the work of Susan and Nathan Isaacs and from the appropriation of aspects of Piagetian psychology for pedagogical purposes during the 1960s and beyond. More particularly, the constructivist requirement to engage the learner actively in learning has been used to justify engaging students in practical and investigative activities of various kinds. In addition, the beguiling, if erroneous, parallels sometimes drawn between the 'construction' of personal knowledge by the learner and the generation of scientific knowledge have been used (Jenkins and Swinnerton 1998) to ally constructivism with 'discovery learning' and with the teaching of science 'by investigation'[4] (Indeed, if knowledge construction is seen as an entirely individual matter, then any distinction between constructivist pedagogy and discovery learning becomes difficult to sustain.) From all these perspectives, 'most types of constructivism are modern forms of progressivism' (Phillips 1995, p. 11).

Some caution is needed, however, to avoid over-estimating the impact of constructivist ideas upon practice in the primary classroom or, indeed, at other levels of education. Sizmur and Ashby (1997) found that few teachers elicited young children's views about natural phenomena in any systematic way when introducing them to scientific concepts, and Larochelle and Bednarz (1998, p. 3) have commented that '... taking students' knowledge into account seems to have scarcely modified the usual teaching modus vivendi at any level of instruction one chooses to examine'. Murphy, following a study of local primary school teachers, has reported that

> Some teachers were not convinced about a constructivist approach, in particular whether it was appropriate for all ages. Others had never heard of constructivism but were nevertheless committed to investigative learning where the children had '*freedom in the practical sense to decide what*

> *they wanted to find out and how to set about doing it*'. Some described a constructivist approach as the children '*doing the doing*'.
>
> Those who had heard of the term were usually strong advocates of the approach and described it as 'children building on their previous experience through practical investigation, learning through investigation, open learning situations within the classroom. (Murphy 1997, pp. 27–28)

A frequent claim of those who invoke constructivist ideas to justify a 'progressive' pedagogy is that other forms of teaching either fail or result in learning that is pejoratively described as superficial, shallow or short-term. A constructivist approach, in contrast, is sometimes said to lead to 'real understanding' or to 'long term' learning. It is not clear what empirical or other evidence exists or might exist to substantiate claims of this kind, but, whatever the evidence, it will remain important to ask whether the classroom activities and practices promoted by those who advocate a constructivist pedagogy to introduce pupils to scientific concepts and ideas can be justified in terms of the time and resources associated with them. As one teacher educator, concerned about the pressures on the science component of the primary curriculum in England and Wales, expressed it, 'When do I tell them the right answer?' (Wadsworth 1997, p. 23). It is not surprising, therefore, that much constructivist writing about school science argues for a reduction in scientific content or demands an assessment technology that is more sympathetic to a constructivist approach.

4. The Issue of Multiple Understandings

Like all who teach science, those who espouse a constructivist approach are faced with the overwhelming evidence that many children retain erroneous 'common-sense' or 'everyday' understandings of a number of scientific phenomena, despite all attempts of science teachers to effect change. For example, students who engage with problems in Newtonian terms in the classroom or laboratory often resort to discredited Aristotelian notions when asked to explain similar problems involving force or motion encountered in an everyday, out-of-school, context. For constructivist, and perhaps for all, teachers, there seems to be both a problem and a challenge here. The problem is the persistence of erroneous ideas among students who, having been well-taught, ought to know better. The challenge seems to lie in helping students develop a more consistent scientific understanding of the natural world. Many constructivists have taken this challenge seriously with some arguing for 'cognitive conflict' i.e., placing a student in a position in which the application of his or her own understanding to a problem leads to cognitive difficulties which the student must then resolve. One of the difficulties of this approach, the problem of 'knowledge in context', is discussed below. A further difficulty arises when, as seems to be the case, all forms of conceptual change are regarded as equally difficult and likely to be effected by some common 'constructivist' pedagogy. Experience in the classroom suggests that, to the contrary, some forms of conceptual change can be brought about much more easily than others, much depending upon the complexity of the scientific ideas and the extent to which they are counter-intuitive. Thus, most secondary school pupils are likely to find greater difficulty with the ideas surrounding the motion of projectiles than they are with the notion that light travels in straight lines and can be reflected by a plane mirror. To the extent that this is the case, 'the constructivist view of teaching and learning' is likely to encounter some difficulty as a 'powerful model for describing how conceptual change in learners might be promoted'.

In addition, the notion that students, and, more generally, adults, should always explain natural phenomena in terms that accord fully with the canon of scientific knowledge presents problems. Phrases like 'the sun rises in the east', 'feed the plants' and 'keep out the cold' persist in everyday use, even though the heliocentric universe makes a nonsense of the first and, in scientific terms, plants make their own food and cold has no scientific meaning save an absence of heat. Moreover, as work in the public understanding of science has revealed, the seemingly straightforward 'application' of scientific knowledge in the world far removed from the laboratory can sometimes be misleading and unhelpful (Irwin and Wynne 1996) and there may be good reasons for its rejection in favour of other, more local or personal knowledge and understanding (Layton et al. 1993). Even among those with a scientific background, outdated scientific ideas may still be used because they adequately serve the purpose in hand. Heating engineers, for example, commonly discuss heat transfer in terms of flow rather than of molecular motion.

What is being suggested here is not that common sense or everyday knowledge should always be valorised over scientific knowledge or that all forms of knowledge are always of equal worth. Common sense or everyday knowledge is sometimes wrong and occasionally dangerously so. The particular point is simply that each of us, in our everyday activities, is usually content to use a model which seems adequate for the purpose we have in mind. The model may draw upon a variety of sources but it will always be tested against experience. This, of course, does not make it 'true', even though, because it works, it may seem so. As noted above, it is on this issue that those who equate constructivist science education with helping students to 'make sense' of the natural world run into some difficulty.

The more general point is captured by Bachelard's notion of a conceptual profile (Bachelard 1968) which acknowledges that individuals have a variety of models of, i.e., ways of thinking about, natural events and phenomena. For example, a physicist who works professionally with quantum models of matter is likely, in other contexts, to invoke the notion of matter as a continuum and, in most everyday practical activities, act on this latter basis. This notion of a conceptual profile, allied with the outcomes of much research into the way in which citizens interact with scientific knowledge, constitutes a direct challenge to the notion that learning science can, or should, be reduced simply to a matter of replacing students' misconceptions/alternative conceptions by more orthodox scientific understandings.[5]

It is also, of course, a challenge to other, more traditional approaches to teaching school science.

5. Conclusion

The preceding paragraphs suggest that constructivism in science education is neither a powerful model for describing how conceptual change in learners may be promoted nor the 'most dangerous intellectual tendency'. Within the literature relating to school science education, the impression is rather one of confusion and often uncritical espousal of a fashionable research paradigm. In addition, 'constructivism' has acquired something of a Humpty Dumpty quality.[6] If the on-going debate about teaching and learning in school science is to become more focused, attention to language would seem to be something of a priority. For example, students' ideas about natural phenomena are too glibly described as 'theories', a description which implicitly, and sometimes explicitly, suggests unhelpful and misleading parallels with scientific theories. Likewise, distinctions often need to be drawn between the guesses which students may make in seeking to explain some phenomenon and scientific hypotheses. Equally important is the need to be clear about the nature of, and the evidence for, the various claims put forward under the umbrella of constructivism. Are these claims, for example, about the nature of knowledge, about how children can be most effectively taught, about the existence of an 'external' reality, or about the locus and nature of scientific or pedagogical authority? Even when these claims are related to each other, important distinctions need to be made, although it would be foolish to suggest that all of the issues involved are capable of unequivocal resolution.

It would also be helpful to know more about the role that theories of learning have played, and might properly be expected to play, in influencing the ways in which science is taught in school. This is a complex and largely unexplored field and one that is likely to reveal significant differences between countries and between primary and secondary schooling. Behaviourist ideas for example, which have exerted a powerful influence in the USA, have had much less impact on teaching in England and Wales, and, in general, the discourse of primary teaching has been more accommodating of theories of learning than is the case at the secondary level where subject disciplinary considerations have held much greater sway.[7] An overview of the position with respect to primary/elementary science education in England and Wales has been provided by Jenkins and Swinnerton (1998), but any historical study is necessarily limited by the difficulty of addressing satisfactorily the gap between pedagogical rhetoric and the reality of classroom practice. In 1967, the Plowden report, usually regarded as a seminal influence on primary education practice in England and Wales, noted of the early twentieth century that

> A considerable body of liberal thinking on the education of children was available to teachers. Rousseau, Pestalozzi, Froebel, Whitehead, Dewey, Montessori and Rachel Macmillan, to mention only a few, had all written

> on lines that encouraged change and innovation. Yet it may be doubted whether the direct influence of these or of any other writers was great. (Central Advisory Council 1967, pp. 189–190).

Only two years earlier, however, the leaders of the Nuffield Junior Science Project were arguing that

> We must be able to justify all that we advocate on psychological grounds. More than this: we need to look at all we plan to do in relation to what is known of children's learning processes. No one who is not prepared to accept this idea should be involved in the scheme! (Nuffield JSP 1965, p. 3).

It is perhaps important to note that any study of the influence of theories of learning on science teachers' practice in the classroom and laboratory will have considerable political resonance since it is likely to be intimately related to the battle between 'progressive' and 'traditional' approaches to teaching.

It is also important to acknowledge that the attention devoted to constructivism in school science teaching may have served to narrow the professional and the research agenda within science education. While recognizing some of the benefits that have flowed from constructivist ideas, notably the emphasis placed upon the learner, Solomon warns of 'tunnel vision' among some researchers within the constructivist tradition, and notes, with O'Loughlin (1992) that 'Mature constructivism tends to abrogate all avenues of research to itself' (Solomon 1994, p. 17). Woolnough, from a different perspective, claims that

> One of the problems about constructivism, and other theories of learning, is not that they get the answers wrong but that they ask the wrong questions! They seek to answer the question 'how do pupils learn?' What we ought to be asking is 'what makes students *want* to learn? (Woolnough 1998, p. 17)

It is difficult to avoid the conclusion that the answer to Woolnough's second question is likely to be of more immediate and practical use to science teachers than any constructivist response to the first.

Notes

[1] Phillips' language is, of course, emotive and pejorative. For a more generous account of the diversity of interpretations of constructivism, together with a scholarly exploration of the 'reach' of constructivist ideas within education, see Larochelle et al. (1998).

[2] It is acknowledged that, for many, perhaps all, constructivists, constructivism requires teachers and learners to devote attention to 'how we know what we know' (Larochelle and Bednarz). While this has an obvious relevance to school science teaching, its implications are somewhat eccentric to the principal focus of this article and are not discussed here.

[3] Science Processes and Concepts Exploration Project.

[4] They have also prompted studies that explore the relationship between the history of scientific ideas and students' contemporary understanding of a range of natural phenomena.

[5] For some constructivists, this issue is one of developing new forms of discourse. See Gee (1996).

[6] 'When *I* use a word', Humpty Dumpty said . . . 'it means just what I choose it to mean – neither more nor less' (Lewis Carroll).

[7] For most of the second half of the 19th century, much of the debate about school science teaching centred on whether, for example, heat should be taught before light, or static electricity before current electricity. When, early in the following century, ideas about children's intellectual development replaced those drawn from faculty psychology to become part of educational thinking, H. E. Armstrong, the advocate of the heuristic method of teaching science, complained that 'The damned boy (*sic*) needs drilling. We forget this and ever twaddle of playing on his interests' (Armstrong 1924, quoted in Brock 1973, p. 145).

References

Bachelard, G.: 1968, *The Philosophy of No*, Orion Press, New York.

Brock, W. H. (ed.): 1973, *H. E. Armstrong and the Teaching of Science 1880–1930*, Cambridge University Press, Cambridge.

Central Advisory Council: 1967, *Children and Their Primary Schools*, Vol.1, HMSO, London.

Claxton, G.: 1986, 'The Alternative Conceivers' Conceptions', *Studies in Science Education* **13**, 123–130.

Devitt, M.: 1991, *Realism and Truth*, Blackwell, Oxford.

Driver, R., Asoko, H., Leach, J., Mortimer, E. & Scott, P.: 1994, 'Constructing Scientific Knowledge in the Classroom', *Educational Researcher* **23**, 5–12.

Driver, R. & Oldham, V.: 1985, 'A Constructivist Approach to Curriculum Development', *Studies in Science Education* **13**, 105–122.

Duit, R. & Treagust, D.: 1998, 'Learning in Science – From Behaviourism Towards Social Constructivism and Beyond', in B. J. Fraser and K. G. Tobin (eds), *International Handbook of Science Education*, Part 1, Kluwer Academic Publishers, Dordrecht.

Fensham, P. J.: 1992, 'Science and Technology', in P. W. Jackson (ed.), *Handbook of Research on Curriculum*, Macmillan, New York, pp. 789–829.

Gee, J.: 1996, *Social Linguistics and Literacies*, Taylor and Francis, London.

Harlen, W.: 1996, *The Teaching of Science in Primary Schools*, David Fulton, London.

Harlen, W.: 1997, 'Ten Years On: The Past and the Future of Research in Primary Science', *Primary Science Review* **47**, 6.

Irwin, A. & Wynne, B. (eds): 1996, *Misunderstanding Science? The Public Reconstruction of Science and Technology*, Cambridge University Press, Cambridge.

Jenkins, E. W. & Swinnerton, B. J.: 1998, *Junior School Science Education in England and Wales Since 1900: From Steps to Stages*, Woburn Press, London.

Keogh, B. & Naylor, S.: 1997, 'Making Sense of Constructivism in the Classroom', *Science Teacher Education* **20**, 12–14.

Larochelle, M. & Bednarz, N.: 1998, 'Constructivism and Education: Beyond Epistemological Correctness', in M. Larochelle, N. Bednarz & J. Garrison (eds), *Constructivism and Education*, Cambridge University Press, Cambridge, pp. 3–20.

Larochelle, M., Bednarz, N. and Garrison, J. (eds): 1998, *Constructivism and Education*, Cambridge University Press, Cambridge.

Layton, D., Jenkins, E. W., Macgill, S. & Davey, A.: 1993, *Inarticulate Science? Perspectives on the Public Understanding of Science and Some Implications for Science Education*, Studies in Education, Driffield.

Locke, J.: 1947, *An Essay Concerning Human Understanding*, Dent, London.

Matthews, M. R. (ed.): 1998, *Constructivism in Science Education. A Philosophical Examination*, Kluwer, Dordrecht.

Murphy, P.: 1997, 'Constructivism and Primary Science', *Primary Science Review* **49**,27–29.

Nuffield JSP: 1965, 'Team Leaders' Meeting', 5 December, Nuffield JSP Archive, King's College, London.

O'Loughlin, M.; 1992, 'Rethinking Science Education: Beyond Piagetian Constructivism Towards a Sociocultural Model of Teaching and Learning', *Journal of Research in Science Teaching* 29, 791–820.

Pfundt, H. & Duit, R.: 1994, *Bibliography of Students' Alternative Frameworks and Science Education*, 4th edn., Institute for Science Education, Kiel.

Phillips. D. C.: 1995, 'The Good, the Bad, and the Ugly: The Many Faces of Constructivism'. *Educational Researcher* 24, 5–12.

Sizmur, S. & Ashby, J.: 1997, *Introducing Scientific Concepts to Children*, NFER, Slough.

Solomon, J.: 1994, 'The Rise and Fall of Constructivism' *Studies in Science Education* 23, 1–19.

Tobin, K. G. (ed.): 1993, *The Practice of Constructivism in Science and Mathematics Education*, AAAS Press, Washington DC.

Wadsworth, P.: 1997, 'When Do I Tell Them the RIGHT ANSWER?', *Primary Science Review* 49, 23–24.

Wolpert, L.: 1992, *The Unnatural Nature of Science'*, Faber and Faber, London.

Woolnough, B. E: 1998, 'Learning Science is a Messy Process', *Science Teacher Education* 23, 17.

Philosophy of Chemistry: An Emerging Field with Implications for Chemistry Education

SIBEL ERDURAN
Department of Science Education, King's College, University of London, Waterloo Road, Franklin-Wilkins Building, London SE1 8WA, UK

Abstract. Traditional applications of history and philosophy of science in chemistry education have concentrated on the teaching and learning of history of chemistry. In this paper, the recent emergence of *philosophy of chemistry* as a distinct field is reported. The implications of this new domain for chemistry education are explored in the context of chemical models. Trends in the treatment of models in chemistry education highlights the need for reconceptualizing the teaching and learning of chemistry to embrace chemical epistemology, a potential contribution by philosophy of chemistry.

1. History and Philosophy of Chemistry and Chemistry Education

Within the last two decades, the overlap of chemistry education research with revived efforts in the application of history and philosophy of science (HPS) in science education has been minimal (Erduran 1997; Kauffman 1989). Brush (1978) has argued that the anti-historical nature of chemistry education is a reflection of chemists' marginal interest in the historical dimensions of their science. Such a claim, however, confuses the status of chemistry education research with the status of the historical and philosophical dimensions of chemistry itself. Many chemists (e.g., Kauffman & Szmant 1984; Partington 1957) have contributed to historical analyses of their discipline. The so-called 'chemist-historians' including Kopp, Thomson, Berthelot, Ostwald and Ihde have maintained a long tradition of interest in history of chemistry (Russell 1985). Furthermore in the United States, for instance, suggestions for the inclusion of history of chemistry in chemistry teaching can be traced back to the 1930s (Jaffe 1938; Oppe 1936; Sammis 1932).

The central argument for the inclusion of history of chemistry in chemistry instruction has been grounded in the need to motivate students' learning (Bent 1977; Brush 1978; Heeren 1990). Often however, history of chemistry, written by chemists from the perspective of the present status of their science consists of 'Whiggish history' (Butterfield 1949): history written from the perspective of contemporary values and criteria. Furthermore, history of chemistry is typically based on the members' account of chemistry (Pumfrey 1989). A member's account extracts from the past what is useful for the present, such as good examples of experimental discovery. What needs to be promoted instead, as Ellis (1989) argues, is

F. Bevilacqua et al. (eds.), Science Education and Culture, 165–177.
© 2001 *Kluwer Academic Publishers. Printed in the Netherlands.*

a stranger's account of history of chemistry: an analysis of historical events without taking for granted what seems self evident to us today.

An implication of the member's versus stranger's issue is that what seems to be self-evident of historical assumptions are not to be taken for granted but should be carefully scrutinised (Shortland & Warwick 1989). For instance, chemists' current criteria may make the Lemery models of acids and bases mythical and the Lewis models more factual. Yet, historical explanations demand more than such classifications. Models of acids and bases proposed by the seventeenth century French chemist Nicolas Lemery were not mythical at his time. The task that today's historians face is precisely to investigate why certain models, explanations and theories in science are taken for granted while others are not. Since the seventeenth century chemists could not anticipate alternative models of acids and bases (namely those based on the electronic configuration of the atom), the historian needs to examine the social, personal and epistemological as well as chemical factors that might provide an account for the route of change in chemists' understanding of acids and bases. Of relevance to chemistry education is that students come to the chemistry classroom not as members of the discipline but as strangers. They are most unlikely to share all of the assumptions that are necessary to see a certain chemical explanation as educators or chemists would see them.

2. Where is Philosophy of Chemistry?

Although history of chemistry has captured the interest of chemists and found its way into the curriculum (Akeroyd 1984; Ellis 1989; Herron 1977; Kauffman 1989), philosophical dimensions of chemistry have not received as much attention (Scerri 1997; van Brakel 1994). Some of the central questions in philosophy of science, such as the distinctive features of science from other endeavours, have been traditionally addressed in terms of what is considered to be the paradigm science: physics. Even though the emphasis on the logical analysis of scientific theories have been challenged by philosophers such as Popper and Kuhn, the legacy of logical positivism as well as physics' dominance in philosophical analyses persist even today.

Reductionism has been regarded as a major factor that has inhibited the development of philosophy of chemistry as a distinct field of inquiry (Primas 1983; van Brakel & Vermeeren 1981). From the logical positivist perspective, chemistry was viewed as being reducible to quantum mechanics. Reduction of one science to another was argued on the basis of correspondence and derivation of laws across these sciences (Nagel 1961; Nye 1993). The argument that chemistry is a reduced science has not gone uncritisized by chemists nor philosophers of science (Scerri 1994a; van Brakel 1994). Roald Hoffmann, the Nobel prize winning chemist for instance, has questioned the credibility of reductionist claims:

> The French rationalist tradition, and the systematization of astronomy and physics before the other sciences, have left science with a reductionist philosophy at its core. There is supposed to exist a logical hierarchy of the sciences, and understanding is to be defined solely in vertical terms as reduction to the more basic science. The more mathematical, the better. So biological phenomena are to be explained by chemistry, chemistry by physics, and so on. The logic of reductionist philosophy fits the discovery metaphor – one digs deeper and discovers the truth. But reductionism is only one face of understanding. We have been made not only to disassemble, disconnect, and analyze but also to build. There is no more stringent test of passive understanding than active creation. Perhaps "test" is not the word here, for building or creation differ inherently from reductionist analysis. I want to claim a greater role in science for the forward, constructivist mode. (Hoffmann & Torrence 1993, pp. 67–78)

Yet the assumption that chemistry is a reduced science has prevailed within the mindset of the HPS community (e.g., Wasserman & Schaefer 1986). Only since the 1990s have some philosophers of science questioned flaws within the reductionism argument. As an influential contributor to the emergence of philosophy of chemistry, Eric Scerri of UCLA has argued that philosophers of science have not been able to demonstrate that laws can be axiomatized in the first place let alone that they can be derived across disciplines (Scerri 1994a). It is further questionable whether or not predictive and explanatory power of laws, conventionally taken to be among the decisive criteria for determining a paradigmatic science, carry the same importance and emphasis, in different sciences. Whereas the history of physics includes numerous dramatic predictions such as the bending of starlight in gravitational fields and the existence of the planet Neptune, chemistry is not known for its predictive successes (Scerri 1994a).

Furthermore, Scerri (1991) argues that chemistry differs from physics generally not in terms of issues of prediction but in terms of classification. Whereas predictions in physics are based on mathematical models, chemical models rely more on the qualitative aspects of matter. Chemistry has traditionally been concerned with qualities such as color, taste and smell. Although both physics and chemistry involve quantitative and dynamic concepts, such concepts are often accompanied by qualitative and classificatory concepts in chemistry, as is also typical in biology. Furthermore, class concepts are used in chemistry as a means of representation. Some examples are 'acid', 'salt', and 'element'. These class concepts help chemists in the investigation and classification of new substances, just as biology is concerned with classification of organisms. Unlike in chemistry and biology, in physics the tendency is towards mathematization, not classification of physical phenomena. Such differences that set apart chemistry from physics as a distinct domain of scientific inquiry have been overlooked within the reductionist framework.

Although chemistry has typically been presented as a branch of physics not capturing sufficient attention within philosophy of science, it is important to note that chemistry demands a particular link to philosophy (Scerri 1997). In posing questions of reduction of one science, such as biology to another, namely physics, one cannot ignore the question of whether or not chemistry can be reduced to physics. If reduction of chemistry to physics fails, then reduction of biology to physics is even more unlikely since chemistry is often been regarded as an intermediary science between physics and biology.

Recent developments in philosophy of science have concentrated on naturalistic analyses of the sciences in which one examines more closely what the practitioners themselves might mean by issues such as reductionism (Kornblith 1985). For chemists and physicists, the attempt to reduce chemistry to physics consists of quantum chemistry which has been developing since the birth of quantum mechanics. Chemists would argue that although some chemical laws relate to physical laws, certain aspects of chemical principles do not necessarily reduce to physical principles. A nice example of this argument is presented by Scerri (1994a) who asks, "Does the periodic law count as a scientific law in the same sense as Newton's laws of motion?" Certainly the arrangement of elements in the periodic table provided some of the most dramatic predictions in the history of science: predictions by Mendeleev of the elements, gallium, germanium and scandium. Such predictions, however, could not have been made at the level of quantum chemistry.

3. Philosophy of Chemistry: An Emerging Field

There is increasing interest in the examination of chemistry as a distinct branch of science. An emergent group of philosophers of science (Green 1993; McIntyre 1999; Scerri 1996) have contributed to the formulation of philosophy of chemistry. The First International Conference on Philosophy of Chemistry was held in 1994. The 1997 annual meeting of the American Chemical Society has devoted a session to issues surrounding the interplay of philosophy and chemistry. The first issue of a new journal, *Foundations of Chemistry*, dedicated to philosophy of chemistry was published in 1999. There is now an on-line journal, *HYLE*, where philosophical analyses of chemistry are reported.

Given that philosophy of chemistry is an emerging field, it is not surprising that literature has barely addressed the applications of this field in chemistry education (e.g., Erduran 2000; Scerri 2000). Philosophy of chemistry, however, has the potential to inform and guide chemistry education particularly through chemical epistemology (Erduran, in press), a line of inquiry that focuses on theories of chemical knowledge. Models and modeling, for instance, provides a crucial and relevant context through which epistemological aspects of chemistry can be promoted in the classroom.

4. Models and Modeling in Chemistry

Chemists have often drawn attention to the significance of models and modeling in chemistry (Suckling, Suckling & Suckling 1978; Tomasi 1988; Trindle 1984).

Chemists model the physical and chemical properties of matter in an effort to explain why matter behaves in certain ways. In the case of acid-base chemistry, for instance, physical and chemical properties of acids and bases are explained with Arrhenius, Brønsted-Lowry and Lewis models (Atkins 1991). The following brief overview of these models will exemplify the role of models in chemistry.

Towards the end of the nineteenth century, Svante Arrhenius classified a compound as acid or base according to its behavior when it is dissolved in water to form an aqueous solution. He suggested that a compound be classified as an 'acid' if it contains hydrogen and releases hydrogen ions, H^+ (see Figure 1). Likewise, a 'base' was defined as a compound that releases hydronium ions, OH^-, in a solution:

$$\underset{\text{Acid}}{HA(aq)} \rightarrow H^+(aq) + A^-(aq)$$

$$\underset{\text{Base}}{BOH(aq)} \rightarrow B^+(aq) + OH^-(aq)$$

Figure 1. Arrhenius model of acids and bases.

In Figure 1, H stands for hydrogen; A and B stand for other element(s) in the compounds; and (aq) stands for aqueous.

At the turn of this century, Johannes Brønsted and Thomas Lowry proposed a broader definition of acids and bases where it is possible to speak of substances as intrinsically acids and bases, independent of their behavior in water. The new model was formulated based on observations that substances could behave as acids or bases even when they were not in aqueous solution, as the Arrhenius model required. In Brønsted-Lowry model (see Figure 2), an acid is a hydrogen donor and a base is a hydrogen acceptor. There is no requirement for the presence of water in the medium:

$$HA(aq) + H_2O \rightarrow H_3O +(aq) + A^-(aq)$$

Figure 2. Brønsted-Lowry model of acids and bases.

The acid-base chemistry took yet another turn when the centrality of hydrogen in both the Arrhenius and Brønsted-Lowry models was challenged. Chemistry, is after all, concerned more with electrons, not hydrogen. Furthermore, the Brønsted-Lowry model did not capture all substances that behave like acids and bases but do not contain hydrogen. Gilbert Lewis formulated yet a broader definition of acid-base behavior. Lewis considered that the crucial attribute of an acid is that it can accept a pair of electrons and a base can donate a pair of electrons (see Figure 3). In Figure 3, : stands for a pair of electrons. In the context of the Lewis model, electron donation results in the formation of a covalent bond between the acid and the base. Lewis hence refocused the definition of acids and bases to something more fundamental about any atom: electrons.

$$\underset{\text{Acid}}{H^+} + \underset{\text{Base}}{:O^{2-}} \rightarrow O\!-\!H^-$$

Figure 3. Lewis model of acids and bases.

What this brief survey of models of acids and bases illustrates is that certain criteria, such as behavior of acids and bases independent of water, shaped the evaluation and revision of each model. The process of model formulation, evaluation and modification is not unique to acid-base chemistry. In chemical kinetics, for instance, the mechanism of chemical change has been explained by various models developed throughout history of chemistry (Justi & Gilbert 1999). The 'anthropomorphic' model described a chemical change in terms of the readiness of the components to interact with each other. The 'affinity corpuscular' model emphasized the chemical change in terms of atomic affinities. 'First quantitative' model introduced the notion of proportionality of reactants for chemical change to occur. The 'mechanism' model began to outline steps in a chemical reaction. The 'thermodynamics' model drew attention to the role of molecular collision (with sufficient energy) in chemical change. The 'kinetic' model introduced the idea of frequency of collisions of molecules. The 'statistical mechanics' model relied on quantum mechanics and identified a chemical reaction as motion of a point in phase space. The 'transition state' model provided a link between the kinetic and thermodynamic models by merging concepts of concentration and rate.

5. Models and Modeling in Philosophy of Chemistry

The role of models in chemistry has been underestimated since the formulation of quantum theory at the turn of the century. There has been a move away from qualitative or descriptive chemistry towards quantum chemistry. Increasingly, chemistry has been projected as a reduced science where chemical models can be explained away by physical theories:

> In the future, we expect to find an increasing number of situations in which theory will be the preferred source of information for aspects of complex chemical systems. (Wasserman & Schaefer 1986, p. 829)

Atomic and molecular orbitals, formulated through quantum chemistry, have been used to explain chemical structure, bonding and reactivity (Luder, McGuire & Zuffanti 1943).

Only recently has an opposition to quantum chemistry (van Brakel & Vermeeren 1981; Zuckermann 1986) begun to take shape with a call for a renaissance of qualitative chemistry. Underlying the emergent opposition is the argument that quantum chemistry has no new predictive power for chemical reactivity of elements that descriptive chemistry does not already provide (Scerri 1994b). Rearrangement of the Periodic Table of elements away from the original proposed by Mendeleev and others, for instance, towards one based on electronic configurations first suggested by Niels Bohr yield no new predictions about chemical or physical behavior of elements. Furthermore, no simple relation exists between the electron configuration of the atom and the chemistry of the element under consideration. In summary, there is no evidence to suggest that new physical and chemical behavior of elements can be explained or predicted by quantum theory.

What the preceding discussion illustrates is that although models have historically been central in the growth of chemical knowledge, in recent years a greater role was granted to quantum theory in chemistry. The purpose of this paper is not to contribute to the philosophical debate surrounding the status of chemical knowledge; this paper is more concerned about promoting the consideration and inclusion in chemistry education of crucial observations and syntheses from philosophy of chemistry. In the following section, the example of models and modeling, a central feature of the chemical sciences (Suckling, Suckling & Suckling 1972) will be proposed as a context worthy of examination by philosophers of chemistry and as a significant educational outcome to be targeted by chemistry educators.

6. Models and Modeling in Chemistry Education

There is substantial evidence that children learn and use models from an early age (Schauble, Klopfer & Raghavan 1991; Scott, Driver, Leach, & Millar 1993; Gilbert & Boulter 2000). Children's learning of models in the classroom has been promoted on the grounds that models can act as "integrative schemes" (National Research Council 1996, p. 117) bringing together students' diverse experiences in science across grades K-12. The *Unifying Concepts and Processes Standard* of The National Science Education Standards specifies that:

> Models are tentative schemes or structures that correspond to real objects, events, or classes of events, and that have explanatory power. Models help scientists and engineers understand how things work. Models take many forms, including physical objects, plans, mental constructs, mathematical equations and computer simulations. (NRC 1996, p. 117).

Science as Inquiry Standards emphasize the importance of students' understanding of *how* we know what we know in science. Taken together, these standards suggest it is not enough that students have an understanding of models as such. In other words, acquisition of declarative knowledge or conceptual information on models is only one aspect of learning models. Students need also to gain an appreciation of *how* and *why* these models are constructed. What is implied with the latter standard is that students need to develop an understanding of procedural knowledge within a domain of science that employs models.

In light of the mentioned standards, it is important to evaluate how models have been conventionally treated in the chemistry classroom. When the use of chemical models in teaching is considered, several trends can be traced that suggest lack of support for students' understanding of models and modeling. First, chemical models have been presented to students as *final* versions of our knowledge of matter: copies of real molecules in contrast to approximate and tentative representations (Grosslight, Unger, Jay, & Smith 1991; Weck 1995). Within the traditional framework of teaching, the motivations, strategies and arguments underlying the development, evaluation and revision of chemical models are overlooked. Classroom teaching typically advances the use of models for conceptual differentiation. For instance, models are used to distinguish weight from density (Smith, Snir & Grosslight 1992), and temperature from heat (Wiser 1987).

Second, textbooks often do not make clear distinctions between chemical models (Glynn, Britton, Semrud-Clikeman & Muth 1989) but rather frequently present 'hybrid models' (Gilbert & Boulter 1997). Carr (1984) provides the following example which illustrates a common model confusion in textbooks:

> Since NaOH is a strong base, Na^+ is an extremely weak conjugate acid; therefore, it has no tendency to react with H_2O to form NaOH and H^+ ion. (p. 101)

The first statement is based on the Arrhenius model of acids and bases. The second statement can be interpreted in terms of the Brønsted-Lowry model although the emphasis on ionization is not consistent with this model. When and why a new model is being used, and how this model differs from another model are not typically explicated in textbooks (Carr 1984).

Third, chemical models have been synonymous with ball-and-stick models which are typically used as visual aids (Grosslight et al. 1991; Leisten 1994). These 'physical models' have been intended to supplement conceptual information taught, and their use has been justified on Piagetian grounds: that students in concrete operational stages, in particular, need concrete models to understand the structure of molecules (Battino 1983). The problem with this perspective is at least threefold:

- The separation of conceptual information about atoms and molecules from physical models that represent them is inappropriate. Physical models *embody* conceptual information. In fact, their very existence is based on conceptual formulations about atoms and molecules.
- The focus on chemical models as physical models underestimates the diversity and complexity of models in chemistry. As illustrated earlier, for instance, models of acids and bases are abstract, and each model is accompanied by different sets of premises about what an acid or a base entails.
- The presumption that students in concrete operational stage *especially* need physical models is simply a weak argument. It is common practice for chemists themselves to use physical models to facilitate their communication and understanding of the structure and function of molecules. What this argument achieves in doing is to stress a deficiency on the part of children's potential to learn.

The fourth trend in the treatment of chemical models in the traditional classroom concerns the shift in emphasis from models to theories since the incorporation of quantum mechanical theories in chemistry. Chemistry and physical science textbooks show a growing tendency to begin with the establishment of theoretical concepts such as the 'atom' (Abraham, Williamson, & Westbrook, 1994). Textbooks often fail to stress the approximate nature of atomic orbitals and imply that the solution to all difficult chemical problems ultimately lies in quantum mechanics (Scerri 1991).

Finally, traditionally chemical knowledge taught in lectures has been complemented by laboratory experimentation which is intended to provide students with the opportunity to experience chemistry as inquiry. Chemical experimentation, however, has rarely been translated in the educational environment as an activity through which models are developed, evaluated and revised. Rather, experimentation is typically confined to data collection and verification of textbook knowledge in the classroom. Evidence suggests, however, that explanatory models may not be generated from data obtained in laboratory activities if explicit construction of such models is not encouraged (Schauble et al. 1991).

Given the trends in the way that models have conventionally been utilized in the classroom, it is not surprising that students' experience difficulties with models (Carr 1984; Gentner & Gentner 1983). Understanding of chemical models has been characterized in terms of three levels in students' thinking (Grosslight et al. 1991). At the first level, students think of models as toys or copies of reality which may be incomplete because they were intentionally designed as such.

At the second level, models are considered to be consciously produced for a specific purpose, with some aspects of reality being omitted, suppressed or enhanced. Here, the emphasis is still on reality and the modeling rather than on the ideas represented, as it is the case with the first level understanding. At the third level, a model is seen as being constructed to develop ideas, rather than being a copy of reality. The modeler is active in the modeling process. Few students demonstrate an understanding of chemical models as characterized by the third level. Many students' conceptions of models as representations of reality persist even after explicit instruction on models (e.g., Stewart, Hafner, Johnson, & Finkel 1992).

It is imperative that more attention is devoted to the effective teaching and learning of chemical models. In particular, omission in the classroom of the heuristics, strategies and criteria that drive generation, evaluation and revision of models, is likely to contribute to chemical illiteracy: a form of alienation where, not fully understanding how knowledge growth occurs in chemistry, students invent mysteries to explain the material world. Concerns have been raised about pseudoscientific interpretations of chemical knowledge (Erduran 1995) and mystification of chemical practices (Leisten 1994). Furthermore, in the classroom, recipe- following continues to be disguised as chemical experimentation – a significant problem often referred to as the 'cookbook problem' (van Keulen 1995). Chemistry, the science of matter, is not driven by recipes, nor by data collection and interpretation alone. Chemists contribute to their science by formulating models to explain patterns in the data that they collect. If effective teaching and learning of chemistry is indeed an intended educational outcome, then classrooms need to manifest what chemists do fundamentally: to model the structure and function of matter.

7. Conclusions

There is a tradition in chemistry education which involves handing down of concepts and principles (e.g., solution, Le Chatelier's principle) to students without engaging them in the processes of chemical inquiry that make possible the generation of these concepts and principles. In particular, rarely are students facilitated in modeling the structure and function of matter themselves. Furthermore, students' experimentation in the chemistry laboratory is conventionally based on rote recipe-following and is not representative of chemical inquiry that underlies what chemists do.

Philosophy of chemistry is a new field that can inform chemistry education about philosophical themes that are crucial aspects of the science of chemistry. Trends in the treatment of models in chemistry education highlights the need for reconceptualizing the teaching and learning of chemistry to embrace chemical epistemology, a potential contribution by philosophy of chemistry.

Acknowledgments

I acknowledge and thank the Spencer Foundation, Chicago for financial and intellectual support throughout the writing of this work.

References

Abraham, M.R., Williamson, V.M. & Westbrook, S.L.: 1994, 'A Cross-age Study of the Understanding of Five Chemistry Concepts', *Journal of Research in Science Teaching* **31**(2), 147–165.

Akeroyd, F.M.: 1984, 'Chemistry and Popperism', *Journal of Chemical Education* **61**(8), 697–698.

Atkins, P.W.: 1991, *Atoms, Electrons and Change*, Freeman, New York.

Battino, R.: 1983, 'Giant Atomic and Molecular Models and Other Lecture Demonstration Devices Designed for Concrete Operational Students', *Journal of Chemical Education* **60**(6), 485–488.

Bent, H.A.: 1977, 'Uses of History in Teaching Chemistry', *Journal of Chemical Education* **54**, 462–466.

Brush, S.: 1978, 'Why Chemistry Needs History and How It Can Get Some', *Journal of College Science Teaching* **7**, 288–291.

Butterfield, H.: 1949, *The Origins of Modern Science*, G Bell & Sons, London.

Carr, M.: 1984, 'Model Confusion in Chemistry', *Research in Science Education* **14**, 97–103.

Ellis, P.: 1989, 'Practical Chemistry in a Historical Context', in M. Shortland & A. Warwick (eds), *Teaching the History of Science*, Basil Blackwell, Oxford.

Erduran, S.: 1995, 'Science or Pseudoscience: Does Science Education Demarcate? The Case of Chemistry and Alchemy in Teaching', in F. Finley, D. Allchin, D. Rhees, & S. Fifield (eds),

Proceedings of the Third International History, Philosophy and Science Teaching Conference, Vol. 1, University of Minnesota,. Minneapolis, pp. 348–354.

Erduran, S.:1997, 'Reflections on the Proceedings from HPSST Conferences: A Profile of Papers on Chemistry Education', in I. Winchester (ed.), *Proceedings of International History, Philosophy and Science Teaching Group North and South America Regional Conference*, University of Calgary, Calgary.

Erduran, S.: 2000, 'Emergence and Application of Philosophy of Chemistry in Chemistry Education', *School Science Review* **81**(297), 85–87.

Erduran, S.: 2000, 'A Missing Component of the Curriculum?', *Education in Chemistry* **37**(6), 168.

Gentner, D. & Gentner, D.R.: 1983, 'Flowing Waters or Teeming Crowds: Mental Models of Electricity', in D. Gentner & A.L. Stevens (eds), *Mental Models*, Erlbaum, Hillsdale, NJ, pp. 99–129.

Gilbert, J. & Boulter, C.: 1997, 'Learning Science Through Models and Modelling', in B. Frazer & K. Tobin (eds), *The International Handbook of Science Education*, Kluwer Academic Publishers, Dordrecht.

Gilbert, J. & Boulter, C. (eds): 2000, *Developing Models in Science Education*, Kluwer Academic Publishers, Dordrecht.

Glynn, S., Britton, B.K., Semrud-Clikeman, M. & Muth, K.D.: 1989, 'Analogical Reasoning and Problem Solving in Science Textbooks', in J.A. Glover, R.R. Ronning & C.R. Reynolds (eds), *Handbook of Creativity*, Plenum Press, New York, pp. 383–398.

Green, J.H.S.: 1993, 'What Is Philosophy of Chemistry?', in E. Scerri (ed.), *The Philosophy of Chemistry*, Report of the meeting held at the Science Museum Library, Royal Society of Chemistry Historical Group Society for the History of Alchemy and Chemistry, London.

Grosslight, K., Unger, C., Jay, E., & Smith, C.: 1991, 'Understanding Models and Their Use in Science: Conceptions of Middle and High School Students and Experts', *Journal of Research in Science Teaching* **29**, 799–822.

Heeren, J.K.: 1990, 'Teaching Chemistry by the Socratic Method', *Journal of Chemical Education* **67**(4), 330–331.

Herron, J.D.: 1977, 'The Place of History in the Teaching of Chemistry', *Journal of Chemical Education* **54**(1), 15–16.

Hoffmann, R., & Torrence, V.: 1993, *Chemistry Imagined: Reflections on Science*, Smithsonian Institute Press, Washington DC.

Jaffe, B.: 1938, 'The History of Chemistry and Its Place in the Teaching of Chemistry', *Journal of Chemical Education* **15**, 383–389.

Justi, R. & Gilbert, J.: 1999, 'A Cause of Ahistorical Science Teaching: Use of Hybrid Models', *Science Education* **83**(2), 163– 177.

Kauffman, G.B.: 1989, 'History in the Chemistry Curriculum', *Interchange* **20**(2), 81–94.

Kauffman, G.B. & Szmant, H.H.: 1984, *The Central Science: Essays on the Uses of Chemistry*, Texas Christian University Press, Fort Worth.

Kornblith, H.: 1985, *Naturilizing Epistemology*, MIT Press, Cambridge, MA.

Leisten, J.: 1994, 'Teaching Alchemy?', *Chemistry in Britain* **30**(7), 552–552.

Luder, W.F., McGuire, W.S. & Zuffanti, S.: 1943, 'Teaching the Electronic Theory of Acids and Bases in the General Chemistry Course', *Journal of Chemical Education*, pp. 344–347.

McIntyre, L.: 1999, 'The Emergence of the Philosophy of Chemistry', *Foundations of Chemistry* **1**(1), 57–63.

Nagel, E.: 1961, *The Structure of Science*, Harcourt New York.

National Research Council: 1996, *National Science Education Standards*, National Academy Press, Washington DC.

Nye, M.J.: 1993, *From Chemical Philosophy to Theoretical Chemistry*, University of California Press, Berkeley.

Oppe, G.: 1936, 'The Use of Chemical History in the High School', *Journal of Chemical Education* **13**, 412–414.

Partington, J.R.: 1957, *A Short History of Chemistry* (3rd edn), MacMillan, New York.

Primas, H.: 1983, *Chemistry, Quantum Mechanics and Reductionism*, Springer, Berlin.

Pumfrey, S.: 1989, 'The Concept of Oxygen: Using History of Science in Science Teaching', in M. Shortland & A. Warwick (eds), *Teaching the History of Science*, Basil Blackwell, Oxford, pp. 142–155.

Russell, C.A.: 1985, *Recent Developments in the History of Chemistry*, Royal Society of Chemistry, Whitstable Litho, Whitstable, Kent.

Sammis, J.H.: 1932, 'A Plan for Introducing Bioraphical Material into Science Cources', *Journal of Chemical Education* **9**, 900–902.

Scerri, E.: 1991, 'Chemistry, Spectroscopy and the Question of Reduction', *Journal of Chemical Education* **68**(2), 122–126.

Scerri, E.: 1994a, 'Prediction of the Nature of Hafnium from Chemistry, Bohr's Theory and Quantum Theory', *Annals of Science* **51**, 137–150.

Scerri, E.: 1994b, 'Has Chemistry Been at Least Approximately Reduced to Quantum Mechanics?', in D. Hull, M. Forbes & R. Burian (eds), *Philosophy of Science Association*, Vol. 1, Philosophy of Science Association, East Lansing, MI, pp. 160–170.

Scerri, E.: 1996: 'Stephen Brush, The Periodic Table and the Nature of Chemistry', in P. Janich & N. Psarros (eds), *2nd Erlenmeyer Colloquium on the Philosophy of Chemistry*, Koningshausen & Neumann, Marburg University, Wurtzburg.

Scerri, E.:1997, 'Are Chemistry and Philosophy Miscible?', *Chemical Intelligencer* **3**, 44–46.

Scerri, E.: 2000, 'The Failure of Reduction and How to Resist the Disunity of Science in Chemical Education', *Science & Education* **9**(5), 405–425.

Schauble, L., Klopfer, L.E. & Raghavan, K.: 1991, 'Students' Transition from an Engineering Model to a Science Model of Experimentation', *Journal of Research in Science Teaching* **18**, 859– 882.

Scott, P., Driver, R., Leach, J. & Millar, R.: 1993, 'Students' Understanding of the Nature of Science', Working papers 1–11, Children's Learning in Science Research Group, University of Leeds, Leeds.

Shortland, M. & Warwick, A.: 1989, *Teaching the History of Science*, Basil Blackwell, Oxford.

Smith, C., Snir, J. & Grosslight, L.: 1992, 'Using Conceptual Models to Facilitate Conceptual Change: The Case of Weight/density Differentiation', *Cognition and Instruction* **9**, 221–283.

Stewart, J., Hafner, R., Johnson, S. & Finkel, E.: 1992, 'Science as Model Building: Computers and High-school Genetics', *Educational Psychologist* **27**(3), 317–336.

Suckling, C.J., Suckling, K.E. & Suckling, C.W.: 1978, *Chemistry Through Models*, Cambridge University Press, Cambridge.

Tomasi, J.: 1988, 'Models and Modeling in Theoretical Chemistry', *Journal of Molecular Structure (Theochem)* **48**, 273–92.

Trindle, C.: 1984, 'The Hierarchy of Models in Chemistry', *Croatica Chemica Acta* **57**, 1231.

van Brakel, J.: 1994, 'On the Neglect of the Philosophy of Chemistry', Paper presented at the First International Conference on Philosophy of Chemistry, London.

van Brakel, J. & Vermeeren, H.: 1981, 'On the Philosophy of Chemistry', *Philosophy Research Archives*, pp. 1405–1456.

van Keulen, H.: 1995, *Making Sense: Simulation-of-research in Organic Chemistry Education*, CD-ß Press, Utrecht.

Wasserman, E. & Schaefer, H.F.: 1986, 'Methylene Geometry', *Science*, p. 233.

Weck, M.A.: 1995, 'Are Today's Models Tomorrow's Misconceptions?', *Proceedings of the Third International History, Philosophy and Science Teaching Conference*, Vol. 2, University of Minnesota, Minneapolis, pp. 1286–1294.

Wiser, M.: 1987, 'The Differentiation of Heat and Temperature: History of Science and Novice-expert Shift', in D. Strauss (ed.), *Ontogeny, Phylogeny and Historical Development*, Ablex, Norwood, NJ, pp. 28–48.

Zuckermann, J.J.: 1986, 'The Coming Renaissance of Descriptive Chemistry', *Journal of Chemical Education* **63**, 829.

Can the Theory of Narratives Help Science Teachers be Better Storytellers?

FRITZ KUBLI
Báulistrasse 26, CH-8049 Zürich, Switzerland

Abstract. The narration of historical details is an art. It can be learned by studying narrative theories which lead to a better understanding of the narrative process. Not every physics teacher is born an expert in storytelling. The analysis of the whole process of story production and its reception by an audience is a precious tool, even in the hand of an inexperienced storyteller. Science teachers can profit from an education in this direction.

1. Plot and Narration Technique in Narrative Theory

The tradition of narrative theory goes back to authors like Quintilian, Horace and even Aristoteles. Modern insights into narrative theory have been developed mainly in the English-speaking parts of Western society. A pioneer was W. Booth (1961, 1974, 1979), other helpful books were written by S. Chatman (1978, 1990), F. Kermode (1967, 1979), R. Alter (1968, 1981) and M. Sternberg (1978, 1985). H. White (1973, 1987) applied these theories to the narration of historical events by historians.

A first insight into modern narrative theory regards the fact that every good story has a consistent and simple plot. Effective storytelling includes a carefully developed structure in the organisation of the narrated facts. In German we use the word 'erzählen' for 'telling a story'. 'Erzählen' contains the word 'zählen' which means counting. The same idea probably led to the English word 'account'. This shows that a logical or even mathematical element is essential for the composition of a good plot. Each story has its proper logic. A simple and elegant plot is always the result of a clear disposition and a clear order of the narrated facts.

2. The Analysis of Short Stories

The simplicity of a good story is deceiving. A simple plot is a piece of art. If we want to go beyond the pure repetition of stories already created, we have to deal with the theory of poetics, that means with the implicit rules of the composition of a plot. Investigations into the poetics of short stories are of particular interest with regard to science teaching, the time available for historical remarks being

F. Bevilacqua et al. (eds.), Science Education and Culture, 179–183.
© 2001 *Kluwer Academic Publishers. Printed in the Netherlands.*

reduced to a few minutes per lesson. German authors like R. Kilchenmann (1967), L. Rohner (1973), K. Doderer (1980), V. Weber (1993) are convincing experts in anecdotes and other short stories. It's worth reading them.

It is no coincidence that the German analysts of short stories are the most interesting to read. Germany has an old tradition of calendar stories. The old-fashioned German calendar stories created by authors like Hebel or Brecht have an educational intention without showing it too openly. We can profit from them because they are laconic, precise and come to the point immediately. If we manage to shape our historical remarks in a similar way we can educate our students rather efficiently.

3. The Effect of the Untold

A story consists of narrated elements and of blanks and gaps (Sternberg 1978). Stories can tell the truth, but never the whole truth. Blanks are necessary, details with no relation to the plot must be excluded to avoid deviating the listener's attention. Gaps, on the other hand, must be left open to the imagination of the audience. A story without gaps to be filled in or even secrets to be guessed can hardly stimulate curiosity. A story without a secret is not a good story (Kermode 1979).

What information do we give to our students and what must be left for them to guess? This selection is essential. It determines the literary quality of the narration. If the narrative is too explicitly formulated, the pleasure of the audience is considerably reduced. A pedantically explicit approach to a subject ignores the intellectual faculties of the listeners. They might even consider this an insult. The same can be true for stories with a simple and penetrating message. In literary theory, these kind of stories are called didactic and must be distinguished from the warm description of life as it is. Becoming too didactic is a frequent mistake made by inexperienced teachers.

This does not mean, however, that there might not be a didactic element in the successful composition of a story. The true didactic element manifests itself by the simplicity of the plot and in the clarity and precision of the descriptions. It avoids any obtrusive messages and leads the reader or listener to the point without moralizing. An unobtrusive author may educate his or her audience far more effectively than a noticeably didactic author.

4. Irony in Narrated Texts

Effective storytelling always includes, so certain authors say, a component of irony (Booth 1974). Even in religious texts, such as the episodes told in the bible, the ironic approach is essential for their effect (Alter 1981, Sternberg 1987). Irony gives the narration a little ease, a playful element. The ironic approach makes it clear that the narrator has a certain freedom to arrange his story, and this freedom creates tension in the listeners.

Irony creates a particular relationship between the narrator and the audience, a kind of complicity. It stimulates the activity of the listeners. If the true meaning of the narrator is not literally spelled out, it must be imagined by a creative process. The stimulated activity leads to a dialectic movement in the minds of the listeners. It generates a sense of superiority to the unwitting objects of the irony, as well as to people who take the ironic assertions at face value.

Irony is a kind of spice that helps to 'digest' the content of the story. It is highly recommended when dealing with the heroes of physical thinking. It helps the students to see them as they really are. This is because, as the theory of narratives puts it: "It is notoriously difficult for novelists to write convincingly about good people" (Alter 1968, p. 80). We do not appreciate being confronted with people who are intellectually or morally superior to us, unless the storyteller helps us to feel equivalent to them. If we can follow their conclusions as if they were our own, we can easily accept their outstanding intellect. A touch of irony makes this easier.

Swiss students are no exception. They don't like historical people to be held up as models, even if strong reasons exist for doing so. A little ironic distance reduces the distance to understanding the genii. An ironic presentation is far from doing harm to the picture we generate of them. It is a condition for the genesis of a friendly image. It even helps to become familiar with their thinking.

5. Implied Author and Implied Addressee

The theory of narrative applies to all sorts of stories and not only to historical remarks. The whole teaching process can be considered as a narrative confrontation. The following statements must not be seen exclusively in connection with short stories.

Stories cannot be told without showing an implied author and an implied addressee (the reader in the case of written stories). A narrative implicitly presupposes partners in the communication. They can be reconstructed from the text of the narrative. The text presupposes an implied reader with a certain faculty to understand the concepts or arguments, and an implied author who talks to this reader. A good text works on a given intellectual level and is also based on a certain system of values which are implicitly supposed to be accepted. We read a text more intensely if we can accept the value system, and if we can accept the role of the implied reader. The implied reader or listener is one of the most important elements of a story.

On the other hand, a story always contains an implied author which must not be confused with the real (or empirical) author. The implied author is a construction. Biographical information about the empirical author is not even necessary and is sometimes not available. The constructed picture of the implied author is often called 'persona', according to a term shaped by the Swiss psychologist Carl Jung.

The distinction of the persona and the empirical author are very helpful in the study of the role of the teacher in class situations. Even in spoken language, the persona or impression we make on our partners is not identical with our empirical being as subjects.

The implicit author can exist in different modes. Most novels show an omniscient author which is excluded from the story itself (comparable to the 'auktorialer Erzähler' in Stanzel 1991). He knows the fate of his protagonists in advance, in contrast with the audience who do not know, but can only suggest from the given information what might happen next. A certain confidence in the reliability of the implicit author is essential, it helps the reader to follow a story even through some complications in the plot.

6. Listeners' Credit and the Narrative Contract

A well told story has a clearly defined logical structure which leads to the expected end. It begins with an implicit promise. The reader must give certain credit to the author. He must be ready to concentrate on the message for a while. On the other hand a sort of contract must be fulfilled by the storyteller in the course of the narration. At the end, all the audience's questions must be answered from the development of the plot. Both the author and the audience are engaged by this contract according to the rules of a narrative process.

The fact that narratives can be translated into foreign languages without loss of information shows that universal laws and general rules of narration exist. Even little children can decide if a certain text is a story or not, if it is complete or if there is something missing. These universal rules are implicitly understood. We realise more easily that they are violated when we are listening to a story rather than when we are telling the story.

7. Final Remarks

In contrast to what has been said, this paper does not completely comply with the rules of narration described earlier. It does not answer all the questions a reader might have asked himself while reading this paper. Its presentation is more of a sort of program for future investigations than a definite answer.

The field of narrative theory is rapidly developing, and more details are given in Kubli (1998).

References

Alter, R.: 1968, *Fielding and the Nature of the Novel*, Harvard University Press, Cambridge, MA.
Alter, R.: 1981, *The Art of Biblical Narrative*, Basic Books, New York.
Booth, W.: 1961 (second edition 1983), *The Rhetoric of Fiction*, University of Chicago Press, Chicago.
Booth, W.: 1974, *A Rhetoric of Irony*, University of Chicago Press, Chicago.

Booth, W.: 1979, *Critical Understanding*, University of Chicago Press, Chicago.
Chatman, S.: 1978, *Story and Discourse. Narrative Structure in Fiction and Film*, Cornell University Press, Ithaca.
Chatman, S.: 1990, *Coming to Terms. The Rhetoric of Narrative in Fiction and Film*, Cornell University Press, Ithaca.
Doderer, K.: 1980, *Die Kurzgeschichte in Deutschland*, Wissenschaftliche Buchgesellschaft, Darmstadt.
Kermode, F.: 1967, *The Sense of an Ending. Studies in the Theory of Fiction*, Oxford University Press, New York.
Kermode, F.: 1979, *The Genesis of Secrecy: On the Interpretation of Narrative*, Harvard University Press, Cambridge, MA.
Kilchenmann, R.: 1967, *Die Kurzgeschichte*, Kohlhammer, Stuttgart.
Kubli, F.: 1998, *Plädoyer für Erzählungen im Physikunterricht – Geschichte und Geschichten als Verstehenshilfen*, Aulis, Cologne.
Rohner, L.: 1973, *Zur Theorie der Kurzgeschichte*, Athenäum, Frankfurt a. M.
Stanzel, F.: 1991, *Theorie des Erzählens*, Vandenhoeck, Göttingen.
Sternberg, M.: 1978, *Expositional Modes and Temporal Ordering in Fiction*, John Hopkins University Press, Baltimore.
Sternberg, M.: 1987, *The Poetics of Biblical Narrative*, Indiana University Press, Bloomington.
Weber, V.: 1993, *Die andere Geschichte*, Stauffenberg, Tübingen.
White, H.: 1973, *Metahistory*, John Hopkins University Press, Baltimore.
White, H.: 1987, *The Content of the Form. Narrative Discourse and Historical Representation*, John Hopkins University Press, Baltimore.

Values in Science: An Educational Perspective

DOUGLAS ALLCHIN

Department of Biological Sciences, University of Texas at El Paso, El Paso, TX 79968-0519, U.S.A.; E-mail: allchin@utep.edu

ABSTRACT. Science is not value-free, nor does it provide the only model of objectivity. Epistemic values guide the pursuit and methods of science. Cultural values, however, inevitably enter through individual practitioners. Still, the social structure of science embodies a critical system of checks and balances, and it is strengthened by a diversity of values, not fewer. Science also exports values to the broader culture, both posing new values-questions based on new discoveries, and providing a misleading model for rational decision-making. Science teachers who understand the multi-faceted relationship between science and values can guide students more effectively in fully appreciating the nature of science through reflexive exercises and case studies.

1. INTRODUCTION

A fundamental feature of science, in most popular conceptions, is that it deals with facts, not values. Further, science is objective, while values are not. These benchmarks often offer great comfort to science teachers, who see themselves as the privileged gatekeepers of the exclusive domain of certain and permanent knowledge. Such views of science are also closely allied in the public sphere with the authority of scientists and the powerful imprimatur of evidence as "scientific". Recently, however, sociologists of science, among others, have challenged the notion of science as value-free and thereby raised questions – especially important for educators – about the authority of science and its methods.

I claim that the popular conceptions – both that science is value-free and that objectivity is best exemplified by scientific fact – are each mistaken. This does not oblige us, however, to abandon science or objectivity, or to embrace an uneasy relativism. First, science does express a wealth of *epistemic values* and inevitably incorporates *cultural values* in practice. But this need not be a threat: some values in science govern how we regulate the potentially biasing effect of other values in producing reliable knowledge. Indeed, a diversity of values promotes more robust knowledge where they intersect. Second, values can be equally objective when they require communal justification and must thereby be based on generally accepted principles. In what follows, I survey broadly the relation of science and values, sample important findings in the history, philosophy and sociology of science from the last several decades, and suggest generally how to address these issues in the classroom.

Values intersect with science in three primary ways. First, there are

F. Bevilacqua et al. (eds.), Science Education and Culture, 185–196.
© 2001 *Kluwer Academic Publishers. Printed in the Netherlands.*

values, particularly epistemic values, which guide scientific research itself (§2). Second, the scientific enterprise is always embedded in some particular culture and its values enter science through its individual practitioners, whether deliberately or not. There are mechanisms, though, for guarding against the bias they might introduce (§3). Finally, values emerge from science, both as a product and process, and may be redistributed more broadly in the culture or society. Also, scientific discoveries may pose new challenges about values in society, though the values themselves may not be new (§4). Ideally, teachers will expose students to the various ways that values apply to science and help them develop skills in distinguishing their differences and in analyzing the role of values in producing particular facts (§5).

2. VALUES OF SCIENCE AND RESEARCH ETHICS

The common characterization of science as value-free or value-neutral can be misleading. Scientists strongly disvalue fraud, error and 'pseudoscience', for example. At the same time, scientists typically value reliability, testability, accuracy, precision, generality, simplicity of concepts and heuristic power. Scientists also value novelty, exemplified in the professional credit given for significant new discoveries (prestige among peers, eponymous laws, Nobel Prizes, etc.). The pursuit of science as an activity is itself an implicit endorsement of the value of developing knowledge of the material world. While few would tend to disagree with this aim, the pursuit of knowledge can become important in the context of costs and alternative values. Space science, the human genome initiative, dissection of subatomic matter through large particular accelerators or even better understanding of AIDS, for instance, do not come free. Especially where science is publicly funded, the value of developing scientific knowledge may well be considered in the context of the values of other social projects.

From the ultimate values of science, more proximate or mediating values may follow. For example, sociologist Robert Merton (1973) articulated several norms or 'institutional imperatives' that contribute to 'the growth of certified public knowledge' (see also Ziman 1967). To the degree that public knowledge should be objective, he claimed, scientists should value 'preestablished apersonal criteria' of assessment. Race, nationality, religion, class, or other personal or social attributes of the researcher should not matter to the validity of conclusions – an ethos Merton labeled 'universalism'. Merton's other institutional norms or values include organized scepticism, disinterestedness (beliefs not biased by authority – achieved through accountability to expert peers), and communism (open communication and common ownership of knowledge). As Merton himself noted, these norms do not always prevail. Still, they specify foundational conditions or proximate values that contribute to the development and certification of knowledge in a community (more below).

Specific social structures (such as certain reward systems or publication protocols) that support these norms thus form the basis for yet another level of mediating values.

Other proximate or mediating values that promote the ultimate goal of reliable knowledge involve methods of evaluating knowledge claims. These *epistemic* values include controlled observation, interventive experiments, confirmation of predictions, repeatability and, frequently, statistical analysis. These values are partly contingent. That is, they are derived historically from our experience in research. We currently tend to discount (disvalue) the results of any drug trial that does not use a double blind experimental design. But such was not always the case. The procedure resulted from understanding retrospectively the biases potentially introduced both by the patient (via the placebo effect) and by the doctor (via observer effects). Each is now a known factor that has to be controlled. The elements of process (both methods of evaluation and institutional norms), of course, are central to teaching science as a process.

While the pursuit of scientific knowledge implies a certain set of characteristically 'scientific' values, the relevance of other values in the practice of science are not thereby eclipsed. Honesty is as important in science as elsewhere, and researchers are expected to report authentic results and not withhold relevant information. Ethics also demands proper treatment of animals and humans, regardless of whether they are subjects of research or not (Orlans 1993). Science is not exempt from ethics or other social values. Knowledge obtained by Nazi researchers on hypothermia and the physiological effects of phosgene, for example, may pass tests of reliability, but the suffering inflicted on the human subjects was unwarranted (Caplan 1992; Proctor 1991). Hence, we may still debate whether it is appropriate to use such knowledge (Sheldon et al. 1989). Similar questions might be asked about U.S. military studies on the effects of radiation on humans. Again, social values or research ethics are not always followed in science (see, e.g., Broad and Wade 1982), but they remain important values. The disparity between the ideal and the actual merely poses challenges for creating a way to achieve these valued ends – say, through a system of checks and balances. Protocols for reviewing research proposals on human subjects, for monitoring the use and care of laboratory animals, or for investigating and punishing fraud each represent efforts to protect wider social values in scientific institutions.

The topics or ends of research, as much as the methods or practice of science, are also the province of ethical concern and social values. Weapons research, even if conducted according to Merton's norms and its results evaluated using scientific standards, is not ethically idle or value-neutral. – Nor is research into better agricultural methods aimed to alleviate hunger or low-cost forms of harnessing solar or wind energy in poor rural areas. In each of these cases, the researcher is an ethical agent responsible for the consequences of his or her actions, good or bad. Again, appeal to science is no escape from ethics. Where the consequences are clear, the frequent distinction in science between 'pure' and 'applied' research is not ethically significant. Many conservation biologists, for example, are well aware of the values inherent in their 'basic' research and sometimes shape and deploy the content of their science in a politically self-conscious way (Takacs 1996). Where debates about research arise – say, about transplanting fetal tissue or gene therapy – there are real conflicts about social values; the question of the ultimate value or ethics of research in these areas can neither be resolved by science alone nor disregarded by scientists in these fields as irrelevant.

3. VALUES ENTERING SCIENCE

Science proceeds through the agency of individuals and – not unexpectedly, perhaps – individual scientists express the values of their cultures and particular lives when they engage in scientific activity. For example, in cultures where women or minorities have been largely excluded from professional activity, they have generally been excluded from science as well. Where they have participated in science, they have often been omitted from later histories (e.g., Rossiter 1982; Kass-Simon and Farnes 1990; Manning 1995). The line demarcating science and society can be fuzzy in practice.

More deeply, however, the conclusions of science at many times and in many places have been strongly biased, reflecting the values of its practitioners (in striking contrast to Merton's norm of universalism). For example, late 19th-century notions of the evolution of humans developed by Europeans claimed that the skulls and posture of European races were more developed than 'Negroes' (Gould 1981). In a progressive view of evolution (adopted even by Darwin himself), persons of African descent were deemed inferior intermediaries on an evolutionary scale–as 'proven' by science. When theories about evolution changed to suggest that 'less-developed' or neotonous (more childlike) skulls were 'more progressive', conclusions from the same data reversed, preserving 'scientifically' the superior status of the scientists' race (Gould 1977). Facts were shaped to fit preexisting judgments and values about race. Likewise, female skulls, skeletal anatomy and physiology were taken by male scientists as evidence of women's 'natural' role in society. The 'scientific' conclusions, which reflected the values of the men, were taken to legitimate social relations that continued to privilege males (Fee 1979; Schiebinger 1990; Smith-Rosenberg and Rosenberg 1973). Perhaps such values should not enter science, but they do.

Values about race and sex, however, have not been the only values to shape science. The phrenology debates in Edinburgh in the early 19th century followed instead class differences (Shapin 1979). Today, notions about biological determinism, especially about the role of genes in governing specific behaviors, follow similar patterns, where some persons appeal to science to try to justify economic disparities as products of nature rather than as the exercise of power (Lewontin, Rose and Kamin 1984). By contrast, disagreement between Boyle and Hobbes over the existence of a vacuum in the late 17th century was guided in part by values about governance and the role of the sovereignty in the state (Shapin and Schaffer 1985). Even natural history museum dioramas of animal groupings designed by researchers have reflected cultural values about nuclear families and patriarchy (Haraway 1989, pp. 26–58). While we may now characterize all these cases as examples of 'bad science', they exemplify how values can and do enter science and shape its conclusions. Moreover, one must always bear in mind that in their own historical context, these examples were considered 'good' science.

While the role of values in these cases can seem obvious from our perspective, it may not be appropriate for us to interpret the scientists as exercising their values deliberately or consciously. To interpret the entry of values into science in cases such as these, one must focus on individual cognitive processes. That is, one must examine the thought patterns of particular agents rather than either abstractly reconstructed reasoning or the influences of a diffusely defined 'culture'. Especially valuable is the notion of cognitive resources: all the concepts, interpretive frameworks, motivations and values that an individual brings from his or her personal experience to scientific activities (Giere 1988, pp.213–21, 239–41). Cognitive resources affect how an individual notices certain things, finds some things as especially relevant, asks questions or poses problems, frames hypotheses, designs experiments, interprets results, accepts solutions as adequate or not, etc. As a set of resources or *tools*, a person's cognitive orientation will both make certain observations and interpretations possible while at the same time limiting the opportunity for others (see also Harding 1991). Succinctly, a person's scientific contributions will be shaped by the domain of his or her resources or values.

An individual's cognitive resources will be drawn from his or her culture, limiting what any one person can contribute to science. Further, because each person's biography and intellectual training are unique, cognitive resources will differ from individual to individual, even within the same culture. Hence, one may well expect disagreement or variation in interpretation in any scientific community. Far from being an obstacle to developing consensus, however, the variation of a community can be a valuable resource. That is, only conclusions that are robust across varying interpretations will tend to be widely perpetuated (Wimsatt 1981).

Indeed, variations in cognitive resources can be critical to isolating and correcting error. For example, in the 1860s through 1890s anthropologists had developed numerous ways to measure skulls and calculate ratios to describe their shapes. In what Fee (1979) described as 'a Baconian orgy of quantification', they developed over 600 instruments and made over 5,000 kinds of measurements. Despite three decades of shifting theories, falsified hypotheses and other unsolved paradoxes, the conclusions of the craniologists – all men – remained the same: women were less intelligent. At the turn of the century, however, two women began work in the field. They showed, among other things, that specific women had larger cranial capacity that even some scientists in the field, and that the margin of error in measurement far exceeded the proposed sex differences – and they strengthened their work with statistical rigor. Here, the women's perspective may have been no less 'biased' or guided by values, but their complementary cognitive resources, with the interests of women, were critical to exposing the deficits in the men's studies. This example illustrates that if science is 'self-correcting', it does not do so automatically. Identifying and remedying error takes work – and often requires applying contrasting cognitive resources or values. The possibly paradoxical conclusion is that one should not eliminate personal values from science – if indeed this were possible. Instead, the moral is: 'the more values, the better'. Contrasting values can work like a system of epistemic checks and balances.

The many cases of bias and error in science have led philosophers to more explicit notions of the social component of objectivity. Helen Longino (1990), for example, underscores the need for criticism from alternative perspectives and, equally, for responsibly addressing criticism. She thus postulates a specific institutional, or social, structure for achieving Merton's 'organized skepticism'. Sandra Harding (1991) echoes these concerns in emphasizing the need for cognitively diverse scientific communities. We need to deepen our standards, she claims, from 'weak objectivity', based merely on notions of evidence, to 'strong objectivity', also based on interpreting the evidence robustly. Both thinkers also point to the role of diversity of individuals in establishing relevant questions and in framing problems, thus shaping the direction of research more objectively. In this revised view, science is both objective and thoroughly 'social' (in the sense of drawing on a community of interacting individuals). Fortunately for science educators, the classroom is an ideal location for modeling this kind of collective activity.

The role of alternative values in exposing error and deepening interpretative objectivity highlights the more positive role of individual values in science. Even religion, sometimes cast as the antipode of science, can be a cognitive resource that contributes positively to the growth of knowledge. For example, James Hutton's theological views about the habitability of the earth prompted his reflections on soil for farming and on food and energy, and led to his observations and conclusions about geological uplift, 'deep time', the formation of coal, and what we would call energy flow in an ecosystem (Gould 1987; Allchin 1994). Likewise, assumptions about a Noachian flood shaped William Buckland's landmark work on fossil assemblages in caves, recognized by the Royal Society's prestigious Copley Medal. Other diluvialists drew attention to the anomalous locations of huge boulders, remote from the bedrock of which they were composed. (While they concluded that the rocks were moved by turbulent flood waters, we now interpret them as glacial erratics.) These discoveries all had origins that cannot be separated from the religious concepts and motivations that made the observations possible. Values entering science from religion – or from virtually any source – can promote good science (Allchin 1996). As suggested above, however, they sometimes also need to be coupled with mechanisms for balancing then with complementary values.

4. VALUES EXPORTED FROM SCIENCE

Just as values of a society can enter science, so, too, can values from the scientific enterprise percolate through society. The most dramatic redistribution of values may be the values of science itself. To the extent that science (and technology) are perceived as successful or powerful, things associated with them can gain authority or value. Commericial advertising, for example, can draw on the images of science to promote certain products as expressions of 'scientific' research or as superior to competing products. The 'scientific' nature of the comparison can even dominate over the values on which the comparison itself rests. The conclusions of science themselves are accorded an image of value. One can see the ethical implications where conclusions that themselves draw on social values (such as those regarding race, sex, class, culture, etc.) are given the imprimatur of scientific authority, thereby reinforcing preexisting distributions of power without justification (see Toumey 1996).

The most dramatic social influence of scientific values, however, may be the image of science itself as a model for all problem-solving. Science (or technology) is sometimes viewed, first, as the panacea for all social problems and, second, as the exclusive or primary means for objectivity, even where other values are involved. Not all problems are amenable to scientific approaches, however, and a narrowly scientific or 'technocratic' view can forestall solving problems in the appropriate realm. Garrett Hardin (1968) noted, for example that 'the population problem has no technical solution'. That is, population pressure is fundamentally an *ethical* challenge about the freedom to bear children in the context of limited global resources. Neither better agricultural efficiency nor reproductive control technology can avert a 'tragedy of the commons'. Instead, we must reach some consensus about the ethics of an individual's use of common resources and how we may enforce such collective judgments about reproductive rights or privileges.

We often need to integrate scientific values with other ethical and social values. Science can help identify unforseen consequences or causal relationships where ethical values or principles are relevant. In addition, individuals need reliable knowledge for making informed decisions. One archetypal hybrid project is risk assessment. Scientists can articulate where, how, and to what degree a risk exists, for example. But other values are required to assess whether the risk is 'acceptable' or not.

Communicating the nature of the risk to non-experts who participate in making decisions can thus become a significant element of science. Where one expects scientists or panels of technical experts to solve the problem of the acceptability of risk, science is accorded value beyond its proper scope – and others abdicate their responsibility in addressing the sometimes more difficult questions of value. Likewise, those who do not address the facts of the matter fail in their responsibility to make an informed decision. Facts and social values function in concert.

As noted above, the values of science may also be applied inappropriately as a model for decision-making. While quantification is often an asset for science, for example, it does not address all the ethically relevant dimensions of technological risk. Cases of risk assessment, in particular, require addressing questions about the distribution of risk among different persons and about the autonomy of accepting risk. Efforts to reduce the problem to a single numerical scale (and then to minimize risk) can obscure the central issues. What matters socially and ethically is the meaning more than the magnitude of the risk (e.g., Sagoff 1992). A 'scientific' approach to solving global warming, for example, might easily focus on cost-effective means of reducing greenhouse gas emissions, diverting attention away from the history of the problem and the ethical need for accountability and remedial justice on the part of industrialized nations. Cases of uncertainty pose special problems for applying scientific values. Scientists generally refrain from advocating claims that cannot yet be substantiated. Ethically, however, one often wishes to hedge against the possibility of a worst case scenario (catastrophic floods, nuclear melt-downs, ozone depletion, etc.) – even if the actual expected consequences are not yet proven. In cases of uncertainty, scientific values about certified knowledge ('assume nothing unproven') and ethical values about communal action ('assume the worst') can diverge (see Shrader-Frechette 1991). One task in teaching is clearly to articulate the limited domain of scientific values and how they integrate with other values.

Controversies over the flouridation of public water supplies exemplify well some of the potential problems and confusions about the role and value of science in social policy (Martin 1991). Both sides of the debate appeal to science as an authority. In each case, however, the science is presented as simple and unproblematic, though complexities and uncertainties exist. In addition, science is treated as the final arbitrer, though research indicates that there are *both* benefits (associated with reducing tooth decay) and risks (associated with flourosis and cancer). In this case, the benefits and risks are not commensurable, and no scientific assessment of the ultimate value of flouridation is possible without the expression of further values. In this case, as in others, the scientific value of empirical certainty can be confused with what science can sometimes actually provide. Even technical certainty does not exclude other, non-scientific values from contributing to resolving disputes about public policy.

Finally, scientific knowledge and new technologies can introduce new ethical or social dilemmas, based on preexisting values. Many medical technologies allow us to express our values in preserving life and health. At the same time, however, they can bring other values into consideration. With the advent of hemodialysis and organ transplants, for example, their limited availability combined with the existing value of fairness in generating a new problem: ensuring fair access to treatment. Subsequently, ethicists developed new solutions for allocating scarce medical resources (e.g., Rescher 1969). Similarly, ecological knowledge – say, about pesticides, heavy metals, toxic chemicals and other pollutants – has educed conventional values about prudence and respect for life in reshaping our values about waste, consumption, modes of production and our relationship to the environment (see, e.g., Des Jardins 1993; Newton and Dillingham 1994). Science does not create these new values. Rather, it introduces novel situations which require us to apply old values in significantly new ways. An awareness that scientific research is typically coupled with new concerns about ethics and values was reflected, for example, in decisions to couple the human genome initiative with funding of research on the humanistic implications of the project.

Some technologies affect values more indirectly. Medical technologies that help sustain life have confounded our traditional definitions of 'life' and 'death' and the values associated with them. New reproductive technologies, likewise, pose challenges for existing concepts of 'parent' and 'family' (Kass 1985); the potential of human cloning forces us to assess more deeply the concept of genetic identity; and the abilities of computers make us reconceive the notion of 'intelligence'. All these innovations challenge us to rethink what it means to be human, just as Copernicus, Darwin and Freud did in earlier centuries. Paradoxically, perhaps, in solving some problems, science and technology can introduce new problems about values that they cannot solve. Yet these consequences are a part of a complete consideration of science and its context in society.

5. TEACHING ETHICS AND VALUES

Many science teachers shy away from addressing values, imagining that they are outside the domain of science or, worse, betray the very core of science. A deeper understanding of science, values, and objectivity (as sketched above), however, supports a mandate for discussing values in the science classroom. This need not seem risky or difficult.

First, the values which guide scientific inquiry may perhaps be best introduced reflexively. That is, in a constructivist setting–where students are engaged in modest scientific activity themselves – they may be asked to reflect on their own process. What are their standards of proof? How has access to multiple investigators influenced their conclusions? The teacher can create opportunities to articulate and illustrate accepted values (perhaps even simulate an instance of fraud!). But if students are to understand the reasoning that supports the epistemic values, they should be able to question and discuss them, like any scientific claim. For example, inviting students both to give and address criticism among their peers can serve to model methods for developing acceptable claims (see Longino and Harding, above). Students should endeavor themselves to articulate what good process is. They may thereby see scientific values as emerging from the scientists' own goals and experiences.

Alternatively, teachers may introduce scientific values through historical case studies. Historical cases offer the advantage that students can clearly see the consequences of certain values, while also seeing how they function in context (e.g., Hagen, Allchin & Singer 1996). It may also be easier to analyze and discuss methods that are not laden with the students' motivations and sense of self.

These approaches may serve as models for introducing other values and ethics. Teaching ethics and values – like teaching science in a constructivist mode – is not centered on teaching specific content, but rather engaging the students in a process. Students should explore values, with the teacher showing how to discuss them collectively. Learning the questions may thus be more important than learning specific answers. Relevant questions for values or ethical discussions include:

- Who are the stakeholders? What are their interests? Are they involved in the decision-making?
- What are the forseeable consequences (possibly remote or hidden)? What are the alternatives? Is the worst case scenario acceptable?
- What intentions or motives guide the choice?
- What are the benefits? What are the costs?
- *Who* benefits? *Who* risks or pays the costs? (Who is upstream, choosing benefits? Who is downstream, experiencing the consequences?) Would you be willing to accept any consequence of this action falling on yourself?
- What would be the outcome if everyone acted this way?

Each of these questions helps foster awareness of some aspect supporting the objectivity of values claims. Working from questions (rather than a list of principles) also allows the teacher to situate discussion at different levels of sophistication, unique to each class group. A teacher may recognize that the potential for ethical reasoning develops with age and experience, just as skills for scientific reasoning do. Lessons need to be designed at a level appropriate to the students' educational maturity.

An important goal is for students to learn that *ethics* and discussion of *public values* require justification just as much as any scientific argument does. Teachers need to emphasize especially that sound ethical conclusions are based on general principles – not on an individual's "feelings" or "personal values". Morals must be publicly endorsed. Ethical principles, in turn, are based on careful reasoning, concrete evidence, and *commonly shared* emotions. The willingness to experience the consequences of one's actions and the ability to universalize a decision (noted above) are two common ways to 'test' whether principles are ethical. A good touchstone for justifying an ethical value (as it is in science) is a good critic: reasons must be strong enough and draw on principles general enough to convince someone with an opposing perspective. Ethics, no less than science, aims at objectivity.

Historical cases, again, can be valuable tools for teaching. One can analyze actions in terms of consequences that are known, but that may not have been obvious to the participants – as in the cases discussed in §3. An essential element, however, is always the active recovery of the context in which actions that may now seem unreasonable were once seen as fully justified (the cognitive interpretive model applies here, as well). To dismiss Nazi research or the Tuskegee syphillis experiment (Jones 1981) as aberrations of a few misguided individuals is to trivialize the deep embeddeness of values in science. One must confront science and values in full context. We must acknowledge that 'blindspots', so obvious in retrospect, are part of the process. Effective ethical inquiry can hopefully make us more aware of potential blindspots, so that we may minimize or reduce them. Other historical cases can help students probe the tensions between context and outcome. The issues of vivisection and the use of animals in research, for example, become sharper when discussed in the context of William Harvey's work on blood flow or Walter Cannon's seminal work on homeostasis (Hagen, Allchin & Singer 1996, pp. 95–103). Students must reconcile the view of common 'textbook knowledge' as value-free with the actual values that were expressed in achieving that knowledge. Engaging in such cases allows students to appreciate the various values that enter and shape the development of scientific knowledge.

REFERENCES

Allchin, D.: 1994, 'James Hutton and Phlogiston', *Annals of Science* **51**, 615–635.

Allchin, D.: 1996, 'Science and Religion' (Special Issue), *SHiPS Newsletter* 6(2).

Broad, W. & Wade, N.: 1982, *Betrayers of the Truth*, Simon and Schuster, New York, NY.

Caplan, A. (ed.): 1992, *When Medicine Went Mad*, Humana Press, Totowa, NJ.

Des Jardins, J.R.: 1993, *Environmental Ethics*, Wadsworth, Belmont, CA.

Fee, E.: 1979, 'Nineteenth-Century Craniology: The Study of the Female Skull', *Bulletin of the History of Medicine* **53**, 415–33.

Giere, R.: 1987, *Explaining Science*, University of Chicago Press, Chicago, IL.

Gould, S.J.: 1977, 'Racism and Recapitulation'. In *Ever Since Darwin*, W.W. Norton, New York, NY, pp.214–21.

Gould, S.J.: 1981, *The Mismeasure of Man*, W.W. Norton, New York, NY.

Gould, S.J.: 1987, *Time's Cycle, Time's Arrow*, Harvard University Press, Cambridge, MA.

Hagen, J., Allchin, D. & Singer, F.: 1996, *Doing Biology*, Harper Collins, Glenview, IL.

Haraway, D.: 1989, *Primate Visions*, Routledge, New York, NY.

Hardin, G.: 1968, 'The Tragedy of the Commons', *Science* **162**, 1243–48.

Harding, S.: 1991, *Whose Science? Whose Knowledge?*, Cornell University Press, Ithaca, NY.

Jones, J.M.: 1981, *Bad Blood*, Free Press, New York, NY.

Kass, L.: 1985, *Toward a More Natural Science*, Free Press, New York, NY.
Kass-Simon, G. & Farnes, P. (eds.): 1990, *Women of Science: Righting the Record*, Indiana University Press, Bloomington, IN.
Lewontin, R.C., Rose, S. & Kamin, L.J.: 1984, *Not in Our Genes*, Pantheon Books, New York, NY.
Longino, H.: 1990, *Science as Social Knowledge: Values and Objectivity in Scientific Inquiry*, Princeton University Press, Princeton, NJ.
Manning, K.: 1995, 'Race, Gender, and Science', In *Topical Essays for Teachers*, History of Science Society, Seatlle, WA, pp.5–34.
Martin, B.: 1991, *Scientific Knowledge in Controversy: The Social Dynamics of the Flouridation Debate*, State University of New York Press, Albany, NY.
Merton, R.: 1973, 'The Normative Structure of Science'. In *The Sociology of Science*, University of Chicago Press, Chicago, IL, pp. 267–78.
Newton, L.H. & Dillingham, C.K.: 1994, *Watersheds*, Wadsworth, Belmont, CA.
Orlans, B.: 1993, *In the Name of Science: Issues in Responsible Animal Experimentation*, Oxford University Press, Oxford.
Proctor, R.: 1991, *Value-Free Science?: Purity and Power in Modern Knowledge*, Harvard University Press, Cambridge, MA.
Rescher, N.: 1969, 'The Allocation of Exotic Medical Lifesaving Therapy', *Ethics* **79**, 173–86.
Rossiter, M.: 1982, *Women Scientists in America: Struggles and Strategies to 1940*, Johns Hopkins University Press, Baltimore, MD.
Sagoff, M.: 1992, 'Technological Risk: A Budget of Distinctions'. In D.E. Cooper and J.A. Palmer (eds.), *The Environment in Question*, Routledge, New York, NY, pp. 194–211.
Schiebinger, L.: 1990, 'The Anatomy of Difference: Race and Gender in Eighteenth Century Science', *Eighteenth-Century Studies* **23**, 387–406.
Shapin, S.: 1979, 'The Politics of Observation: Cerebral Anatomy and Social Interests in the Edinburgh Phrenology Disputes'. *Sociological Review Monographs* **27**, 139–78.
Shapin, S. & Schaffer, S.: 1985, *Leviathan and the Air Pump*, Princeton University Press, Princeton, NJ.
Sheldon, M., Whitely, W.P., Folker, B., Hafner, A.W. & Gaylin, W.: 1989, 'Nazi Data: Dissociation from Evil. Commentary', *Hastings Center Report* **19**(4), 16–18.
Shrader-Frechette, K.S.: 1991, *Risk and Rationality*, University of California Press, Berkeley, CA.
Smith-Rosenberg, C. & Rosenberg, C.: 1973, 'The Female Animal: Medical and Biological Views of Woman and Her Role in Nineteenth-Century America', *Journal of American History* **60**, 332–56.
Takacs, D.: 1996, *The Idea of Biodiversity: Philosophies of Paradise*, Johns Hopkins University Press, Baltimore, MD.
Toumey, C.: 1996, *Conjuring Science*, Rutgers University Press, Rutgers, NJ.
Wimsatt, W.C.: 1981, 'Robustness, Reliability and Overdetermination'. In M. Brewer and B. Collins (eds.), *Scientific Inquiry and the Social Sciences*, Jossey-Bass, San Francisco, CA, pp. 124–63.
Ziman, J.: 1967. *Public Knowledge*, Cambridge University Press, Cambridge.

Which Way Is Up? Thomas S. Kuhn's Analogy to Conceptual Development in Childhood[1]

ALEXANDER T. LEVINE

Department of Philosophy, Lehigh University, 15 University Drive, Bethlehem, PA 18015, USA

ABSTRACT. In the Preface to his *Structure Of Scientific Revolutions*, Thomas S. Kuhn let it be known that his view of scientific development was indebted to the work of pioneering developmental psychologist Jean Piaget. Piaget's model of conceptual development in childhood, on which the child passes through several discontinuous stages, served as the template for Kuhn's reading of the history of a scientific discipline, on which mutually incommensurable periods of normal science are separated by scientific revolutions. The analogy to conceptual change in childhood pervades Kuhn's corpus, serving as the central motif in his well-known essays, 'A Function for Thought Experiments' and 'Second Thoughts on Paradigms'. But it is deeply problematic. For as a careful student of Piaget might note, Piaget, and the developmental psychologists he inspired, relied on the same analogy, *but with the order of epistemic dependencies reversed*. One begins to worry that Kuhn's use of the analogy, and its subsequent re-use by developmental psychologists, sneaks a vicious circularity into our understanding of important processes. This circularity is grounds for some concern on the part of science educators accustomed to employing such Kuhnian notions as 'incommensurability' and 'paradigm'.

1. KUHN'S DEVELOPMENTAL ANALOGY

Early on in the *Structure of Scientific Revolutions* (hereafter *Structure*), we learn that Kuhn's understanding of the history of science was profoundly influenced by such thinkers as W. V. O. Quine, B. L. Whorf, Ludwig Fleck and, of course, the great developmental psychologist, Jean Piaget.[2] Since this early mention of Piaget already points the way to the problem to be discussed in the present paper, it merits close attention. Kuhn tells us,

> A footnote encountered by chance led me to the experiments by which Jean Piaget has illuminated both the various worlds of the growing child and the process of transition from one to the next.[3]

In a footnote to this passage, Kuhn cites two works of Piaget's as having played a role: his *The Child's Conception of Causality* (1930) and *Les Notions de Mouvement et de Vitesse Chez l'Enfant* (1946). Nearly ten years later, Kuhn would elaborate as follows:

> Almost twenty years ago I first discovered, very nearly at the same time, both the intellectual interest of the history of science and the psychological studies of Jean Piaget. Ever since that time, the two have interacted closely in my mind and in my work. Part of what I know about how to ask questions of dead scientists has been learned by examining Piaget's interrogations of living children . . . . [I]t was Piaget's children from

F. Bevilacqua et al. (eds.), Science Education and Culture, 197–212.
© 2001 *Kluwer Academic Publishers. Printed in the Netherlands.*

> whom I . . . learned to understand Aristotle's physics . . . I am proud to acknowledge the ineradicable traces of Piaget's influence.[4]

Given Kuhn's stated interest in 'the various worlds of the growing child and the process of transition from one to the next', we may, in light of what we know about his understanding of the history of science,[5] offer a reasonable extrapolation as to what he saw in Piaget's results.

Kuhn was surely taken by the most celebrated of Piaget's explanatory hypotheses, the claim that conceptual development in childhood proceeds through a number of discrete, discontinuous stages, at each of which the concepts whereby the child copes with a given domain are profoundly different both from adult concepts, and from the child's own concepts at other developmental stages, both earlier and later. If Piaget's model accurately describes our childhoods, it follows that we are possessed, at least during our maturation, of mental machinery which allows us to undergo rapid and dramatic conceptual transformations. There seems no reason to expect, a priori, that this capacity for radical conceptual change disappears at the onset of adulthood and, thus, it would not be surprising to find the same or similar mechanisms at work beyond the supposed end-point of our developmental trajectory. Furthermore, Kuhn would argue in *Structure*, such mechanisms *must* operate in adult scientists, for when we read the history of science attentively we discover marked discontinuities, or 'incommensurabilities', analogous to the discontinuities found by Piaget in his work on conceptual development in childhood.

If the above sketch accurately portrays the manner in which Kuhn was influenced by Piaget then, turning to the larger issue of the relation between Kuhn's understanding of psychology and his understanding of scientific development, the following picture emerges: Kuhn's understanding of scientific development, along with its trademark notions of incommensurability and scientific revolution, is based at least in part on an analogy to conceptual change in childhood. This analogy serves as a source of inspiration not unlike that Rutherford derived from his understanding of the Solar System. But whereas atomic nuclei are not stars, adult scientists *are* former children, and while their conceptual tools may be in some sense more sophisticated than those employed by children at any stage of development, they are not different in kind. Evidence in support of Piaget's account of conceptual change is thus, derivatively, at least, evidence in support of Kuhn's theory of scientific development (provided numerous standard conditions on the treatment of data as positive evidence are also met).

Whether Kuhn really relied on the results of Piaget and other psychologists to this degree or in precisely this way is, of course, impossible to tell from the evidence of *Structure* alone. But that he conceived of the analogy between conceptual change in childhood and scientific development along the lines sketched above is made amply clear in Kuhn's later essays. Two of the papers included in Kuhn (1977), 'A Function for Thought

Experiments' and 'Second Thoughts on Paradigms', prove especially illuminating in this regard.[6]

Before we take a closer look at these essays, it should be noted that they are by no means isolated examples. These two papers are merely those in which the analogy presently under scrutiny is most clearly developed, but others might have been chosen with almost equal justification. Following the footnote in *Structure* and 'A Function for Thought Experiments', originally published in 1964, Kuhn's next public reliance on an analogy to conceptual change in childhood appears to be a paper originally given at a conference in 1965, and eventually published in 1970.[7] In a lecture originally given in 1968, and published in Kuhn (1977) as 'The Relations Between the History and the Philosophy of Science', Kuhn offers a related analogy, this time likening the interpretive task of the historian of science to the learning task performed by a child in mastering the similarity relations constitutive of adult concepts.[8] The main analogy, and the associated debt to Piaget, play a central role in a lecture delivered to a group of Piagetians in 1970, republished in Kuhn (1977) as 'Concepts of Cause in the Development of Physics'. It is in this essay, cited above, that Kuhn most fervently acknowledges Piaget's influence. As originally published in 1974, 'Second Thoughts on Paradigms' appeared along with transcripts of commentaries and rejoinders. Of particular interest for present purposes is Kuhn (1974, pp. 505ff), where he replies to objections regarding the relevance of his analogy to questions of theory change in science. Finally, the analogy reappears in Kuhn (1983, p. 682).[9]

In 'A Function for Thought Experiments', Kuhn sets out to refute and improve upon a widely accepted conception of the role of thought experiments in science, the view on which a thought experiment is said to serve not to illuminate nature, but rather aspects of the conceptual apparatus of the scientist who performs it. Toward this end, Kuhn begins with a reflection on Piaget (1946):

> The historical context within which actual thought experiments assist in the reformulation or readjustment of existing concepts is inevitably extraordinarily complex. I therefore begin with a simpler, because nonhistorical, example, choosing for the purpose a conceptual transposition induced in the laboratory by . . . Jean Piaget . . . Piaget dealt with children, exposing them to an actual laboratory situation and then asking them questions about it. In slightly more mature subjects, however, the same effect might have been produced by questions alone in the absence of any physical exhibit. If those same questions had been self-generated, we would be confronted with [a] pure thought-experimental situation . . . .[10]

Kuhn then proceeds to sketch one of Piaget's more celebrated experiments, in which subjects come to distinguish claims of the form 'A is faster than B' from claims of the form 'A reaches its goal before B'. The process underlying this transformation is said to be essentially the same one whereby a Galileo much older than any of Piaget's subjects arrived at the distinction between average and instantaneous velocity. Since Galileo's insight is a paradigm case for our understanding of the role of thought experiments in science, an analysis of Piaget's experiments can be expected to illuminate this role.

The details of Kuhn's treatment of thought experiments, and its relation to his better-known accounts of paradigm shift and revolutionary scientific change, need not concern us here. Of primary importance for present purposes is what the supposed parallel between the Piagetian subject and Galileo tells us about Kuhn's analogy between conceptual change in childhood and conceptual change in science. We will return to this question shortly, after first considering a second essay, 'Second Thoughts on Paradigms'.

In 'Second Thoughts on Paradigms', Kuhn attempts to remedy a serious defect in *Structure*, the ambiguity of the key term 'paradigm'. Toward this end he argues that his uses of the term fall into two broad classes. The first class includes references to paradigms as sets of commitments shared by all members of a scientific community; a paradigm, in this first sense, will henceforth be called the 'disciplinary matrix' of its associated community.[11] Among the commitments which comprise a community's disciplinary matrix may be found commitments to exemplary scientific problems and their solutions. Such problem/solution pairs are paradigms in the second sense of the word, the only sense Kuhn wishes to allow in the 'Second Thoughts' essay.[12] In this sense, for example, the problem of calculating a pendulum's period given its length, along with the classic solution to that problem, would constitute a paradigm for the community of Newtonian physicists.

With regard to paradigms, in this narrower sense of the word, the following question arises: how do they help constitute the scientific community of whose disciplinary matrix they form a part? The answer, for Kuhn, lies in the role they play in inducting new members into the community. In discovering the exemplary solutions to exemplary problems, Kuhn asserts, the student masters the set of similarity relations which will enable him or her, as a mature scientist, to identify the conceptual tools appropriate to the solution of novel problems he or she may encounter. This claim surely requires some explanation and defense, as Kuhn is well aware. One might expect Kuhn to illustrate his account by showing, for a *given* exemplary scientific problem, how discovering its solution furthers the education of a scientific novice. Interestingly enough, Kuhn begs off on this sort of illustration, announcing instead his intention to

> Return now to my main argument, but not to scientific examples. *Inevitably the latter prove excessively complex*. Instead I ask that you imagine a small child on a walk with his father in a zoological garden. The child has previously learned to recognize birds . . . . During the afternoon now at hand, he will learn for the first time to identify swans, geese, and ducks. Anyone who has taught a child under such circumstances knows that the primary pedagogic tool is ostension.[13]

Kuhn has thus made a conscious decision to defend an account of the role of paradigms in establishing membership in a scientific community with examples not from science, but from childhood. Underlying this decision, one would suppose, is the assumption that the latter sort of examples are appropriate to the issue at hand; indeed, they are perhaps the most appropriate examples, given the 'excessive complexity' of cases drawn from science.[14] This assumption is made explicit later on in the 'Second Thoughts' essay, when Kuhn asks, rhetorically,

> Need I now say that the swans, geese, and ducks which Johnny encountered during his walk with his father were what I have been calling exemplars? . . . they were solutions to a problem that the members of his prospective community had already resolved . . . Johnny is, of course, no scientist, nor is what he has learned yet science. But he may well become a scientist, and the technique employed on his walk will still be viable . . . . [T]he same technique, if in a less pure form, is essential to the more abstract sciences as well.[15]

Explicating the role of paradigms in science by examples from childhood is justified, for Kuhn, because the child and the mature prospective scientist both employ 'the same technique' in mastering the concepts of their respective communities. Paradigms emerge as universal conceptual tools, with applications far beyond the bounds of science.

Having canvassed two of Kuhn's arguments in which the analogy between conceptual development in childhood and conceptual change in science looms large, it is time to take stock of such features of Kuhn's analogy as may be derived from these sources. Kuhn appears prepared to assent to all of the following:

(1) Evidence for the operation, in childhood, of a mechanism for radical conceptual change is, all other things being equal, also evidence for the operation of a similar (or identical) mechanism in science.
(2) The mechanisms supporting conceptual change in science are not fundamentally different from those supporting conceptual change in childhood.
(3) Some of the cognitive processes which, in childhood, require direct interaction with the world, also take place in the mind of the adult scientist, but without the need for interaction.
(4) Instances of conceptual change in childhood are, on the whole, less complex and more easily grasped than instances of conceptual change in science.

The two essays sketched above provide ample evidence that Kuhn would have agreed to all of (1–4) – at least when those essays were written. But it should be noted that there is absolutely no reason, a priori, to think that any of (1–4) is true. There is also no reason, a priori, to think that (1–4) are false; whether they are true or false is presumably the sort of fact discoverable only by empirical research, research Kuhn certainly never conducted. But what follows if they are assumed to be true? The following is an immediate consequence of (1–4):

(5) The analogy between conceptual development in childhood

> and conceptual change in science permits inferences in one direction: from childhood to science.

I shall call (5) Kuhn's 'unidirectionality principle'. By itself, the truth of (2) establishes an analogy permitting inferences in both directions; if the mechanisms underlying conceptual change in childhood and science are stipulated to be the same or similar, there is no reason not to make use of new insights into the history of science in devising explanatory hypotheses for conceptual development in childhood.[16] But it is clear that, for Kuhn, the class of inferences sanctioned by the analogy is much narrower. Certain processes become progressively internalized with age, and are thus easier to study in children. Furthermore, the conceptual dimensions of episodes in the history of science are sometimes prohibitively complex, while their analogues in childhood development are manageably simple. Finally, while not all children become scientists, all scientists were at one time children, and what worked for the child may well work for the scientist.

With the parameters of Kuhn's analogy in mind, let us return to his initial acknowledgment of Piaget's influence in the Preface to *Structure*. There, as we have seen, Kuhn recalls the impact of the 'experiments by which Jean Piaget has illuminated both the various worlds of the growing child and the process of transition from one to the next'.[17] So far, Kuhn's use of Piaget seems consistent with the unidirectionality principle derived from his later work. However, a footnote to this same passage appears to tell a somewhat different story: 'Because they displayed concepts and processes that also emerge directly from the history of science, two sets of Piaget's investigations proved particularly important (Piaget 1930, 1946)'.[18]

The footnote is ambiguous. On the one hand, it might be interpreted as an expression of Kuhn's pleasure at having found a convergence between Piaget's studies of development and his own work on the history of science. But to those familiar with Piaget, a somewhat different interpretation suggests itself. For Piaget was himself an avid student of the history and philosophy of science, to which he frequently alluded in explaining his choice of developmental hypotheses. Now, one might expect Piaget, and psychologists in general, to conceive of the relationship between childhood and science along lines similar to those proposed by Kuhn. After all, Kuhn's unidirectionality principle makes their discipline, psychology, epistemically prior to any philosophically informative interpretation of the history of science. And indeed psychologists are sometimes found to speak approvingly of this order of epistemic dependencies. But when we canvass the ways in which Piaget and other developmental psychologists have conceived of the relationship between childhood and science (Section 2), we generally find the order of epistemic dependencies reversed, with the history and philosophy of science serving as a source for explanatory hypotheses in developmental psychology, rather than the other way around.

And so, when we consider the consequences of employing Kuhn's unidirectional analogy in the history and philosophy of science, while at the same time the *reverse* analogy is being employed in developmental psychology, we begin to worry about the potential for vicious circularity (Section 3).

2. TURNING THE TABLES ON KUHN'S ANALOGY

As suggested above, Kuhn's is not the only way of conceiving of the analogy between childhood and science. Given what we have already seen of the influence of Piaget on Kuhn's thinking, it will be instructive to consider Piaget's views on the respective roles of the history and philosophy of science and developmental psychology. Piaget's (1970) *Genetic Epistemology* provides a useful, brief illustration of these views.[19]

A casual reading of Piaget (1970) suggests an epistemic relationship between childhood and science on which, as in Kuhn, the former helps us understand the latter. 'Genetic epistemology', Piaget tells us,

> attempts to explain knowledge, and in particular scientific knowledge, on the basis of its history, its sociogenesis, and especially the psychological origins of the notions and operations on which it is based.[20]

The 'psychological origins of the notions and operations' on which scientific knowledge is based are presumably the province of cognitive developmental psychology, on which Piaget spent most of his career. Nothing about the project of genetic epistemology, as Piaget here defines it, demands an *analogy* between childhood development and conceptual change in science. The picture of the genetic explanation of scientific knowledge that emerges over the first few pages of Piaget (1970) evokes more the idea of a *reduction* of the study of scientific change to the study of psychology than it does an analogy. Yet as in Kuhn, psychology is epistemically prior to our understanding of scientific change.

This initial impression appears to find confirmation later on in Piaget's discussion, when he asserts,

> The fundamental hypothesis of genetic epistemology is that there is a parallelism between the progress made in the logical and rational organization of knowledge and the corresponding formative psychological processes . . . . Of course the most fruitful, most obvious field of study would be reconstituting human history . . . . Unfortunately . . . [s]ince this field of biogenesis is not available to us, we shall do as biologists do and turn to ontogenesis. Nothing could be more accessible to study than the ontogenesis of these notions. There are children all around us.[21]

Here, as in Kuhn, we find hints that it is easier to study transformations in childhood than it is to study transformations in science. Furthermore, the genesis of human knowledge in general and the ontogenesis of the human epistemic subject are thought to be similar enough that studying the latter will, at least in some sense, serve as a stand-in for the far more difficult task of studying the former. Finally, in this passage, the relationship between science and development appears clearly to be one of *analogy*, rather than *reduction*.

But of course Piaget *does* study the history and philosophy of science, often directly, without prior recourse to developmental psychology. In other work, Piaget claims to be pursuing developmental and historical studies in parallel, comparing childhood and science in search of 'common characteristics between the history of science and psychological development . . .'.[22] Piaget's attention to both sides of the equation has led some, including Bärbel Inhelder, to read him as follows:

> [Piaget's and Garcia's] purpose in studying these generalized mechanisms was not to describe term-by-term correspondences [between science and childhood], and even less to propose a recapitulation of phylogenesis by ontogenesis, nor even to demonstrate the existence of analogies in sequencing. Instead, they wished to see if the mechanisms mediating the transition from one historical period to the next . . . are analogous to those mediating transitions from one developmental stage to the next.[23]

If Inhelder is right, then for Piaget, any analogy between science and childhood is the *product* of research into both, and a *presupposition* of research into neither.

However, close reading of Piaget's analyses of cases from the history of science sometimes tells a different story. Consider Piaget's treatment of one aspect of the move toward a relativistic physics at the beginning of the twentieth century, the revision in the concept of simultaneity:

> . . . how is it that Einstein was able to give a new operational definition of simultaneity at a distance? How was he able to criticize the Newtonian notion of universal time without giving rise to a deep crisis within physics? . . . A few metaphysicians . . . were appalled by this revolution in physics, but . . . among scientists themselves it was not a very drastic crisis . . . . It was not a crisis because simultaneity is not a primitive . . . concept, and it is not even a primitive perception.[24]

Of interest in this passage is the fact that Piaget treats the physical community's relatively placid acceptance of Einstein's definition of simultaneity at a distance as evidence that the Newtonian conception of simultaneity, for all that it is in some sense easier to grasp, is not psychologically primitive. To be sure, Piaget also has independent, psychological evidence for this claim. But the fact that he treats his interpretation of this episode in the history of science as evidence in support of a developmental thesis *at all* shows that, for Piaget, the relationship between science and psychology is one in which the order of epistemic dependencies is at least sometimes reversed from that observed by Kuhn.

For a younger generation of cognitive developmental psychologists, all influenced to one or another degree by both Piaget and Kuhn, the order of epistemic dependencies almost always runs from science to childhood. For ease of exposition, I shall call this the 'inverse ordering', so as to distinguish it from the direction of inference we observed in Kuhn.

One excellent example of the inverse ordering is Susan Carey's (1988) 'Conceptual Differences Between Children and Adults'. In this paper, Carey, a developmental psychologist, asks whether, or not, the conceptual schemes of children and adults can be characterized as 'locally incommensurable'.[25] The notion of local incommensurability was developed by Kuhn (1983) in an effort to clarify his interpretations of key episodes in the history of science, and further explicated by another philosopher of science, Philip Kitcher (1988). The purpose of Carey's paper is clearly to explain data gleaned from the study of children by recourse to a device developed to explain conceptual change in science. Carey, like Kuhn, relies on an analogy between science and childhood, but for her, science is epistemically prior.

Another developmentalist, former Piaget student Annette Karmiloff-Smith, clearly exhibits the inverse ordering in an essay suggestively titled, 'The Child is a Theoretician, Not an Inductivist' (Karmiloff-Smith 1988). In this essay, as elsewhere, she asserts,

> [C]hildren go about their spontaneous discovery task by behaving like the typical scientist. Kuhn . . . was right in his view that only on the very rare occasions when scientists must actually choose between competing theories do they reason like philosophers or logicians! Both for the child and the adult researcher, scientific progress does not stem from the use of logical criteria on the basis of rational induction from observations.[26]

Like Carey, Karmiloff-Smith applies lessons learned from the history and philosophy of science toward understanding conceptual development in childhood. Also like Carey, she acknowledges that she is indebted for these lessons to Kuhn, whose argument (in *Structure*) that scientists make fewer appeals to logic or the canons of good reasoning than one might expect, is well known.

More recently, Allison Gopnik has acknowledged the extent to which, in the past few years,

> . . . cognitive and developmental psychologists have invoked the analogy of science itself. They talk about our everyday conceptions of the world as implicit and intuitive theories, and about changes in those conceptions as theory changes.[27]

Gopnik, however, is eager to distance herself from this trend, of which she herself was a sometime standard-bearer.[28] In Gopnik (1996), she announces her intention

> to argue that the analogy cuts both ways: specifying the parallels between cognitive development and science not only can help us to understand cognitive development, it can also help us to understand science itself.[29]

By now such pronouncements should strike us as ironic. The inverse ordering reading of the analogy between science and childhood, which Gopnik now seeks to reverse, was championed by psychologists who, like Carey, Karmiloff-Smith, and Gopnik herself, were inspired by Kuhn.[30] But for Kuhn, the whole point of the analogy *was* to help us understand *not conceptual change in childhood*, *but 'science itself'*. And finally Kuhn, for his part, was inspired by Piaget, who as we have seen was himself often historically inclined. We begin to sense a circle: a historical circle, certainly, and perhaps an epistemic circle, as well.

3. THE THREAT OF CIRCULARITY

The threat of circularity becomes clear when we imagine the following two claims to both be true:

(1) Philosophers and historians of science look to the results of developmental studies for well-confirmed explanatory hypotheses suitable for use in interpreting scientific change.

(2) Developmental psychologists look to the history and philosophy of science for well-confirmed explanatory hypotheses suitable for use in interpreting data on child development.

In Sections 1 and 2, we have seen some evidence for the truth of both (1) and (2). But as Kuhn pointed out, in a different context, not every circularity is vicious.[31] Whether the apparent circularity is vicious or not, in this case, turns on the degree to which, for both developmental psychologists and philosophers of science, their respective analogies are the source of *substantive evidence*, rather than merely informal inspiration.

A definitive answer to *this* question would, it seems, require a general account of the role of analogies in scientific reasoning, and of their evidential status in particular. Clearly this is not the place to pursue such an account, even if there were one in the offing. But even in the absence of a general theory of analogy, I think a persuasive case can be made that the analogy between childhood and science, as employed both by Kuhn and the psychologists, has some significant evidential standing. A review of this case is in order.

As I hope to have established by now, Kuhn himself ascribed great importance to the analogy between childhood and science. But, it might be objected, this analogy (and perhaps analogies in general) serve only as inspiration, as motivation for looking elsewhere for hard evidence. In short, so the objection goes, the analogy is relevant only in what has been called the *context of discovery*, but *not* in the *context of justification*.[32]

There are two replies to this objection, either of which suffices, in my view, to show why it fails to apply. The first simply recalls that Kuhn himself saw the rejection of the distinction between context of discovery and context of justification as a natural outgrowth of his views on the philosophy of science.[33] For Kuhn, a whole range of subjective factors not only *does* play a role in scientific theory choice, such factors *must*, and indeed *should* weigh in.[34] Thus, if both Kuhn and the post-Kuhnian psychologists discussed in Section 2 endorse enough of Kuhn's philosophy of science to undermine the distinction between context of justification and context of discovery, the objection misses its mark.

For those less favorably inclined toward Kuhn's philosophy of science, the second reply will doubtless prove more compelling. Let us assume that the distinction between context of discovery and context of justification is valid, and that to have evidential standing, our analogy must function not only in the former, but also in the latter. I believe it can be shown that, both for Kuhn and the post-Kuhnian developmentalists, this condition is met.

There are at least two ways (and doubtless many more) to use an analogy in scientific reasoning. The first appears to be exemplified by the Rutherford's analogy, alluded to in Section 1, between the structure of the atom and the structure of a solar system. Consider a set with four members: two different solar systems and two different atoms, perhaps atoms of different elements. Each member will differ from the other three in various respects. But on the whole, the differences between the two solar systems will be differences only in *degree*, as will the differences between the two atoms. The differences between either of the atoms and either of the solar systems, on the other hand, will be differences in *kind*. Furthermore, the difference between the two homogeneous pairs (solar systems and atoms) will also be differences in kind. Now, were we to have discovered, shortly after the First World War, that there are certain systematic differences between kinds of solar system, that discovery would have been of *absolutely no consequence* for the development of atomic theory. The reason is that Rutherford's analogy functioned only in the context of discovery. The Solar System offers a useful model for thinking about a system which, experiment had shown, must be made up mostly of empty space. But there is no basis whatsoever for treating the solar system as a useful research tool in the quest for insights into atomic structure.

By contrast, consider Charles Darwin's (1859) analogy between domestic breeds and naturally occurring species. This analogy, the development of which occupies the first two chapters of Darwin (1859) in their entirety, is central to the higher-level analogy between domestic and natural selection and, thus, a cornerstone of Darwin's evolutionary theory. Indeed, as far as *Origin* is concerned, this analogy constitutes *the main reason* for believing in Darwin's evolutionary theory, and it is difficult to see how any of Darwin's contemporaries would have been persuaded in its absence. Certainly the fossil record could be explained in non-Darwinian terms, as Lyell and others have shown.

But as Hull has demonstrated, though most naturalists were quickly won over by *Origin*, contemporary philosophers of science, including Whewell, Herschel and Mill, resisted, all of them apparently on methodological grounds.[35] Mill's objection, in particular, is interesting, for it amounts to the claim, made in nineteenth century terms, that Darwin has violated the distinction between context of discovery and context of justification.[36] If Darwin's book *did* provide Darwin's scientific colleagues with a good reason to believe in his evolutionary theory, it follows that

Mill and the others must have been missing something. What they were missing, I believe, is the crucial distinction between analogies like Darwin's and analogies like Rutherford's.

Consider a second four-member set, this time consisting of two breeds of a given domestic species, and two species of a given genus. A pair of organisms belonging to the two breeds will likely differ from one another less than a pair selected from the two species. But now compare the two sorts of difference. At the pair level, we observe that the two homogeneous pairs, breed-pair and species-pair, *differ only in degree, not in kind*. In other words, a breed-pair differs from a species-pair only in that the two members of the former are more alike than the two members of the latter. But there is no difference in kind, i.e., no *categorical* difference, between breeds and species. And here is the crux of Darwin's analogy. To paraphrase, breeds or varieties are incipient species.

Darwin's analogy succeeds in the context of justification, and that success is legitimate. Its legitimacy is perhaps explicable by recourse to the following principle: *An analogy may serve as evidence* (*function in the context of justification*) *to the extent to which it approximates the literal truth*. If breeds or varieties *really are* incipient species, lessons learned about the origins of breeds are evidence in the quest for the origin of species. Returning to Kuhn: if children *really are* incipient scientists (or scientists grown-up children), then knowledge of the former constitutes evidence about the latter. And as for atoms and solar systems, atoms *really aren't* solar systems.

That Kuhn sees his analogy this way is made amply clear by a passage cited earlier from 'Second Thoughts'. We recall,

> Johnny is, of course, no scientist, nor is what he has learned yet science. *But he may well become a scientist, and the technique employed on his walk will still be viable*. [*T*]*he same technique*, if in a less pure form, is essential to the more abstract sciences as well.[37]

Not all children become scientists, but all scientists were once children. Evidence about how children perform certain cognitive tasks is thus, prima facie, evidence about how scientists perform analogous tasks.

My argument that an analogy between childhood and science can, *in principle*, function in the context of justification, applies equally well to the inverse ordering. That it *in fact* functions in this way is, I think, even more clear for the psychologists cited in Section 2 than it is for Kuhn. But even if one were to question the evidential status of the analogy between childhood and science as employed, say, by Carey or Karmiloff-Smith, that status is surely beyond question in Paul Thagard's *Conceptual Revolutions*:

> Most comparisons of scientists and children by developmental psychologists and science educators have been limited by the inadequacy of available theories of conceptual change in science. If the much more detailed account of conceptual change in earlier chapters is acceptable, then richer comparisons of scientists and children become possible. We can contrast conceptual change in scientists during revolutionary periods in the history of science with conceptual change in children and in students learning science.[38]

Kuhn's influence is plain to see, as is what I have called the inverse ordering of the analogy between childhood and science. But more importantly, for Thagard, an account of conceptual change in science serves as the evidential foundation for an account of conceptual change *tout court*.

> In developmental psychology, "ontogeny recapitulates phylogeny" means that the development of knowledge in individual children corresponds to the development of knowledge in the history of science. Support for the developmental version of the thesis has been drawn from analogies between, for example, the child's transition from holding a naive, Aristotelian view of motion to, eventually, with enough education, a Newtonian view.[39]

Such 'support' is unintelligible unless an analogy to science is presupposed; without the analogy, it would clearly be impossible even to identify a child's view of motion as 'Aristotelian' or 'Newtonian'.

Clearly, all of the historians, philosophers and psychologists cited above also have independent reasons for believing that conceptual change in science, or conceptual development in childhood, operates as they think it does. Countless historical scientific documents have been dusted off and analyzed, and countless experimental studies of children conducted and replicated. No one relies exclusively on arguments from analogy. And yet, if the argument of this section is sound, the analogy between childhood and science constitutes real evidence, whichever way it is taken. When it comes to contemporary psychologists, sociologists, and educators, one is struck not only by the magnitude of Kuhn's influence, but also by the nearly universal inattention to Kuhn's own evidential employment of the analogy. The potential for circularity is very real, and must be deliberately avoided.

Awareness of this threat should also inform the way we teach science. Summarizing current consensus on the importance of a historical dimension in science education, Robert Carson asserts,

> The use of history as an organizing framework appears to have psychological validity for several reasons. Piaget argued that the construction of scientific understanding in individuals often parallels the historic evolution of scientific theories (Piaget 1970, p. 13). Sequencing the approach to a scientific concept by means of history takes the student through progressively more current viewpoints, dignifies students' interim conceptions, and shows how one set of ideas gets overthrown by another, more valid.[40]

Our inquiry into the interaction between the study of scientific change and the study of conceptual development should have taught us, at a minimum, to be very careful about assuming parallels between historical and psychological transformations. No one, Piaget included, has offered a convincing argument that, in the emergence of scientific knowledge, ontogeny recapitulates phylogeny.

As educators, we ought to be equally careful about metaphorically extending some of Kuhn's more celebrated notions. We should ask, for example, whether we can meaningfully (i.e., noncircularly) describe the conceptual schemes of a science teacher and his or her novice student as 'incommensurable'. To adopt the incommensurability metaphor for a conceptual gap is, after all, to buy into an account of what can be done to bridge that gap, and what can't. Or again, should we talk, as we so often do, of the 'paradigm' of the freshman physics student? Is it meaningful to claim that children master an 'Aristotelian paradigm' of motion before progressing to a 'Newtonian paradigm?' And bearing in mind the second Kuhnian use of the word, how should we conceive of the psychological role played by 'paradigms', exemplary problems and their solutions, in the minds of our students? I hope to have shown that all such metaphors ought to be treated as such, and that when we lose track of their analogical component, we are on very shaky ground.

NOTES

[1]I am indebted to Mark Bickhard and three anonymous referees for their insightful comments and suggestions. Any remaining errors are mine.
[2]Kuhn (1970a, pp. vi–vii).
[3]Ibid.
[4]Kuhn (1977, pp. 21–22).
[5]For a comprehensive exposition of Kuhn's views, along with a review of important criticisms, see Hoyningen-Huene (1993).
[6]The 'Second Thought' essay largely overlaps with the 'Postscript' to Kuhn (1970a).
[7]See Kuhn (1970b, pp. 17–19).
[8]See Kuhn (1977, pp. 16–17). It could be argued, of course, that the analogy invoked in this essay is, at bottom, the same analogy between science and childhood. For it is clear that Kuhn believed that his account of scientific revolutions enjoyed rather broad generality, applying not only to the natural sciences from which most of his examples were drawn, but also to the social sciences, and more specifically, to history. To claim that the conceptual chores of the historian are in many respects like those of the child is thus to offer a special case of the analogy to which the present paper is devoted.
[9]To my knowledge, this is the latest explicit appearance of the analogy to conceptual change in childhood in Kuhn's published work. But the *implicit* importance of the analogy only increased as, throughout the 1980s and 1990s, Kuhn came to rely on *language learning* as a model for understanding how disputes between parties on either side of a revolutionary divide might be resolved short of the extinction of one or another party. Our paradigmatic language learners are, of course, children.
[10]Kuhn (1977, p. 243).
[11]Kuhn (1977, pp. 296–297). For a careful analysis of the notion of a disciplinary matrix, and a more thorough treatment of Kuhnian paradigms in general, see Hoyningen-Huene (1993, especially Chap. 4).
[12]Kuhn (1977, pp. 306–307).
[13]Kuhn (1977, p. 309), emphasis added.
[14]Of course, one must not overstate the case. In the 'Postscript' to the second edition of *Structure*, Kuhn does develop examples from the history of science to serve this purpose.
[15]Kuhn (1977, p. 313).
[16]Indeed, for many psychologists, the primary interest of any analogy between science and childhood lies in the utility of the former in helping us understand the latter. I will touch on such efforts later.
[17]Kuhn (1970a, loc. cit).
[18]Kuhn (1970a, p. vi, fn. 2).
[19]Other works in which the relationship between conceptual development in childhood and conceptual change in science is a primary focus include Piaget (1950), and Piaget and Garcia

(1989). A systematic survey of Piaget's views on this subject would exceed the scope of the present paper.
[20]Piaget (1970, p. 1).
[21]Piaget (1970, p. 13).
[22]Piaget and Garcia (1989, pp. 28).
[23]Inhelder (1989, p. x).
[24]Piaget (1970, p. 6).
[25]See e.g. Carey (1988, p. 168).
[26]Karmiloff-Smith (1988, p. 183).
[27]Gopnik (1996, p. 485).
[28]See e.g. Gopnik (1994).
[29]Gopnik (1996, p. 486).
[30]The inverse ordering is also employed by Philip Kitcher, a philosopher of science, in his 1988 article. Kitcher believes that the analogy between science and development is really symmetrical, but insists, 'one has to begin somewhere, and I start with the area where my ignorance is less'. Since Kitcher's area of expertise is the same as Kuhn's, it is interesting that Kuhn chooses the other starting-point.
[31]Kuhn (1970a, p. 208).
[32]See Popper (1968, Chap. 1) for a canonical exposition of this distinction.
[33]See *Structure* (pp. 8–9); Kuhn (1977, pp. 325–330). For a thorough exposition and defense of Kuhn's attack on the distinction between context of discovery and context of justification, see Hoyningen-Huene (1993, pp. 245–252).
[34]See Hoyningen-Huene (1993, pp. 250–251).
[35]Hull (1989, Chap. 2).
[36]Hull (1989, p. 31).
[37]Kuhn (1977, p. 313), emphasis added.
[38]Thagard (1992, p. 246).
[39]Thagard (1992, p. 259).
[40]Carson (1997, p. 233).

REFERENCES

Carey, S.: 1988, 'Conceptual Differences Between Children and Adults', *Mind and Language* 3(3).

Carson, R.: 1997, 'Science and the Ideals of Liberal Education', *Science and Education* 6(3).

Darwin, C.: 1859, *On the Origin of Species*, John Murray, London.

Gopnik, A.: 1994, 'How We Know Our Minds: The Illusion of First-Person Knowledge of Intentionality', *Brain and Behavioral Sciences* 16.

Gopnik, A: 1996, 'The Scientist as Child', *Philosophy of Science* 63(4).

Hoyningen-Huene, P.: 1993, *Reconstructing Scientific Revolutions: Thomas S. Kuhn's Philosophy of Science*, A. T. Levine (trans.), University of Chicago Press, Chicago.

Hull, D.: 1989, *The Metaphysics of Evolution*, State University of New York Press, Albany.

Inhelder, B.: 1989, 'Foreword', in Piaget & Garcia (1989).

Karmiloff-Smith, A.: 1988, 'The Child is a Theoretician, Not an Inductivist', *Mind and Language* 3(3).

Kitcher, P.: 1988, 'The Child as Parent of the Scientist', *Mind and Language* 3(3).

Kuhn, T. S.: 1970a, *The Structure of Scientific Revolutions*, 2nd edn., University of Chicago Press, Chicago.

Kuhn, T. S.: 1970b, 'Logic of Discovery or Psychology of Research', in Lakatos & Musgrave (eds.), *Criticism and the Growth of Knowledge*, Cambridge University Press, Cambridge.

Kuhn, T. S.: 1974, 'Discussion', in F. Suppe (ed.), *The Structure of Scientific Theories*, University of Illinois Press, Urbana.

Kuhn, T. S.: 1977, *The Essential Tension*, University of Chicago Press, Chicago.

Kuhn, T. S.: 1983, 'Commensurability, Comparability, Communicability', in Asquith & Nickles (eds.), *Proceedings of the 1982 Biennial Meeting of the Philosophy of Science Association*, Philosophy of Science Association, East Lansing, Michigan.

Piaget, J.: 1930, *The Child's Conception of Causality*, M. Gabain (trans.), London.

Piaget, J.: 1946, *Les Notions de Mouvement et de Vitesse chez l'Enfant*, Presses Universitaires de France, Paris.

Piaget, J.: 1950, *Introduction a l'Epistemologie Genetique*, Presses Universitaires de France, Paris.

Piaget, J.: 1970, *Genetic Epistemology*, E. Duckworth (trans.), Columbia University Press, New York.

Piaget, J. & Garcia, R.: 1989, *Psychogenesis and the History of Science*, H. Feider (trans.), Columbia University Press, New York.

Popper, K.: 1968, *The Logic of Scientific Discovery*, Harper & Row, New York.

Thagard, P.: 1992, *Conceptual Revolutions*, Princeton University Press, Princeton, NJ.

Saving Kuhn from the Sociologists of Science

ROBERT NOLA

Department of Philosophy, University of Auckland, Private Bag 92019, Auckland, New Zealand. E-mail: r.nola@auckland.ac.nz

ABSTRACT. For many in the science education community Kuhn is often closely identified with a sociological approach, as opposed to a philosophical approach, to matters raised in his book *The Structure of Scientific Revolutions*. This paper is an attempt to liberate Kuhn from too close an association with the sociology of scientific knowledge. While Kuhn was interested in some sociological issues concerning science, e.g., how to individuate communities of scientists, many of his other interests were not sociological. In fact in later writings he was quite hostile to the claims of the Strong Programme. This difference in his post-*Structure* writings is explored, along with his model of weighted values as an account of theory choice. This model has little in common with the model of theory choice advocated by Strong Programmers and much more in common with traditional philosophical concerns about theory choice.

Some commentators have declared that the idea of a theory of scientific method, long advocated by philosophers and scientists, is this century's stale debate and that it has already been replaced by a quite new outlook before the new century has even begun. Taking an even longer view in which science has dominated at least the second half of the second millennium, they might add that the idea of a scientific method was one of our millennium's long lasting but nevertheless illusive, because non-existent, ideals. Some sociologists of science herald this new view; and some in the community of philosophers have been thought to do so as well, for example Paul Feyerabend – another being Thomas Kuhn.

What I hope to do in this paper is at least cast Kuhn in a different light in order to show that he was deeply preoccupied by issues to do with methodology and even proposed a methodology of his own and a justification for it, though in the long run neither are fully defensible. I will also show that Kuhn distanced himself from the claims of sociologists of scientific knowledge, especially those of the Strong Programme. This is a significant issue not least for the field of science education. Both Kuhn and the sociology of science are commonly invoked together by constructivists and postmodernists in science education as a new 'referent' or 'paradigm' for the discipline. But this is to play down the serious differences between Kuhn and the sociologists, as will be seen.

David Bloor, and advocate of the Strong Programme adapts a phrase of Wittgenstein's when he says of the sociology of knowledge that its practitioners 'are the heirs to a subject that used to be called philosophy' (Bloor 1983, pp. 182–183). Though Kuhn is not mentioned specifically in this context, at least a new outlook is heralded in which the role of

F. Bevilacqua et al. (eds.), Science Education and Culture, 213–226.
© 2001 *Kluwer Academic Publishers. Printed in the Netherlands.*

philosophy, with is compartments of epistemology and methodology, is severely diminished. However many sociologists of science see Kuhn as having been one of the forerunners of the new point of view. Thus Brante, Fuller and Lynch tell us: 'The publication of Thomas Kuhn's *The Structure of Scientific Revolutions* in 1962 pointed the way toward the integrated study of history, philosophy and the sociology of science (including technology) known today as science and technology studies (STS)'. Kuhn set a good example they say, adding of his book:

> It alerted STS practitioners to the mystified ways in which philosophers talked about science, which made the production of knowledge seem qualitatively different from other social practices. In the wake of STS research, philosophical words such as *truth*, *rationality*, *objectivity*, and even *method* are increasingly placed in scare quotes when referring to science – not only by STS practitioners, but also by scientists themselves and the public at large. (Brante et al. ibid., p. ix)

Perhaps all is not well in the STS camp because the authors confess: 'the field [of STS] has yet to articulate aspirations that go beyond this deflation of philosophical pretensions'. Part of the problem appears to be with aspects of Kuhn's book; as path breaking as it might have been, there are shackles from which STS ought to free itself given its too close an adherence to some of Kuhn's less acceptable positions. Though there are several ways in which Kuhn used sociological notions in his book, I will discuss only one appeal he made to the sociology of science, and then one appeal that he did not make to the sociology of scientific knowledge, the latter setting Kuhn apart from advocates of the Strong Programme.

KUHN ON SCIENTIFIC COMMUNITIES AND 'PARADIGMS'

One of the central ideas of Kuhn's book that has excited much comment is that of a paradigm. In the 'Postscript' to the 1970 edition of his book, Kuhn tries to tackle the individuation of paradigms through an appeal to the notion of a scientific community. The idea of a community of scientists appears prominently both at the beginning and end of the 'Postscript'; for Kuhn it is a central notion to be articulated within the sociology of science, but not within the sociology of scientific *knowledge*. Kuhn recognises the circularity in the claim: 'A paradigm is what the members of a scientific community share, and conversely a scientific community consists of men who share a paradigm' (Kuhn 1970, p. 176). So he sets out to independently individuate scientific communities and then use that notion to individuate paradigms, rather than the other way around: 'Scientific communities can and should be isolated without prior recourse to paradigms; the latter can then be discovered by scrutinising the behaviour of a given community's members' (loc. cit.).

Many might think that the task of individuating such communities without appeal to the actual content of some science believed by members of the community is a hopeless task. Some of the criteria Kuhn mentions for individuating communities are: having 'undergone similar educations and professional initiations'; having 'absorbed the same technical literature and drawn many of the same lessons from it'; pursuing 'a set of shared goals' (ibid., p. 177). But it is hard to see how the last criterion, or even the second, can be employed without appeal to the content of science. If the goal were to obtain, say, an hypothesis which was more fruitful than its rivals in the sense that it has more successful predictions, then some appeal is made to some science, and its methods, in order to establish this – and these are important aspects of paradigms.

A similar problem arises when Kuhn tries to individuate not the world-wide community of scientists, but each sub-group such as radio-astronomers, optical telescope astronomers, and so on. Such groups may be individuated by 'their attendance at special conferences' or their informal networks of communication and 'linkages amongst citations' (ibid., pp. 177–178). But in a review of Kuhn's book, Musgrave (1980, p. 40) points out that citations are not a good criterion for membership of the same community because rival communities often cite one another. Finally Kuhn tells us: 'Communities . . . are the units that this book has presented as the producers and validators of scientific knowledge. Paradigms are something shared by the members of such groups' (ibid., p. 178). But consider some claim p which is said to be validated scientific knowledge by a community. If this is to be much more than the mere assent given to 'p' by each member, then this takes us towards the very content of some science, and some paradigm with its criterion of what is to count as 'validated scientific knowledge'; and this was something that the appeal to independently individuated communities was meant to underpin. Though the enterprise of individuating scientific communities is a useful task for various sociological purposes, what is doubtful is that the communities will be so finely individuated that they can then be used in turn to individuate paradigms as that which each finely individuated community shares. Rather elements of paradigms will enter into the individuation of the communities themselves; paradigm individuation then becomes a separate problem.

At the same time that Kuhn was attempting to individuate the notion of a community independently from that of a paradigm on sociological grounds, he also elaborated upon and changed the notion of a paradigm that he had employed in the first edition of his book. A paradigm was replaced by the notion of a disciplinary matrix which in turn comprised exemplars, symbolic generalisations (or laws), model and values (ibid., pp. 182–187). Kuhn recognised that he had not been specific enough in his account of a paradigm and in subsequent writings on the history of science did not make any significant use of the notion. In an interview published in Italian in 1991 Kuhn said: 'I'm not sure I would use the term paradigm in such a wide sense anymore' (Borradori, 1994, p. 166). But alert readers of the 1970 'Postscript' would have already noticed that Kuhn effectively abandoned the notion of a paradigm in favour of the more articulated notion of a disciplinary matrix. As will be seen, the values of a disciplinary matrix can be shared by different communities, and may well be widely endorsed across most scientific communities. This also underlines the way in which the enterprise of individuating scientific communities in order to individuate paradigms falters when it come to a disciplinary matrix's values which may be shared across communities.

Independently of its success or failure, the above shows one way in which Kuhn took an interest in a sociological investigation. It also introduces Kuhn's notion of values and the role they play in his account of scientific method, especially that of theory choice; this is an important theme which will be taken up in the third section of the paper. But first we need to set out some of the views held by advocates of the Strong Programme in order to see in what way Kuhn distances himself from their sociologically based accounts of theory choice. Even though sociology of *science* can play a role in the individuation of scientific communities, for Kuhn sociology of *scientific knowledge* plays very little role in theory choice and none in his account of the justification of his principles of theory choice.

THE STRONG PROGRAMME AND KUHN'S OBJECTIONS

What is the Strong Programme (SP) in the sociology of scientific knowledge? David Bloor expresses SP in the form of four tenets, of which only the first, the Causality Tenet (CT), need concern us: 'It would be causal, that is, concerned with the conditions which bring about belief or states of knowledge. Naturally there will be other types of causes apart from social ones which will cooperate in bringing about belief' (Bloor 1991, p. 7). Sociologists do not take the care that philosophers do over the distinction between knowledge and belief; so we will go along with Bloor and understand CT as pertaining to *beliefs* held on the part of some individual x. Since beliefs can be either true or false, then we can immediately incorporate Bloor's second 'Impartiality Tenet' into CT, viz., '[SP] would be impartial with respect to truth or falsity, rationality and irrationality, success or failure' (loc. cit). Thus CT is quite broad with respect to the beliefs within its scope. It is also quite broad with respect to the people it applies to; we will consider it only with respect to individual scientists.

What causes a belief p in the mind of some scientist x? It is x's social condition (call this 'S_x') in cooperation with other conditions of x which are not social (for convenience call these 'N_x'). Thus CT, in its full generality with respect to all beliefs and all scientists says:

CT: for all (scientific) beliefs p, and all scientists (and perhaps others) x, there are social conditions S_x, and there are other non-social conditions N_x, which cooperate together such that (S_x & N_x) cause x's belief that p.

The non-social causes N_x can be taken broadly to include items such as our biology, our cognitive structures which we have inherited through the processes of evolution, our inherited sensory apparatus, and our history of sensory inputs (but not our reports of them). While the non-social is constant for most of humanity, the history of our individual sensory input, as well as our individual (or our group's) social condition, does vary from person to person, or from group to group. Thus though the factor N_x must enter into the causal nexus producing our beliefs, it will not explain, apart from appeal to variable sensory input, the variation in our beliefs. What will do this will be the varying social factor S_x. The list of variable social factors is open-ended, but it includes at least: the language we learn and the way we thereby express our beliefs and report our experiences; the beliefs we acquire through acculturation and education; our social circumstance including the class into which we fall. It is the variation in social factors such as these which is alleged to be the main cause of variation in belief. We are to appeal to these factors in offering causally-explanatory accounts of why, say, some person x believes that p.

If the social factors are quite external to us (e.g., our class status), how do they manage to causally produce beliefs in us in a way which is not overtly behaviouristic? It is unclear how such external factors, such as class status, manage to penetrate our brains to cause beliefs in something as remote as science without at least our *awareness* of our class interests entering into the casual chain. This suggests that cognitive factors enter into the causal story and that it is cognitive attitudes to social factors, such as beliefs or interests, which are causally active and not the social factors themselves. If so, CT has been improperly expressed since what has been left out are cognitive attitudes to social factors. This omission is remedied in what might be called the 'interests' version of CT (ICT).

ICT: for all (scientific) beliefs p, and all scientists x, there are interests of x, I_x, such that I_x cause x's belief that p.

Here we can drop reference to the non-social factors since the causal chain is alleged to run independently of them at the cognitive level from interests to beliefs. The task of SP is to flesh out the above two schema by finding, for each person x and their particular belief p, either the social conditions S_x that cause their believing p, or some interest they have, I_x, that causes their believing p. The radical challenge that SP with its causality tenet presents to philosophical theories of method is this. Scientists do not believe their theories for the reasons philosophers suppose, e.g., that their theories might explain more, or be highly confirmed, or any of this ilk.

Rather, as CT and ICT show, the causes of their belief in their theories is social (omitting reference to the common non-social causes).

As an illustration of the above schema consider Paul Forman's often cited study of the emergence of the belief in acausality in physics in Weimar Germany in the decade after the end of World War I. Forman says of this new widely spread belief in physics of the time that he 'must insist on a causal analysis' (Forman 1971, p. 3) thus bringing his analysis into conformity with SP. So let 'p' stand for the set of beliefs in acausality and restrict the scope of x in CT to the dozen physicists of Weimar Germany that Forman investigates, and in particular, the physicist Richard von Mises. We can accept Forman's analysis of the cultural milieu of the time in which post World War I German society was hostile to science and to its notions of causality and embraced the neo-romantic, even 'existentialist', doctrines such as those to be found in Spengler's book *Decline of the West* in which a causal view of the world was condemned. However it is not sufficient merely to claim that the dozen physicists lived in a milieu of this sort; this could be the case whether they were aware of their cultural milieu or not. For the milieu to have an affect on the physicists' beliefs we need to add that they were *aware* of doctrines that were embraced in their cultural milieu. So CT needs to be reformulated as a thesis not about some S_x causing x's belief in acausality (how could it?), but the significantly different thesis: x's *awareness* of S_x causes x's belief in acausality.

Being socially aware people, most of the dozen physicists Forman investigates were aware of their cultural milieu and its hostility to notions of causality. The important questions to ask are now these. Was their awareness merely a mental accompaniment to their belief in acausality – this belief being caused in other ways due to the internal development of physics? Or was their awareness of their milieu's hostility the very cause of their belief in acausality? Advocates of SP have to show the latter. And if they can, the result would be rather shocking. Major beliefs in science are just 'social imagery', to quote the title of Bloor's book; they are not acquired on grounds to do with the sciences themselves and their accompanying principles of scientific method.

However the sociologists do not correctly employ the methodology of causal analysis to establish the following claims. Let 'A' be the physicists' awareness of their hostile cultural milieu, 'B' their belief in acausality, and P an internalist story of how, on the basis of issues in physics alone, acausality came to be believed. Then the following needs to be shown about what went on in the mind of each physicist: (1) A accompanies B; (2) P accompanies B; (3) P does not cause B; (4) A causes B. All can agree with (1) and (2); but the sociologists do not do show the crucial causal claims (3) or (4). Historical research can show that A, B, and P accompany one another; but it is another matter altogether to show that A, not P, is the cause of B.

Of the physicists Forman discusses, perhaps Richard von Mises comes nearest to being his best example as he was a convert to the *Weltschmerz* of Spengler. Did Von Mises' Spenglerism cause him to believe in acausality rather than matters internal to the physics with which he was acquainted? Forman does not show that it was Von Mises' awareness of his cultural milieu rather than issues to do with physics which caused this belief. In fact Forman's investigations of Von Mises' mental states get no better than this when he says of his research into Von Mises' writings on Quantum Mechanics at the time: 'Admittedly, Von Mises has invoked the quantum theory as the occasion for the repudiation of causality' (ibid., p. 81). Maybe; but this does not show that it was Von Mises' *awareness* of his cultural milieu, rather than his physics, that caused him to believe in acausality. To invoke other reasons or rationalisations based on Spenglerian considerations would be to attribute a massive amount of self-delusion to Von Mises, given his avowed reasons based in physics.

Critics are aware of the methodological shortcomings in the studies alleged to support SP, even those who take the view that scientific belief needs both a externalist (sociological) and internalist (using some methodological principles) explanation:

> when we come down to the content of physics, we must of necessity take into account internal as well as external considerations. ... Forman has succeeded in demonstrating that physicists and mathematicians were generally aware of the values of the milieu .... But when we come to the crucial claims, that there was widespread rejection of causality in physics, and that there were no internal reasons for the rejection of causality, then the weakness in his argument also becomes crucial. For there were strong internal reasons for the rejection of causality ... (Hendry 1980, p. 160)

Let us now turn to Kuhn's critical remarks about SP (which differ from those just given above). During the more than thirty years after he published *Structure* Kuhn worked at modifications of its views, projecting even another book which remained unpublished at the time of his death. The new book, he said, would return to the philosophical problems *Structure* bequeathed to him by the incomplete account he had given of them in his earlier work. The list of philosophical problems include: 'rationality, relativism and, most particularly realism and truth ... [and a notion of] incommensurability ... [which] is far from being the threat to rational evaluation of truth claims that it has frequently seemed'. (Kuhn 1991, p. 3).

Surprisingly for a book whose historical examples were largely culled from physics and chemistry, *The Structure of Scientific Revolutions* has been taken up by the social sciences and lauded by sociologists of scientific knowledge as one of the works which inaugurated the new sociological approach in STS. What was Kuhn's attitude to the burgeoning literature of sociological studies of science? Though he says that there is a lot to learn from such studies, he expresses his view of its underlying philosophical stance with uncharacteristic harshness. His projected book was to provide an account of incommensurability consistent with the notion of rational evaluation of truth as a corrective to the sociological stance:

'Rather, it's what is needed, within a developmental perspective, to restore some badly needed bite to the whole notion of cognitive evaluation. It is needed, that is, to defend notions like truth and knowledge from, for example, the excesses of post-modernist movements like the strong program' (ibid., pp. 3–4). Even though Kuhn would not have adopted a realist correspondence theory of truth, his unfinished project was going to conserve some of the traditional ideas of scientific method.

In a 1991 address Kuhn said: 'I am among those who have found the claims of the strong program absurd: an example of deconstruction gone mad' (Kuhn 1992, 9). He says in agreement with advocates of SP:

> Interest, politics, power and authority undoubtedly do play a significant role in scientific life and its development. But the form taken by studies of 'negotiation' has, as I've indicated, made it hard to see what else may play a role as well. Indeed, the most extreme form of the movement, called by its proponents 'the strong program', has been widely understood as claiming that power and interest are all there are. Nature itself, whatever that may be, has seemed to have no part in the development of beliefs about it. Talk of evidence, of the rationality of claims drawn from it, and of the truth or probability of those claims has been seen as simply the rhetoric behind which the victorious party cloaks its power. What passes for scientific knowledge becomes, then, simply the belief of the winners. (ibid., 8–9)

With these remarks in mind, David Bloor, in an obituary notice for Kuhn, thought it was still possible to co-opt him to the sociologists' cause. In relation to Kuhn's remarks about the role of nature Bloor says: 'How nature is to be described is not predetermined, or independent of the traditional resources that are bough to bear on it. That is why, despite some uncharacteristically ill-focused remarks of his own to the contrary, Kuhn is properly called a 'social constructivist' (Bloor 1997, p. 124). Setting aside the obscure matter as to whether Kuhn was a 'social constructivist', his model of consensus through individually weighted shared values hardly makes theory choice 'social constructed'. What Kuhn also requires is 'cognitive evaluation with rational bite'.

Kuhn clearly agrees with the sociologists' claim that interests, power and politics do play a significant role in *scientific life* and its development. This is almost a banal truth with which no one would disagree. The extent to which interests of power, money and politics support, say, modern scientifically based agribusinesses, or timber-felling companies, over those who wish to retain elements of biodiversity in the countryside is an empirical matter to decide – and one in which it is very clear who is the winner in a number of cases. But it is a very different matter to allege that power, politics, money, interests and authority influence not only *scientific life* (as Kuhn says) but also the *acceptance* of *scientific hypotheses* or *observations* in scientific investigations. For the latter Kuhn would apply the 'cognitive bite' of his model of weighted values – and this says little of power and politics amongst its list of values.

This highlights Kuhn's worry. If the power politics of negotiation is the only, or dominant, factor in determining what we are to accept as scientific facts, or to accept as hypotheses; or accept as what evidence supports what hypothesis, then Kuhn can see no role for nature as that which enters into our very experimental interventions, or that to which our theories and observations are a response, or that which our theories are about. Nature appears to be irrelevant while society is paramount. For Kuhn SP so marginalises the role of nature in our theory choices, or in our accounts of what science is about, that nature seems to drop out of the picture all together. In sum, despite some of the more radical claims of the first edition of *Structure* which subsequently underwent refinement, Kuhn clearly sees himself as a member of the long tradition which still has a use for the notions of rationality, truth and knowledge and does not jettison them.

KUHN'S VALUES AND THE METHODOLOGY OF THEORY CHOICE

Those who view Kuhn as holding either an irrationalist or anti-methodology stance, or endorsing a paradigm-relative account of method, can find passages in the Kuhn of 1962 that support these views. Using a political metaphor to describe scientific revolutions, Kuhn says of scientists working in different paradigms that 'because they acknowledge no supra-institutional framework for the adjudication of revolutionary difference, the parties to a revolutionary conflict must finally resort to the techniques of mass persuasion, often including force' (Kuhn 1970, p. 93). Continuing the metaphor, there is also a suggestion that the methods of evaluation in normal science do not carry over to the evaluation of rival paradigms:

> Like the choice between competing political institutions, that between competing paradigms proves to be a choice between incompatible modes of community life. Because it has that character, the choice is not and cannot be determined merely by the evaluative procedures characteristic of normal science, for these depend in part upon a particular paradigm, and that paradigm is at issue. When paradigms enter, as they must, into a debate about paradigm choice, their role is necessarily circular. Each group uses its own paradigm to argue in that paradigm's defence. (Kuhn 1970, p. 94)

Later he speaks of paradigms 'as the source of the methods . . . for a scientific community' (ibid., p. 103). These and other passages tell us that methodological principles might hold within a paradigm but that there are no paradigm transcendent principles available. And they show that Lakatos is justified when he says that the theory choice involved in paradigm change is a matter of 'mob psychology' (Lakatos & Musgrave 1970, p. 178).

But by the time he came to write the 'Postscript' for the 1970 edition of his book Kuhn, as already noted, effectively abandoned talk of paradigms in favour of talk of exemplars and disciplinary matrices. Values are one of the elements of a disciplinary matrix; and contrary to the impression given above, they are 'widely shared among different communities' (ibid., p. 184). That is, scientists in different communities, and so working in different 'paradigms', value theories because of their following features: they yield predictions (which should be accurate and quantitative rather than qualitative); they permit puzzle-formation and solution; they are simple; they are self-consistent; they are plausible (i.e., are compatible with other theories currently deployed); they are socially useful.

In a 1977 paper 'Objectivity, Value Judgement, and Theory Choice' (Kuhn 1977, chapter 13), Kuhn re-endorses these values and adds to them: scope (theories ought to apply to new areas beyond those they were designed to explain), and fruitfulness (theories introduce scientists to hitherto unknown facts). Kuhn initially thinks of these as rules of method. But owing to the imprecision which can attach to their expression as rules, and the fact that they are open to rival interpretations and can be ambiguous in application or can be fulfilled in different ways, Kuhn prefers to think of them as values to which we could given our general assent. Thus Kuhn adopts a methodology which avoids talk of the rules which we ought to follow as means to realise values, and instead focuses on the values we do, or ought to, adopt in our choice of theories.

Kuhn's list of values does not mention several other important values that have been endorsed by other methodologists. Thus inductivists and Bayesians put store on high degree of support of hypotheses by evidence. Constructive empiricists put high value on theories which are empirically adequate. Realists wish to go further and value not only this but also truth, or increased verisimilitude, about non-observational claims. Yet others value high explanatory power. And so on. Kuhn rejects the idea that our theories do approximate to the truth about what is "really there" (Kuhn 1970, p. 206); so even though we might not view him as endorsing the realists' value of truth, he could adopt the value of empirical adequacy (i.e., truth at the observational level) espoused by constructive empiricists such as Van Fraassen. Also there is a strong anti-inductivist strain in Kuhn which would take him away from notions of high degree of support, though this is linked to the Kuhnian values of scope and fruitfulness. Finally some methodologists would downplay some of the values Kuhn endorses, such as external consistency or social utility.

The position of Kuhn on methodology after the first edition of *Structure* yields the following picture of a model of weighted values. (1) There is a set of values which can vary over time, and can vary from methodologist to methodologist. A sub-set of these values could comprise a cluster of central values which hold for most sciences and throughout most of their history. For Kuhn's own model, the values he mentions fall into this subset with few, if any, falling outside (but within the larger set). (2) These values are used to guide and inform theory choice across the sciences and within the history of any one science, including its alleged 'paradigm' changes. That is, the cluster of values are science and paradigm transcendent. (3) The model may be either descriptive or normative. Kuhn does not make it clear whether his model is to be understood as a description of how scientists do in fact make their choices, or whether it is to understood normatively in that it tells us how we ought to make choices. If the latter then there is a need for a justification of the norms it embodies. (4) Kuhn says that the values may be imprecise and be applied by different scientists in different ways. While this is not the case for the value inconsistency (there are fairly precise criteria for internal and external inconsistency for any theory), or for any given degree of accuracy of predictions, some values do exhibit imprecision. Thus accuracy could differ in the required degree. Simplicity might be taken in different ways (simplicity in equations versus simplicity in ad hoc assumptions) so that different aspects of a theory might be deemed simple; or the notion of simplicity itself might be taken in different ways or in different degrees. Such imprecision in the interpretation of values can, however, be readily overcome by precisification so that there need not be the wide divergence over the interpretation of values that Kuhn alleges. (5) Different scientists do, as a matter of fact, give different weightings to each of the values.

Following from (4) and (5) there are two aspects to theory choice – an objective aspect in shared values and a subjective aspect in idiosyncratic weightings of values (and interpretation where this arises). In the light of this, Kuhn claims that there is no general 'algorithm' for theory choice – though there is hardly any methodologist who has required that methodological principles should be algorithmic. This allows that different scientists can reach different conclusions about what theory they should choose. First, they might not share the same values; but where they do share the same values they might interpret them differently or give them different weightings. Shared values (with the same interpretation) and shared weightings of these values will be sufficient for sameness of judgement within a community of scientists. However this might not be necessary; it might be possible for scientists to make the same theory choices yet to have adopted different values and/or have given them different weightings. Thus there is the possibility that consensus might be a serendipitous outcome despite lack of shared values and different weightings. However it is more likely that, where values and weightings are not shared, different theory choices will be made, and there is no consensus.

Whether scientists do or do not make theory choices according to Kuhn's model is a factual question to answer. But what does the model say about what we ought to do, and what is its normative/rational basis? In particular why, if T_1 exemplifies some Kuhnian value(s) while T_2 does not, should we adopt T_1 rather than T_2? Kuhn's answer to the last metamethodological question is often disappointingly social and/or 'intuitionistic' in character. In his 1977 paper Kuhn refers us to his earlier book saying: 'In the absence of criteria able to dictate the choice of each individual, I argued, we do well to trust the collective judgements of scientists trained in this way. "What better criterion could there be", I asked rhetorically, "than the decision of the scientific group"'; (Kuhn 1977, pp. 320–321). As to why we ought to follow the model, Kuhn makes a convenient is-ought leap when he says in reply to a query from

Feyerabend: 'scientists behave in the following ways; those modes of behaviour have (here theory enters) the following essential functions; in the absence of an alternative mode that would serve similar functions, scientists should behave essentially as they do if their concern is to improve scientific knowledge' (Lakatos & Musgrave (eds) 1970, p. 237). The argument is not entirely clear, but it appears to be inductive: in the past certain modes of behaviour (e.g., adopting Kuhn's model of theory choice) have improved scientific knowledge; so in the future one ought to adopt the same modes of behaviour if one wants to improve scientific knowledge. What is surprising about this line of argument is its dependence on meta-induction from the past success of the use of Kuhn's model to its future success. However Kuhn later offers other reasons for adopting his model.

Much later comments from Kuhn (in a paper entitled 'Rationality and Theory Choice') on the status of his methodology arise in a 1983 symposium on 'The Philosophy of Carl G. Hempel' with Salmon and Hempel. In subsequent reflection on that symposium, Salmon argues that it is possible to reconstrue the features of Kuhn's model of weighted values in terms of subjective Bayesianism (Salmon 1990; see also Earman chapter 8 for a further attempt by a Bayesian to incorporate Kuhn's model of theory choice into a Bayesian decision framework). Bayes' Theorem is able to account for a large number of our central methodological principles, including accuracy, fruitfulness, scope, and so on (but not social utility unless set in a decision-theoretic context). If Salmon's project in which 'Tom Kuhn meets Tom Bayes' is able to account for the theory choices of Kuhn's model, then the independent status of the model is undercut as it is incorporated into a more wide ranging theory of method. Whatever may be the case here, let us focus on the new features of Kuhn's 1983 account of his methodology.

In his symposium paper Kuhn addresses a point that Hempel had made in an earlier paper about his position, viz., that Kuhnian values are goals at which science aims, and not means to some goal such as puzzle-solving (Hempel 1983). Both Kuhn and Hempel take puzzle-solving, accuracy, simplicity, etc, to be values (ends) rather than rules (means), and throughout this paper this position has been adopted. Given that theories are judged by the values they exemplify, Kuhn takes up a further point that Hempel makes, viz., that rationality in science is achieved through adopting those theories that satisfy these values *better*. Hempel thinks that this criterion of rational justification is near-trivial. However Kuhn turns Hempel's near-triviality into a virtue by proposing that the criterion is analytic, thereby adopting as his meta-methodology a theory of analyticity concerning the term 'science'. In developing his views in this paper Kuhn tells us that he is 'venturing into what is for me new territory' (Kuhn 1983, p. 565). So we can take it that the meta-methodological justification developed here is not one that Kuhn had had in mind before.

Kuhn's account of analyticity is based on what he calls 'local holism'. This is the view that the terms of any science can not be learned singly but must be learned in a cluster; the terms have associated with them generalisations that must be mastered in the learning process, and the cluster of terms form contrasts with one another that can only be grasped as a whole. If 'learning' is understood as 'understanding the meaning', then analyticity becomes an important part of the doctrine of 'local holism' for the central terms of each sufficiently broad scientific theory. In Kuhn's view the doctrine applies not only to specific theories such as Newtonian Mechanics with its terms 'mass' and 'force' which must be learned together holistically. It also applies to quite broad notions signified by the terms 'art', 'medicine', 'law', 'philosophy', 'theology', and so on; the central terms associated with these notions must also be learned holistically. Importantly 'science' is another such broad notion to be learned holistically since 'science' is in part to be understood in contrast to these terms.

Kuhn recognises that not every science we adopt should possess every value since the values are not necessary and sufficient conditions for theory choice; rather they form a cluster associated with the local holism of the term 'science'. But what he does insist is that claims such as 'the science X is *less* accurate than the non-science Y' is a violation of local holism in that 'statements of that sort place the person who makes them outside of his or her language community' (Kuhn 1983, p. 569). For Kuhn, Y's being more accurate is just one of the things that the local holism of the word 'science' makes Y scientific; Y cannot be non-scientific. For Kuhn, Hempel's near-triviality is not breached because a convention has been violated; nor is a tautology negated. Rather 'what is being set aside is the empirically derived taxonomy of disciplines' (loc. cit) that are associated with terms like 'science'. Like many claims based on an appeal to analyticity, meaning or taxonomic principles, one might feel that the later Kuhn has indulged in theft over honest toil. However in linking his model of weighted values to the alleged local holism of the term 'science', Kuhn comes as close as any to adopting the meta-methodological stance that his theory of method has an analytic justification for its rationality. In this respect Kuhn's position has close affinities with that of Strawson (1952, Chapter 9, part II) who tried to justify the rationality of induction in much the same way.

CONCLUSION

Since Kuhn's later position on methodology is a far cry from the Strong Programme of the sociologists, there are no good grounds for co-opting him into the sociologists camp as far as theory choice is concerned. His philosophical account of values and the status of his model all fall within the traditional concerns of rationality in science, the very rationality that sociologists of scientific knowledge set out to undermine.

ACKNOWLEDGEMENT

The comments of Howard Sankey and Paul Hoyningen-Huene on an earlier draft of this paper were appreciated.

REFERENCES

Borradori, G.: 1994, *The American Philosopher: Conversations with Quine, Davidson, Putnam, Nozick, Danto, Rorty, Cavell, MacIntyre and Kuhn*, The University of Chicago Press, Chicago.

Bloor, D.: 1983, *Wittgenstein: A Social Theory of Knowledge*, Macmillan, London.

Bloor, D.: 1991, *Knowledge and Social Imagery*, The University of Chicago Press, Chicago, second edition; first edition 1976.

Bloor, D.: 1997, 'The Conservative Constructivist', *History of the Human Sciences* **10**, 123–125.

Brante, T., Fuller, S. & Lynch, W. (eds): 1993, *Controversial Science: From Content to Contention*, State University of New York Press, Albany, NY.

Earman, J.: 1992, *Bayes or Bust?: A Critical Examination of Bayesian Confirmation Theory*, The MIT Press, Cambridge, MA.

Forman, P.: 1971, 'Weimar Culture, Causality, and Quantum Theory 1918–27: Adaptation by German Physicists and Mathematicians to a Hostile Intellectual Environment', in R. McCormmach (ed), *Historical Studies in the Physical Sciences* **3**, University of Pennsylvania Press, Philadelphia, pp. 1–116.

Gutting, G. (ed): 1980, *Paradigms and Revolutions*, University of Notre Dame Press, Notre Dame, IN.

Hempel, C.: 1983, 'Valuation and Objectivity in Science', in R. S. Cohen and L. Laudan (eds), *Physics, Philosophy and Psychoanalysis*, Reidel, Dordrecht, pp. 73–100.

Hendry, J.: 1980, 'Weimar Culture and Quantum Causality', *History of Science* **18**, 155–180.

Kuhn, T.: 1970, *The Structure of Scientific Revolutions*, The Chicago University Press, Chicago, second edition; first edition 1962.

Kuhn, T.: 1977, *The Essential Tension*, The Chicago University Press, Chicago.

Kuhn, T.: 1983, 'Rationality and Theory Choice', *The Journal of Philosophy* **80**, 563–570.

Kuhn, T.: 1991, 'The Road Since Structure', in A. Fine, M. Forbes and L. Wessels (eds), *PSA 1990, Volume Two*, Philosophy of Science Association, East Lansing, MI, pp. 3–13.

Kuhn, T.: 1992, 'The Trouble With the Historical Philosophy of Science', Robert and Maurine Rothschild Distinguished Lecture, Department of the History of Science, Harvard University, Cambridge, MA, 3–20.

Lakatos, I. and Musgrave, A. (eds): 1970, *Criticism and the Growth of Knowledge*, Cambridge University Press, Cambridge.

Musgrave, A.: 1980, 'Kuhn's Second Thoughts', reprinted in G. Gutting (ed).

Salmon, W.: 1990, 'Rationality and Objectivity in Science, *or* Tom Bayes meets Tom Kuhn', in C. Wade Savage (ed), *Scientific Theories: Minnesota Studies in the Philosophy of Science volume XIV*, University of Minnesota Press, Minneapolis.

Strawson, P.: 1952, *Introduction to Logical Theory*, Methuen, London.

PART THREE
History and Philosophy in Physics Education

The papers in Part Three elaborate the role of history and philosophy in physics education. The history and philosophy of physics can address two levels of problems in physics education: at the first level, there are problems of how best to promote subject matter competence, knowledge and understanding; at the second level, there are problems about the purposes, aims and goals of physics instruction.

It is recognised that physics education shares all the difficulties of science education, and then some more. There has been a flight from physics by both students and teachers. In only a few thousand of the 24,000 or so high schools in the United States is physics taught by a physics graduate. The situation in Australia is not much better, with approximately one-tenth of graduates going into science teaching having a physics degree. These teacher-supply problems are societal and structural, and only peripherally amenable to educational solutions.

But there are important aspects of the physics crisis that are amenable to educational effort. One aspect is the unfortunate trend to reduce school physics to the 'physics of gadgets'. The physics of gadgets - sometimes called 'applied physics' - is one reaction to the daunting formalism and higher mathematics so pervasive in curricula and texts. There are some progressive aspects to this trend.

But competence in the physics of gadgets is surely not the only goal to which a physics programme can aspire. David Goodstein, in his Oersted Medal Address to the American Association of Physics Teachers, correctly said of the physics profession that 'What we have is nothing less than the wisdom of the ages. It's that vast body of knowledge, the central triumph of human intelligence, our victory over mystery and ignorance' (Goodstein 1999, p. 186). Without the history and philosophy of physics, this 'wisdom of the ages' can hardly be appreciated. Copernicus on the heliocentric solar system, Galileo on falling bodies, Newton on the unity of celestial and terrestrial physics, Volta on electricity, Einstein on relativity, Planck on quantum effects, and so on. This history is a rich storehouse, the walking through of which should leave students with some sense of the wisdom of the ages and the personal and social requirements whereby it was acquired.

Igal Galili and Amon Hazan here present research on the effect of a history-based course in optics on students' views about science. Encouragingly, students doing the history-based course performed equally as well as students who studied a more standard, professional, course. Their subject matter knowledge and comprehension was not diminished in virtue of their historical curriculum. But their comprehension of the nature of science, and its place in the big picture of human intellectual and practical endeavour, was considerably advanced compared to the professional group.

Nahum Kipnis outlines one of the great and engaging controversies in the history of science, that between the two Italian professors Alessandro Volta and Luigi Galvani concerning the nature of 'animal electricity'. This late eighteenth century debate is an exemplary model of rival theory development and appraisal. Kipnis shows how its main features can be reproduced in classrooms, thus shedding light on the understanding of the basic scientific concepts, and importantly on scientific methodology and the place of experimental results in the resolution of theoretical controversy.

Roberto de Andrade Martins and Cibelle Celestino Silva address one of the central issues concerning the use of history in science teaching: How much can the history of science be deformed to suit pedagogical purposes? This question was raised by Martin Klein, a research physicist turned historian in 1970 at an MIT conference sponsored by the International Commission on Physics Education. Klein's argument was basically: teachers of science select and use historical materials to further contemporary scientific or pedagogical purposes, such selection is contrary to the canons of good history, and therefore 'in trying to teach physics by means of its history, or at least with the help of its history, we run a real risk of doing an injustice to the physics or to its history - or to both' (Klein 1972, p. 12).

Michael Matthews uses the prehistory of the metre length standard as an example of how history in the classroom can shed light on the 'big picture' of science, and enable students to appreciate some basic methodological issues in science, as well as enabling them to see how political considerations can bear upon its conduct.

Olivia Levrini discusses a topic that is tailor made for historical and philosophical elaboration in the classroom, namely Einstein's Theory of General Relativity. Among the issues she delineates is the long-standing debate about the reality of space. Students of relativity theory can, with benefit, relive this debate, and other basic philosophical disputes with which the subject blossoms.

Fanny Seroglou and Panagiotis Koumaras provide perhaps the most comprehensive extant review of research on the theoretical, curricular and pedagogical aspects of using history of physics in physics education. Their conclusions are encouraging for those advocating such usage.

Goodstein, D.L.: 1999, 'Now Boarding: The Flight from Physics', *American Journal of Physics* 67(3), 183-186.

Klein, M.J.: 1972, 'Use and Abuse of Historical Teaching in Physics'. In S.G. Brush & A.L. King (eds.), *History in the Teaching of Physics*, University Press of New England, Hanover

The Effect of a History-Based Course in Optics on Students' Views about Science

IGAL GALILI and AMNON HAZAN
Hebrew University of Jerusalem, The Science Teaching Center, Givat Ram, Levin Building, Jerusalem 91904, Israel

Abstract. In light of the convincing claims extolling the multifaceted merit of the "genetic" (historical) approach in designing learning materials (Matthews 1994), we developed an experimental course in optics. We tested the new materials and determined their effectiveness in a year long course given in several 10th grade high school classes. A special feature, which soundly contrasted our course from a typical one, was its essential incorporation of historical contents: the ideas, views and conceptions which constituted the early understanding of light and vision. In this report, we present that part of the assessment which concerns the course's impact on the students' views about science and some related technological and cultural issues. In our analysis, we used a special hierarchical organization to represent pertinent data. Strong differences were found between the views elicited in the experimental group and parallel data regarding students in the control group. In our view, this demonstrated the advantage of utilizing historical materials in an aspect which is additional to our first intention – to improve students' disciplinary knowledge. Such materials naturally address issues of much broader general interest, appropriate for physics education as opposed to physics training. Touching on a variety of features of science the materials positively effect students' views about science.

1. Background of the Study

Besides "installing" the fundamentals of disciplinary knowledge, school introductory physics courses should aid in the development of knowledge about science: a capacity for understanding of what is science, how it works, its features, methods of activity and interaction with its cultural environment. The importance of such knowledge, especially in a society rapidly being saturated with science and technology, was widely discussed during the last century (e.g., Dewey 1916; Jaffe 1938; Conant 1951; Schwab 1964; Bybee et al. 1991; Matthews 1994; Driver et al. 1996) and even declared as policy (Nuffield Advanced Science 1988; AAAS 1990, 1993; NAS 1996). We likewise expect this knowledge to be in the possession of educated non-scientists.

It has been shown that besides a rich variety of conceptions of scientific content (e.g., Wandesee et al. 1994), students hold a variety of views about science (e.g. Aguirre et al. 1990). Naive objectivism fused with naive pragmatism and naive ideas of accurate unambiguous scientific knowledge are often pronounced

F. Bevilacqua et al. (eds.), Science Education and Culture, 229–254.
© 2001 *Kluwer Academic Publishers. Printed in the Netherlands.*

in these views, and in fact, it is seemingly encouraged by the traditional instruction, which seeks simplicity (Rowell and Cawthron 1982; Abell and Smith 1994). The simplified image of science and scientific knowledge, often framed within a positivistic philosophy, displayed in a great many of the available curriculum materials (DeBoer 1991), is often far from being adequate or realistic. Such an image poorly facilitates the intellectual growth in the individual perception of science, and may negatively affect students' understanding of science in general (Loving 1991; Meichtry 1993; Thomsen 1998).

Despite existing pluralism, many share a dissatisfaction with students' and teachers' views about science (e.g., Meichtry 1993; Abell and Smith 1994; Matthews 1997; Lederman et al. 1998). Misinterpretations of scientific method by learners, and awareness of cultural factors in science have been discussed with regard to science education (Lederman and O'Malley 1990; Solomon et al. 1992; Griffiths and Barman 1995; Aikenhead 1997; McComas et al. 1998). Thus, Lederman et al. (1998) found no empirical support for the belief that teachers' competence in disciplinary knowledge and "hands on science" teaching strategy bring students to an adequate understanding of the nature of science. At the same time, Halloun and Hestenes (1998) found a positive correlation between the profiles in students' views about science and their achievements.

Utilization of elements of the history and philosophy of science (HPS) in introductory science courses has been argued for quite a long time (e.g., Conant 1951; Schwab 1963; Klopfer 1969; Brush 1989; Niedderer 1992; Duschl 1994; Matthews 1994; Monk and Osborn 1997). A broad cultural approach to science which accompanies HPS based introduction to teaching science, replaces the traditional focus on the correct-for-now scientific contents and problem solving training. As such, it reflects a cardinal change in the philosophy of education. History-based instruction uncovers the non-linear process by which current scientific knowledge was attained. It presents a number of views, ideas, and theories regarding the same subject, along with the complex interaction in which they may replace one another as well as co-exist in the course of history. This process of collective knowledge construction in science is, to a certain extent, similar to the process of individual construction of knowledge (widely adopted interpretation of learning – Glasersfeld 1995). Therefore, the exposure to science history may help students in their making sense of scientific claims, and the reconstruction of scientific ideas (Bettencourt 1993; Galili and Hazan 2000a).

Moreover, historically based teaching cannot ignore the social settings of scientific activity, inherent in history. The various human, philosophical, cultural and social aspects of scientific enterprise, which make the picture of science more colorful, realistic and educationally important in a broad sense: "History is necessary to understand the nature of science, ... [it] counteracts the scientism and dogmatism that are commonly found in science texts and classes. ... examining the life and times of individual scientists humanizes the subject matter of science making it less abstract and more engaging for students" (Matthews 1994, p. 50).

Another merit of intergrating historical materials emerges when one realizes that school teaching faces an extremely heterogeneous population, with a great span of interests, skills and abilities, ambitions and motivations. Many of them (if not the majority) will never pursue the learning of science, and a school course in physics may be their first and last physics course.[1] These circumstances imply the need to not overlook such cultural values as "integrity, diligence, fairness, curiosity, openness to new ideas, skepticism, and imagination" which obtain a very special meaning with regard to science (AAAS 1990). This, even at the expense of attention paid to the procedural ("instrumental") knowledge of problem solving which for the majority of students remains useless and is quickly forgotten.

Klopfer and Cooly (1963), Lochhead and Dufresne (1989), Johnson and Stewart (1991), Aikenhead (1992), Solomon et al. (1992) suggested that HPS-based learning materials could improve views about science in a manner meaningful and interesting to students. However, widely known attempts to teach HPS based materials, and through them to reach this goal seemingly have been abandoned, i.e., the Harvard Project Physics course of the seventies. One of the ways to interpret such an outcome was to question the competency of the teachers, and to study their views on, and knowledge of, the subject (Rowell and Cawthron 1982; Brickhouse 1989; Aguirre et al. 1990; Koulaidis and Ogborn 1989, 1995; Mellado 1997; Tobin and McRobbie 1997). These studies revealed the necessity for specific pre-service training of teachers.

The implementation of HPS based learning emphasizes the problem of a shortage of appropriate teaching materials, since historical primary sources require special knowledge to be used. For instance, the texts on optics by Descartes (1637/1965), Newton (1704/1952) and even Mach (1913/1926), as they are, do not present materials appropriate for direct use by students or teachers. Thus, construction of special learning materials is required for those who want to teach and learn physics by means of history. We invested our efforts in this direction, and prepared a special textbook on optics for 10th grade students. The text presented the disciplinary knowledge in its historical evolution. The present study considers the impact of the innovative materials on students' views about science.

2. Curriculum and Sample

Optics was chosen for our educational experiment, after discovering that its extremely rich history of 2,500 years is representative for scientific progress in general (e.g., Mach 1913/1926; Cohen and Drabkin 1948; Lipson 1968; Ronchi 1970; Lindberg 1976). This history can also elucidate many aspects of the nature of science and its features, through an exposure to the rise and fall of scientific theories and conceptions, along with their adoptions, refinements and replacements. A specially prepared textbook served as the main learning resource for the course. In its preparation, we used a variety of resources including primary ones (Descartes 1637/1965; Newton 1704/1952; Huygens 1690; Priestley 1772), secondary, as quoted above, and educational texts (e.g., Finegold and Olson 1972; Kipnis 1993).

In designing the new course we were guided by our preliminary study, which explored the content and structure of students' knowledge in optics (Galili and Hazan 2000b). No explicit effort was made in the textbook to present aspects of scientific literacy, epistemology, or cultural values in science. This agenda was inherent in the context of the presented scientific views, theories and activities of Aristotle, Alhazen, Kepler and many others, which we selected as appropriate, and hopefully appealing, in light of the documented views of students.

Although our course preserved the standard menu of topics, it greatly differed in several aspects from a regular curriculum. The most pronounced difference was the exposure of the learner to the historical growth of the understanding of vision, interwoven within the nature and behavior of light (Ronchi 1970; Lindberg 1976). Besides the historical correctness of this dichotomy, both these topics are of equal subjective importance for learners. The two trends (light and vision) kept the relationship within the triad – object, light and eye (the main participants in the vision process) in permanent focus throughout the instruction.

If in the past, history of science was often used to inform the learner about the history of the discipline, in our approach in choosing historical materials, the schemes of students' alternative knowledge with respect to vision, nature of light, optical imaging and shadow guided us. Many students could recognize their own thoughts in the history of ideas and theories raised and refuted, in the legendary attempts to account for vision, light and related phenomena. Table Ia specifies examples of conceptual parallelism in optics knowledge used in the design of the experimental course. Among the materials incorporated into the curriculum were ancient Greek, medieval Arabic and early modern theories of vision, ideas regarding light nature and expansion, light rays, shadows, reflection and refraction of light, mirror, lens and pin- hole images, and the speed of light.

The subject matter contents of the course were fortified with social and technological background. This provided a potential for teachers to add to the specific issues with opinions about science in general, when either they themselves, or students, wished to address such an issue. To illustrate this potential, we listed some possible associations between the particular subject matter contents of the curriculum and features of science provoked by them for a discussion in a classroom (Table 1b). It is important to stress that the issues related to the nature of science were not isolated topics within the curriculum (which would make addressing them obligatory), but were naturally interwoven in the context of the subject matter presentation, as they were not rejected by the teachers as non relevant when spontaneously arising in the class. Furthermore, since the history of science cannot be reduced solely to epistemology, the features of science as they appeared in the course, expand beyond the epistemology of science to a variety of features of scientific enterprise.

Table Ia. Examples of conceptual parallelism in optics knowledge used in the experimental course

Historical conception (practiced in past science)	Student's conceptions (practiced when learning present physics courses)
• Pythagorean conception of vision	• "Active" vision*
• Euclidean visual and light rays	• Rays of sight, rays of light, (rays reification)
• Atomists' conception of "Eidola"	• Image holistic scheme
• Biblical – medieval dichotomy of light as an entity and perception (Lumen-Lux)	• Static light located in/around light sources, halos, bright sky, illuminated surfaces (light reification as static entity)
• Al-Hazen conception of vision by means of light rays	• Image projection scheme

*The exact meaning of students' conceptions such as "Active vision", "Image holistic scheme", and "Image projection scheme", are provided in Galili and Hazan (2000a).

Importantly, the teachers who volunteered to deliver the experimental course, although guided and encouraged throughout the year-long experiment, lacked experience or formal training in HPS. Likewise, as we were only interested in the influence of the innovative learning materials, we did not monitor the instructional pedagogy, and no special teaching strategy was suggested to the teachers regarding how to deliver the experimental course. We served only as supporters for the teachers to clarify, on request, some unfamiliar contents. Thus, the difference between the classroom setting in the experimental and control groups was only designed with regard to the contents of the learning materials, which, despite their novelty, kept with the indisputable requirement - to cover the same subject matter topics of the optics curriculum.

The topics of the optics curricular covered the disciplinary contents one can find in any regular physics course of high-school (college) level: the nature of the light, light expansion, reflection and refraction of light, lenses and mirrors, optical instruments etc. The experimental course covered the same topics but in a different historical ("genetic") perspective, within which, the commonly presented knowledge about light and vision was fortified with historical models and alternative theories that preceded the currently adopted scientific knowledge of the subject, and are considered incorrect in the perspective of our modern understanding of light and vision.

Our student sample comprised two groups. The experimental group included four 10th grade high-school classes ($N = 141$) who were the subject of a specially designed history-based course. The control group, consisted of three 10th grade classes ($N = 93$). The instruction in the classes of both groups was four weekly hours, and both groups were tested simultaneously at the end of the academic year.

Table Ib. Associations of subject matter contents and features of science which could arise in class discussion in the course of an HPS-based instruction in optics

Subject matter issues	Associated features of science
Ancient theories of sight and vision (Atomists' holistic eidola, Pythagorean phenomenological internal fire. Euclid's mathematical ray, Aristotelian role of medium, Galen's physiological model)	• driving forces to understand nature • conflict between different ideas in science • features of scientific explanation • co-existence of rival scientific theories • temporal nature of scientific theories • limited validity of scientific theories
The history of light rays (the birth of the concept in Euclid's optics, advantages and disadvantages of light rays in the explanation of vision and visual perspective, transition to flux)	• the role of modeling in science • limitation of models • abstract concepts in scientific theories • exchange of models (ray/flux)
The medieval optics of Alhazen (isotropic emission/reflection of light from each point of the source/reflector)	• falsity and truth in the development of science • synthesis of scientific knowledge • merging different areas of research • empirical versus theoretical scientific exploration
Rectilinear propagation of light (exploration of reflection and refraction, empirical account for reflection and refraction, theoretical account for refraction)	• scientific exploration, regularity, elicitation of a rule • empirical laws precede theoretical ones • the problem of superiority of a theoretical law over empirical data • the role of a theoretical account
The history of optical "technology" (lenses, eyeglasses, and telescope, attempts to measure the speed of light)	• inter-relationship between science and technology • the role of technology in scientific progress • technology and crucial experiments in science
Competitive scientific theories (corpuscular versus undulatory theories of light)	• the role of authority in science • crucial experiments • co-existence of rival theories

The classes participated in the study were taken from three types of schools: urban, rural and boarding, which are representative of the secular Israeli educational system. They were randomly chosen, and each school contributed to both the experimental and control groups. The students were of equivalent potentials in learning, as was inferred from school records and accumulated preliminary opinions of the schools' academic supervisors and teachers.

3. Instruments of Assessment

The data regarding the students' views about science was collected in tests administered simultaneously to both groups in their natural classroom environment, during 45-minute sessions.

The importance of both a reliable and valid assessment of students' knowledge is extreme, when one wishes to measure the effect of an innovative instruction. A special difficulty within our study was its subject, views about science, which cannot be evaluated in simple terms and are highly dependent on the interpretation of texts which are often of ambiguous language. The major source of our data was a written questionnaire, the same given for both the experimental and control groups, comprising 18 items, ten open-ended and eight multiple choice questions. They were adopted from several previously performed studies where they were validated for probing the specific subject of our study (Rowell and Cawthron 1982; Yoshida 1989; Koulaidis and Ogborn 1989; Milgrom 1989; Aguirre et al. 1990; Lederman and O'Malley 1990; Tamir 1994; Aikenhead 1997; Tobin and McRobbie 1997). The choice of the distractors in the mutiple choice questions reflected the experience accumulated in this field of research with regard to students' pertinent knowledge. Inversely, the answers to open ended questions often may be statements of multiple ideas which relate to more than one topic thus complicating the identification of distinct views. Yet, this same feature may be seen as an advantage, providing complementary data on the items of inquiry elicited from the interwoven contexts. It thus helps in building a more realistic, and reliable picture, constructed from a variety of its contextual manifestations. Students were encouraged to explain their answers as fully as they could. The combination of both formats was used to complement each other making the data more reliable.

To fortify our collected information with illustrative examples providing further details regarding affective aspects of the views of the experimental group, we also interviewed a few students and teachers of the experimental group about their perception of the new course, after its completion. However, the derivation of the features, categories and their distribution in the resultant knowledge in which the comparison between the groups was performed, was based *solely* on the results of the final (end of the year) written test, identically applied to both groups.

The following six general conceptual topics were addressed in the questionnaire. They represent some important dimensions of what is commonly considered to be the framework of one's perception of science, as summarized by McComas et al. (1998) to reflect "a consensus view of the nature of science objectives":

1. relationship of science and technology;
2. attitude to the science of the past;
3. influence of external factors on science and its products;
4. conception of scientific theories;
5. reliability of scientific knowledge;
6. critical perception of science in the classroom.

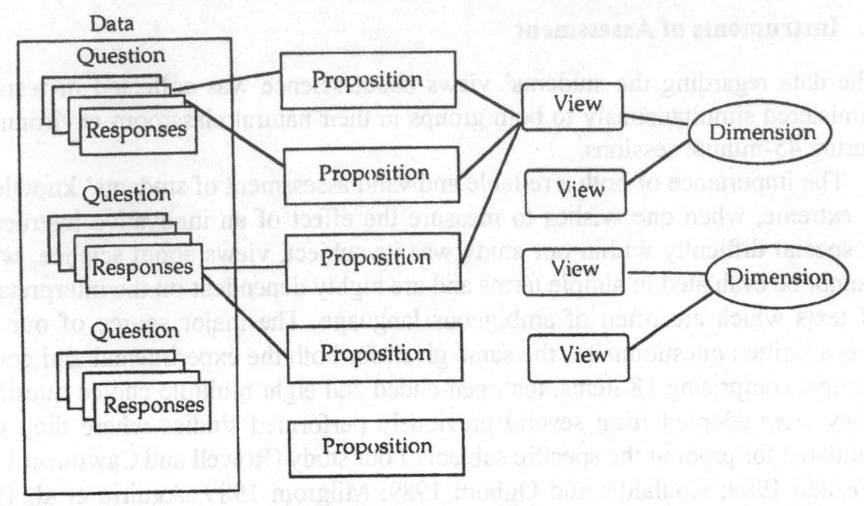

Figure 1. Process of categorizing data towards its propositions-view organization. Only representative connections are shown.

3.1. QUALITATIVE ANALYSIS

To elicit students' views on the subject, a step by step categorization, increasing in level of generality, was applied (Figure 1). In the first step, the responses to each question were categorized into "propositions". Each proposition represented a group of students' expressions of ideas similar in meaning (differing mainly in wording). Within a given question, a particular proposition normally represented more than one answer. To compose a proposition, we tried to use actual student answers. In case of extended answers, wherever possible, we considered the most *inclusive* version to represent the whole group. Next, after examining the propositions accumulated from responses to all 18 questions, "views" were identified and formulated by the researchers. Each of the views represented a group of propositions, and served as their generalization. Obviously, the two steps of generalization, to *propositions* (more concrete), and then into *views* (more abstract), enhanced the representative power of the elicited structure. "Views" represented a claim of value, the ideology of the individual. The obtained "views" were then sorted into corresponding theoretical topics of inquiry, one of the six dimensions mentioned above. We illustrate the procedure of propositions-view derivation in the Appendix.

In the process of data evaluation, both researchers first independently created the initial categorization of the answers in the form of representative propositions. To a large extent (about 90%), there was agreement between the researchers, and the final form of the propositions was determined by common consent. The elicitation of views , which to a large extent is a matter of personal value (never totally coinciding in two persons), was performed by discussion between the researchers, thus providing a greater reliability for the inferences made.

3.2. QUANTITATIVE ANALYSIS

To include a quantitative aspect in the analysis, four coefficients were calculated as percentages. As we were only interested in group differences, the answers were analyzed per sample rather than per student. First, Proposition Abundance (PA) is calculated in a straightforward manner, as the frequency (percentage) of which it appears in the responses to a particular question. The second coefficient measures View Abundance (VA), which might draw on a number of propositions, stemming from either the same question, or from different ones.

When the view is comprised only of propositions given in response to a single question, the VA's score is computed as the sum of the PAs, since no individual student can contribute more than one proposition. To compute the contribution of propositions that appeared in responses to different questions, but contributed to the same "view", one must average the percentages as had been calculated in the first stage, regarding the responses to each question. Averaging can only cause a *moderation* of the effect of the results. It also suppresses the influence of unbalanced (too frequent or extremely rare) responses which could occur due to some unexpected factor (association) which is foreign to the goals of the study.

The third and forth coefficients are Proposition Difference (PD) and View Difference (VD). They measure the absolute differences in the frequency of the proposition and the view respectively, between the experimental and the control groups. Their magnitudes were calculated in a straightforward way: $PD = PA_e - PA_c$, and $VD = VA_e - VA_c$. The calculations of relative differences in views (VD_r) and propositions (PD_r) were obvious ($VD_r = VD/VA_e$, $PD_r = PD/PA_e$). The significance of differences in frequencies of appearance between the groups were statistically tested according to the "Test for Significance of Difference Between Two Proportions" (Bruning and Kintz 1977).

Table II presents an example illustrating the derivation of the VA coefficient of the view: "Science of the past was primitive, its quality was of a low standard", who's qualitative elaboration was presented in the Appendix. As was shown, this view generalizes five propositions that emerged from the responses to two of the test questions, the first three propositions – to one, and the following two – from the other. After all five propositions were interpreted as manifestations of the same view, the calculation of the VA was as follows. The initial set of PAs in the control group was [43, 18, 16, 5.4, 6.5]. In the first stage, the first three PAs, originating from the responses to the same question, are substituted by their sum, 77, while the following two by 11.9, for the same reason. This provided an intermediate set [77, 11.9], which yields the resultant value of VA = 44%.

4. Findings and Discussion

Students' views elicited with regard to the mentioned six conceptual dimensions present our major findings in this study, and the differences in their frequencies between the experimental and control groups allowed us to evaluate the influence

Table II. Example of propositions-view relationship for the view "Science of the past was primitive, its quality was of a low standard"

#	Propositions	PA_C(%)	PA_E(%)	PD(%)
P-1	In the past scientists used primitive tools and very limited information. Therefore, they understood science differently.	43	26	−17
P-2	In the past scientists were very often wrong in their knowledge and reasoning. Scientists today understand nature correctly. This is why people today know more and better than then.	18	3.3	−14.7
P-3	In the past, scientists understood science differently because they could not conduct the complex experiments which are performed today.	16	6.6	−9.4
P-4	The main problem of scientists in the past, which made them often wrong in their views, is the lack of precise and sophisticated instruments and advanced laboratories.	5.4	0	−5.4
P-5	In the past, science was simple and straightforward, and so it could be worked out with simple tools and instruments. Today we must use hi-tech equipped laboratories.	6.5	17	10.5
	The view: "Science of the past was primitive, its quality was of a low standard"	VA_C (%) 44	VA_E (%) 26	VD (%) −18

of the innovative learning materials more specifically. Along with the views, we will display some representative propositions that will illustrate the more prominent views. Importantly, all quotations in this report are translations from the colloquial Hebrew, which caused some loss of the original "flavor" in favor of a more "scientific" look.

4.1. DIMENSION 1 – RELATIONSHIP OF SCIENCE AND TECHNOLOGY

Table III presents students' views with regard to the relationship between science and technology. The four shown views, apparently represent two pairs of opposites. The first pair (1 and 2) illustrates the views regarding which of the two, science or technology, is the historical "leader" of the other. Here we observe a clear difference in the opinions of the students. Those of our sample who were taught in the regular manner tended more to think that science had to lead any technological progress, by providing specific knowledge of how to construct tools and devices.

Table III. Students' views concerning the relationship of science and technology

#	Views	VA_c (%)	VA_e (%)	VD (VD_r) (%) z(statistics)
1	Technology never precedes science	66	30	−36 (−0.6) $z = 5.43^{**}$
2	Sometimes technology precedes scientific progress	22	70	48 (2.2) $z = -7.197^{**}$
3	Progress in modern science is dependent on advanced technology and sophisticated tools	88	47	−41 (−0.5) $z = 6.38^{**}$
4	Scientific progress can be attained without advanced technology	4	45	41 (10.3) $z = -6.79^{**}$

VA – view abundance, VD – view abundance difference ($VD = VA_e - VA_c$).
VD_r – relative differences in views ($VD_r = VD/VA_e$).
$^{*}p < 0.01$; $^{**}p < 0.001$; (ns) – statistically Non-Significant difference ($p > 0.05$).

For example, one of the students reasoned his answer as follows (corresponds to View 1, Table III):

> No one can invent instruments such as a TV or telephone without knowledge of scientific laws which govern their functioning.

However, students in the experimental group more significantly tended to think that scientific progress neither always, nor solely, draws on technological development (View 2, Table III). Thus, among the propositions affiliated to this view we found:

> Sometimes technology precedes theories, such as when people developed lens and eyeglasses without knowing how they worked. ($PA_c = 0$ and $PA_e = 25\%$)

The other pair, Views 3 and 4, represent the perception of the role of technological equipment in facilitating scientific progress. Here the views were also different, with the experimental group displaying, to a greater extent, the view that scientific progress is not solely dependent on the utilization of sophisticated tools and equipment. The view that technology is the unique (and maybe necessary) basis for scientific growth, was significantly reduced by the HPS materials, in favor of a greater appreciation of intellectual and cultural factors. Thus, a student of the experimental group wrote as response:

> Scientific development may occur based on thoughts and ideas, not necessarily involving sophisticated experiments. Ancient scientist discovered so much about light, they measured the universe almost without tools, look how much Alhazen knew. If it happened then, why can't it happen today? [corresponds to View 4, Table III]

Among the propositions found only in the experimental group was:

> Science and technology are interdependent. The invention of the telescope promoted science, but it was science that enabled the development of the microscope. ($PA_c = 0$, $PA_e = 18\%$) [affiliated with View 4, Table III]

Addressing the same topic, when interviewed, one teacher of the experimental group stated:

> Students were surprised to learn that technology may precede accurate scientific knowledge. During the lessons, when we considered the history of eyeglasses, their invention and use, we had a chance to discuss the complexity of the realistic relationship between science and technology.

Although half the experimental group remained convinced that in our time no research activity can take place without using highly sophisticated technology and extremely complex equipment, the frequency of this view was still less than in the control group. The increase in awareness of factors other than technical equipment which may determine scientific research, was mentioned by all teachers in the experimental classes.

4.2. DIMENSION 2 – ATTITUDE TO THE SCIENCE OF THE PAST

Table 4 presents the views elicited regarding students' attitude to the science of the past. As was expected, students in the experimental group had a significantly more positive view of past science. A significant growth of the opinion that it is beneficial to know about scientific knowledge of the past, both to students (+31%), as well as to scientists (+27%), indicates that students recognized the importance of this knowledge beyond just curious historical facts. Thus the following proposition:

> Scientists can learn from history that theories, views and beliefs could, and were, replaced or changed over time as happened with the corpuscular theory of light or the belief about an infinite velocity of light. ($PA_c = 0\%$; $PA_e = 19\%$) [affiliated with View 2, Table IV]

appeared only in the experimental group (19%). In another proposition, students explained the validity and importance of historical knowledge:

> When we know how scientific knowledge came about, we understand more of what it is about. ($PA_c = 11\%$; $PA_e = 25\%$) [affiliated with View 3, Table IV]

Some experimental group students mentioned that learning about attempts to account for light refraction in the old theories of vision, caused them to appreciate the non-obvious phenomenon of the human eye.

As was stated in an interview:

> I: Why did you say it is so important to learn about science of the past?
>
> S: Science develops based on previous knowledge, eliminating false understanding. But new theories are synthesized from former ones, and new knowledge from what had been known before. One sees it in what Alhazen

Table IV. Students' attitude to the science of the past

#	Views	VA_c (%)	VA_e (%)	VD (VD_r)(%) z(statistics)
1	Old science is invalid today and its knowledge is unnecessary.	39	14	−25 (−0.6) $z = 4.39^{**}$
2	Scientists should be aware of former theories even those which appear to them incorrect.	55	82	27 (0.5) $z = -4.47^{**}$
3	Learning about science in the past is beneficial for students.	50	82	31 (0.6) $z = 5.70^{**}$
4	Scientists, as well as students, should not study former theories, refuted as incorrect, but focus on the correct scientific knowledge of today.	45	12	−33 (−0.7) $z = -5.20^{**}$
5	Science of the past was primitive, its quality was of a low standard.	44	26	−18 (−0.4) $z = 2.87^{*}$
6	In the past, there were bits of "good" science which were of a high quality.	13	48	35 (2.7) $z = -5.53^{**}$
7	Former scientific theories are important regardless of their correctness in our modern view.	31	66	35 (1.1) $z = -5.25^{**}$

VA – view abundance, VD – view abundance difference ($VD = VA_e - VA_c$).
VD_r – relative differences in views ($VD_r = VD/VA_e$).
$^{*}p < 0.01$; $^{**}p < 0.001$; (ns) – statistically Non-Significant difference ($p > 0.05$).

did in order to account for vision in his theory. I think that in order to really understand some theory, we should also know the way it came about.

The increased respect and appreciation given to old science by students of the experimental group, incorporated a more mature attitude toward scientists of the past and their enterprises.

> We know more facts that they did in the past, but scientists then were not less intelligent. ($PA_c = 23\%$; $PA_e = 64\%$) [affiliated with View 6, Table IV]

In contrast to the position:

> In the past scientists used primitive tools and very limited information. Therefore, they understood science differently. ($PA_c = 43\%$; $PA_e = 26\%$) [affiliated with View 5, Table IV]

Likewise, the often observed disdain to the quality of the scientific research in antiquity and the middle ages, a certain disparagement with regard to its products, became apparently weaker in the experimental group. This idea was seen in their explanations regarding the ignorance about vision in early history, that the lack of knowledge about the way eyes function and the nature of light, could be objective factors impeding understanding of things. Such knowledge provides us today with a practical advantage.

The feeling of the importance of old theories, despite their obviously limited correctness and accuracy, grew in the experimental group. This was contrasted with some views in the control group. The claim:

> Scientist should focus only on the correct knowledge since they should establish new theories. ($PA_c = 14\%$; $PA_e = 3\%$) [affiliated with View 4, Table IV]

in referring to the old science, was rare in the experimental group.

Only in the experimental group did students show a wider perspective by responding:

> In studying theories of the past we learn that scientists can make mistakes. ($PA_c = 0\%$; $PA_e = 31\%$) [affiliated with View 7, Table IV]

or, to quote a student interview protocol:

> Scientists of the past were even more clever than their colleagues today, because they had to use logic and could not help themselves with sophisticated instruments. The difference in knowledge between them and us is that we know a lot more. But they were smart and could see much around them. They often were good in making astute inferences from what they observed.

The contrary view, discrediting the validity of historical knowledge of the subject matter, was not given in the experimental group. The following proposition was registered only in the control group:

> It does not make sense for students to base their knowledge on wrong theories [of the past], as they are misleading. ($PA_c = 13\%$; $PA_e = 0\%$) [affiliated with View 4, Table IV]

4.3. DIMENSION 3 – INFLUENCE OF EXTERNAL FACTORS ON SCIENCE AND ITS PRODUCTS

Table V represents students' views on whether science and its products are subject to the influence of external factors, or are autonomous in their development. A great majority of the students in the experimental group (82%) shared the view that scientists in their research activity may be influenced by a variety of external factors, as expressed in the following proposition:

> Scientists may be biased. They are affected by prejudices and beliefs, and may be mislead by intuition and claims of other scientists. ($PA_c = 6\%$; $PA_e = 43\%$) [affiliated with View 2, Table V]

This, compared to 57% of the control group who were convinced of the opposite, that scientists in their research are strictly objective and independent of their social, "non-scientific" environment when making decisions, as exemplified by the following:

Table V. Students' views on the influence of external factors on science and its products

#	Views	VA_c (%)	VA_e (%)	VD (VD_r)(%) z(statistics)
1	Products of scientific research, and scientists' activities, are autonomous and not affected by factors external to science.	57	15	−42 (−0.7) $z = 6.77^{**}$
2	Scientific research and scientists' activities may be influenced by external factors, not directly scientific.	37	82	45 (1.2) $z = -7.03^{**}$

VA – view abundance, VD – view abundance difference ($VD = VA_e - VA_c$).
VD_r – relative differences in views ($VD_r = VD/VA_e$).
$^{*}p < 0.01$; $^{**}p < 0.001$; (ns) – statistically Non-Significant difference ($p > 0.05$).

> Science is objective and exact, it refutes any subjective considerations, social environment or politics. ($PA_c = 53\%$; $PA_e = 12\%$) [affiliated with View 1, Table V]

Some students in the experimental group exemplified their views by the fact that Newton's authority for many years shielded the corpuscular theory of light from the critique of others (e.g., Huygens) who argued for its wave nature.

> Scientists belong to a cultural and social environment, which affects their views and activities. Therefore, science in different times was so different. ($PA_c = 22\%$; $PA_e = 70\%$) [affiliated with View 2, Table V]

One teacher commented:

> We had many discussions with students about the effect of external factors on science. There were many opportunities during the course, to see the way scientific ideas were born, and we connected them to the actual political and religious issues of those times.

Students pursued and exemplified:

> One can see how scientists were affected by the conditions in which they lived. Like when life in Greece became hard, science moved to the countries of Islam. The influence of life conditions and surroundings on scientists was very strong then, as it is today.

> The Church put Galileo on trial and convicted him. It greatly changed his life and activity.

> We see in the confrontation between the corpuscular and the wave theories of light, how scientists can be affected also by prejudices. They may prefer a theory because of some invented belief or a religious idea, or because of the authority of great scientists they knew.

Table VI. Students' conceptions of scientific theories

#	Views	VA_c (%)	VA_e (%)	VD (VD_r)(%) z(statistics)
1	Rival scientific theories never co-exist, since scientists decide which of them is correct to hold.	25	18	−7 (−0.3) $z = 1.29$ (ns)
2	It may happen that different scientists, at the same time, hold and practice rival scientific theories.	22	56	34 (1.5) $z = -5.16^{**}$
3	Knowledge about rejected invalid scientific theories is useless.	46	15	−31 (−0.7) $z = 5.21^{**}$
4	Some knowledge about invalid (old) scientific theories can be beneficial and helpful in research.	28	61	33 (1.8) $z = -4.95^{**}$
5	In addition to the experimental data, a good way to resolve controversies in science is to refer to theoretical arguments.	29	62	33 (1.1) $z = -4.95^{**}$
6	In addition to the experimental data, a good way to decide in favor of a scientific statement is to rely on the opinion of an expert.	32	14	−18 (−0.6) $z = 3.30^{*}$

VA – view abundance, VD – view abundance difference ($VD = VA_e - VA_c$).
VD_r – relative differences in views ($VD_r = VD/VA_e$).
$^{*}p < 0.01$; $^{**}p < 0.001$; (ns) – statistically Non-Significant difference ($p > 0.05$).

4.4. DIMENSION 4 – CONCEPTION OF SCIENTIFIC THEORIES

Table 6 presents students' views on scientific theories. The first four views apparently represent two contradictory pairs. The recognition of a possibility that rival theories may coexist in science was very pronounced in the experimental group (56%). Although 18% of the group believed that scientists can always decide which theory is "correct", they seemingly perceived science as choosing between the theories and interpretations. Students' propositions were:

> Sometimes scientists cannot make the choice between different theories, as they could not conduct an appropriate experiment. It would require technology not available at that time. This happened when there was no way to resolve whether light comprised waves or small particles. ($PA_c = 0\%$; $PA_e = 28\%$) [affiliated with View 2, Table VI]

> Rival theories regarding the same reality can exist at the same time, where there is still not enough knowledge accumulated. ($PA_c = 2\%$; $PA_e = 19\%$) [affiliated with View 2, Table VI]

Students of the experimental group specifically referred to the contradictory views regarding the speed of light and the nature of vision. The following is an illustrative excerpt from an interview:

Rival, even contradicting, scientific theories can exist at the same time, and it can happen even in our times. There are many reasons for this. Scientists' assumptions may come from different sources, as in the ancient theories of vision, or later, when there were no technological means to conduct proper experiments to measure light speed in water and a vacuum, in order to infer whether the wave or corpuscular theory is correct.

A naive view was expressed in:

Science provide unique answers, because science operates with precise tools. (PA_c = 26%; PA_e = 5%) [affiliated with View 1, Table VI]

Every natural phenomenon must be explained correctly in one particular way. That is why it is always possible to resolve between rival scientific theories. (PA_c = 13%; PA_e = 11%) [affiliated with View 1, Table VI]

The second pair of views (3 and 4, Table VI) dealt with the replacement and refutation of scientific theories. Students argued that even dismissed theories possess a certain validity (examples referring to Alhazen's theory of vision or incorrect versions of the law of refraction, were given in the experimental group).

Scientists might use, if not the whole theory then some of its parts, in forming new theories or when incorporating old knowledge in current research. (PA_c = 9%; PA_e = 25%) [affiliated with View 4, Table VI]

Important ideas can be learnt even from incorrect old theories, and be reconsidered in a new way. (PA_c = 13%; PA_e = 20%) [affiliated with View 4, Table VI]

In contrast, some students mentioned the knowledge of replaced theories as useless. Among the propositions stating such an attitude was:

If a theory failed, its knowledge is incorrect. It cannot be utilized in a beneficial way anymore. (PA_c = 43%; PA_e = 15%) [affiliated with View 3, Table VI]

The other views (5 and 6, Table VI) reflect the perception of the role of empirical data in making scientific judgments and arguments. Considerably more students in the experimental group thought that theoretical arguments are highly useful in making judgments and raising new ideas and scientific theories. Likewise, they were less in favor of solely relying on the authority of experts, a view found to be stronger in the control group.

4.5. DIMENSION 5 – RELIABILITY OF SCIENTIFIC KNOWLEDGE

Views concerning this dimension are presented in Table VII. Apparently, the refutable nature of scientific knowledge was more appreciated in the experimental group. At the same time, many students in both groups considered scientific knowledge to be accurate and reliable, although in the experimental group this view

Table VII. Students views on the reliability of scientific knowledge

#	Views	VA_c (%)	VA_e (%)	VD (VD_r)(%) z(statistics)
1	Results of scientific research are well founded and not refutable.	42	19	−23 (−0.5) $z = 3.83^{**}$
2	Scientific knowledge is refutable.	33	72	39 (1.2) $z = -5.90^{**}$
3	Scientific knowledge is temporary, and correct only for the time being.	15	27	12 (0.8) $z = -2.16^{*}$
4	Scientific knowledge is accurate, certain and reliable.	67	30	−37 (−0.5) $z = 5.58^{**}$

VA – view abundance, VD – view abundance difference ($VD = VA_e - VA_c$).
VD_r – relative differences in views ($VD_r = VD/VA_e$).
$^{*}p < 0.01$; $^{**}p < 0.001$; (ns) – statistically Non-Significant difference ($p > 0.05$).

was less prevalent (30%), compared to 67% in the control group. The following propositions may be illustrative:

> Scientists argue by employing precise proofs/methods/instruments. (PA_c = 25%; PA_e = 0%) [affiliated with View 1, Table VII]
>
> Science is accurate and objective, its laws were tested and proven for a long time. (PA_c = 32%; PA_e = 11%) [affiliated with View 1, Table VII]
>
> Science of nature is an objective and exact realm, free of subjective features common in social sciences or politics. (PA_c = 53%; PA_e = 12%) [affiliated with View 4, Table VII]
>
> Scientists rely solely on the accurate data they obtain in the designed experiments. (PA_c = 24%; PA_e = 0%) [affiliated with View 4, Table VII]

These, as opposed to:

> Sometimes scientists rely on wrong presumptions. (PA_c = 0%; PA_e = 11%) [affiliated with View 3, Table VII]

Addressing the temporal nature of scientific theories, students of the experimental group referred to the theory of vision rays, Newton's particles of light and Alhazen's mistaken understanding of visual images. A teacher in the experimental group said:

> Some students revealed for themselves the fact that scientists can be wrong, and that scientific theories may change. However, most students still considered science to be very precise in nature, and lack any inaccuracy. Those students were surprised to learn about incorrect theories which were held for a very long time. Some of the students realized that this could also happen in contemporary science, and some modern theories might appear false in future.

Table VIII. Students' attitudes toward critical perception of science in the classroom

#	Views	VA_c (%)	VA_e (%)	VD (VDr)(%) z(statistics)
1	It is pointless to criticize scientific claims	57	23	−34 (−0.6) $z = 5.29^{**}$
2	Critical discussions about scientific claims are worthy, important and beneficial for the learners.	26	76	50 (1.9) $z = -7.55^{**}$

VA – view abundance, VD – view abundance difference ($VD = VA_e - VA_c$).
VD_r – relative differences in views ($VD_r = VD/VA_e$).
$^*p < 0.01$; $^{**}p < 0.001$; (ns) – statistically Non-Significant difference ($p > 0.05$).

4.6. DIMENSION 6 – CRITICAL PERCEPTION OF SCIENCE IN THE CLASSROOM

Views on whether it is worthwhile to criticize science in the classroom are presented in Table VIII, and can be condensed into a pair of opposite views regarding the importance of critical classroom discussions about science. The majority (76%) of the experimental group thought such discussions are valid and were helpful to them, whereas 57% of the control group thought such discussions to be pointless.

Students who rejected the idea of critical discussions on science, reasoned that the accuracy and objectiveness of scientific knowledge had been repeatedly checked, ever since the laws were established. Some stated that the correct knowledge established by scientists, should not be further discussed or criticized by the learners:

> Science is precise and objective, and its laws were checked and proven many times. ($PA_c = 32\%$; $PA_e = 11\%$) [affiliated with View 1, Table VIII]
>
> We should learn scientific knowledge and not criticize it because it is correct and valid. ($PA_c = 16\%$; $PA_e = 0\%$) [affiliated with View 1, Table VIII]

The opposite can be seen in the following propositions:

> Discussions are important because they help us to be more critical in considering claims, even if they are scientific. We know that there were many theories in science which were incorrect. ($PA_c = 0\%$; $PA_e = 26\%$) [affiliated with View 2, Table VIII]
>
> It was exciting to know what were the thoughts of the scientists inventing theories and how they arrived at the laws. We can discuss the same ideas as they did when faced with the same problems. This way we learn how science acts. ($PA_c = 0\%$; $PA_e = 12\%$) [affiliated with View 2, Table VIII]

As well as in quotes from two interviews:

> S: "To discuss the advantages and disadvantages of scientific theories means to know them better. It teaches you a lot of things about the way scientists think and how to be critical of other theories we will learn in the future".

T: "It was very interesting to see students in this class keep asking "why" questions, and not accepting new material as a given truth, as usually happens in regular classes. The fact they were encouraged to criticize scientific theories helped them to discover that they could think critically themselves, as scientists did and do."

5. Cumulative Presentation of Views About Science

Based on the elicited views and their frequencies in both groups, we could construct a schematic conceptual profile of each group with regard to their views about science. In doing so, we grouped those views in which each group scored higher, and listed the frequency ratio of occurence (vs. the opposite group) in each view (Table IX). These profiles summarize our interpretation of the data, illustrating the noticeable differences between the groups as affected by the types of instruction. In particular, they can facilitate analysis of the use of HPS materials, and illustrate its results regarding students' views about science.

The profile of the control group (traditional instruction) confirms feelings common in science educators with regard to students' image of science (Carr et al. 1994). Students, often hold inadequate views on what science is, its activities and products. Considering this image of science, one finds belief in dogmatic truth, deprived of features inherent in true science: dynamics, uncertainty, controversy, plurality of views, limited validity and accuracy. Observing the profile of the experimental group, one can detect a definite shift in the desirable direction. The information about the latter in the educational context may be drawn from the studies on scientific literacy (Bybee 1997) and declarations of position on science education (AAAS 1993; NAS 1996). These profound positive results in the experimental group cannot be isolated from the difference between the instruction of the groups. The nature of the instructional materials used in the experimental group, together with a reduced emphasis on standard problem solving, introduced a general scientific, cultural and philosophical background into the discussion on vision, the role of the observer and the nature of light.

Another feature of the data was that among students sharing the same view, those in the experimental group often responded by illustrating their statements with concrete examples from the history of optics, whereas the others frequently remained only declarative. Views elicited in the control group about the nature of science, commonly were naive and oversimplified. Some of them – regarding the relationship with technology, the nature of the old science (as "primitive and mistaken"), the theories in science (as never controversial), the nature of scientific knowledge (as always objective and accurate), the refutation of discussions on scientific subjects in the instruction, should concern science teachers regardless their pedagogical standpoint.

Table IX. Conceptual profile of students with regard to their views about science

#	Conceptual dimension	Control group (#) = frequency ratio vs. the experimental group	Experimental group (#) = frequency ratio vs. the control group
1	The relationship of science and technology	Technology never precedes science (2.2) Progress in modern science is stipulated by advanced technology and sophisticated tools (1.9)	Sometimes technology precedes scientific progress (3.2) Scientific progress can be attained without advanced technology (11.3)
2	Attitude to the science of the past	Old science is invalid today and its knowledge is unnecessary. (2.8) Science majors should not study former theories refuted as incorrect, but focus on the correct scientific knowledge of today. (3.8) Science in the past was primitive, its quality was of a low standard (1.7)	Scientists should be aware of the former theories even those which appear to them incorrect. (1.5) Learning about science in the past is beneficial for students. (1.6) In the past, there were bits of "good" science which were of a high quality (3.7) Former scientific theories are important regardless of their correctness in modern view. (2.1)
3	Influence of external factors on science and its products	Results of scientific research and scientists' activities are autonomous and not affected by factors external to science. (3.8)	Scientific research and scientists' activities may be influenced by external factors, not directly scientific. (2.2)
4	Conception of scientific theories	Rival scientific theories never co-exist, as the scientists decide which of them is correct to hold. (1.5) Knowledge about rejected invalid scientific theories is useless. (3) In addition to the experimental data, a good way to decide in favor of a scientific statement is to rely on the opinion of an expert. (2.3)	It may happen that different scientists, at the same time, hold and practice rival scientific theories. (2.5) Some knowledge about invalid (old) scientific theories can be beneficial in research. (2.8) In addition to the experimental data, a good way to resolve controversies in science is to refer to theoretical arguments. (2.1)
5	The reliability of scientific knowledge	Results of scientific research are well founded and not refutable. (2.2) Scientific knowledge is accurate, certain and reliable. (2.2)	Scientific knowledge is refutable. (2.2) Items of scientific knowledge are temporary correct are subject for changes (1.8)
6	Critical perception of science in the classroom	It is pointless to criticize scientific claims. (2.5)	Critical discussions about scientific claims are important and beneficial for the learners. (2.9)

6. Conclusion

We generally view helping students to construct a more realistic view of science as extremely important. Specifically, with regard to the course under discussion, it was even more important in light of the fact that for the majority of Israeli students, the tenth grade course presents their last institutionally organized encounter with physics.

In this study, we have considered students' views on science in an empirical-data-supported manner, while comparing between those taught along a traditional curriculum with those taught using an HPS-based program.

The constructed assessment tool, presented the obtained knowledge in terms of a hierarchical, view-propositions structure. Such an organization allowed us to express the versatile qualitative (textual and descriptive) and voluminous data (usual in this type of studies), in a succinct, yet representative, manner.

Our results provided pronounced evidence of the beneficial use of the HPS-based learning materials in regular school instruction. The advantage of the experimental group, in the type of students' views about various aspects of science, was evident in the constructed conceptual profiles and statistically significant differences of the quantitative parameters. Students became more aware of a variety of human, cultural and historical issues; this in contrast to the currently employed and highly formal instruction, which leaves such subjects to the realm of self-education. Importantly, this effect was not on the expense of the disciplinary knowledge of the subject (Galili and Hazan 2000a).

Appendix

Here we illustrate the derivation of the view: "Science of the past was primitive, its quality was of a low standard", within dimension-2, "attitude to the science of the past" (Table IV).

Analyzing students' answers to the questions

- What do you think are the reasons that scientists of the past explained nature differently than they do today?

and

- Consider the following statement: "In our times one cannot expect a scientific development which would not be the result of using highly sophisticated technology and complex experiments." Explain your opinion.

we first gathered the answers which expressed similar ideas that differed one from the other mainly by variations in the wording used, and provided each such group with a unified textual form. These constructs were labeled as *propositions*. Among the propositions representing students' responses to the first of the above questions, were (Table II):

P-1. In the past, scientists used primitive tools and very limited information. Therefore, they understood science differently.

P-2. In the past, scientists were very often wrong in their knowledge and reasoning. Scientists today understand nature correctly. This is why people today know more and better than then.

P-3. In the past, scientists understood science differently because they could not conduct the complex experiments which are performed today.

Similarly, the following propositions appeared in response to the second question:

P-4. The main problem of scientists in the past, which made them often wrong in their views, is the lack of precise and sophisticated instruments and advanced laboratories.

P-5. In the past, science was simple and straightforward, and so it could be worked out with simple tools and instruments. Today we must use hi-tech equipped laboratories.

By further considering these five propositions, it is apparent that they all share students' belief (a *view*) that "science of the past was primitive, its quality was of a low standard". This value statement was formulated by the researchers to represent an important, in their opinion, idea characterizing students' "attitude to the science of the past", one of the dimensions of knowledge as it was presented in the study. Other views of the same dimensions are presented in Table IV.

Notes

[1] For instance, this is the case in Israel, where science courses are compulsory for tenth grade students (age 15–16). In eleventh and twelfth grades, science courses are elective. They are taken only by a minority of the students, and less than 10% are enrolled in physics classes.

References

AAAS (American Association for the Advancement of Science): 1990, *Project 2061: Science for All Americans*, Oxford University Press, New York, pp. 183–194.

AAAS (American Association for the Advancement of Science): 1993, *Benchmarks for Science Literacy*, AAAS Press, Washington, DC.

Abell, S. & Smith, D.: 1994, 'What is Science? Preservice Elementary Teachers' Conceptions of the Nature of Science', *International Journal of Science Education* **16**(4), 475–487.

Aguirre M., Haggerty, S. & Cedric, L.: 1990, 'Student-Teachers' Conceptions of Science, Teaching and Learning: A Case Study in Preservice Science Education', *International Journal of Science Education* **12**(4), 381–390.

Aikenhead, G.: 1992, 'How to Teach the Epistemology and Sociology of Science in a Historical Context', in S. Hills (ed.), *The History and Philosophy of Science and Science Education*, Vol. 1, Kingston University, Ontario, Canada, pp. 23–34.

Aikenhead, G.: 1997, 'Students' Views on the Influence of Culture on Science', *International Journal of Science Education* **19**(4), 419–428.

Bettencourt, A.: 1993, 'The Construction of Knowledge', in K. Tobin (ed.), *The Practice of Constructivism in Science Education*, Laurence Erlbaum, Hillsdale, NJ, pp. 39–50.

Brickhouse, N.: 1989, 'The Teaching of Philosophy of Science in Secondary Classrooms: Case Study of Teachers' Personal Theories', *International Journal of Science Education* **11**(4), 437–449.

Bruning, J.L. & Kintz, B.L.: 1977, *Computational Handbook of Statistics*, Scott & Foresman, Glenview, IL, pp. 220–224.

Brush, S.J.: 1989, 'History of Science and Science Education', *Interchange* **20**(2), 60–70.

Bybee, R., Powell, J., Ellis, J., Giese, J., Parisi, L. & Singleton, L.: 1991, 'Integrating the History and Nature of Science and Technology in Science and Social Studies Curriculum', *Science Education* **75**(1), 143–155.

Bybee, R.: 1997, 'Toward an Understanding of Scientific Literacy', in W. Graber & C. Bolte (eds), *Scientific Literacy*, IPN, Kiel, Germany, pp. 37–68.

Carr, M., Barker, M,. Bell. B., Biddulph, F., Jones, A., Kirkwood, W., Pearson, J. & Symington, D.: 1994, 'The Constructivist Paradigm and Some Implications for Science Content and Pedagogy', in P. Fensham, R. Gunstone and R. White (eds.), *The Content of Science*, Falmer Press, London, pp. 147–160.

Cohen, M.R. & Drabkin, I.E. (eds): 1948, *A Source Book in Greek Science*, McGraw Hill, New York.

Conant, J.B.: 1951, *Science and Common Sense*, CT Yale University Press, New Haven.

Conant, J.: 1951, *On Understanding Science: An Historical Approach*, New America Library, New York.

DeBoer, G.: 1991, *A History of Ideas in Science Education: Implications for Practice*, Teachers College Press, New York.

Descartes, R.: 1637/1965, *The Discourse on Method, Optics, Geometry, and Meteorology*, Bobbs-Merrill, New York.

Dewey, J.: 1916, *Democracy and Education*, Free Press, New York

Driver, R., Leach, J., Miller, A. & Scott, P.: 1996, *Young People Image of Science*, Open University Press, Bristol, PA.

Duschl, R.: 1990, *Restructuring Science Education: The Importance of Theories and Their Development*, Teachers College Press, New York.

Duschl, R.: 1994, 'Research in History and Philosophy of Science', in D.L. Gabel (ed.), *Handbook of Research on Science Teaching and Learning*, MacMillan, New York, pp. 443–465.

Finegold, M. & Olson, J.: 1972, *An Enquiry into the Development of Optics: Conception of Light and Their Role in Enquiry*, Department of Curriculum, the Ontario Institute for Studies in Education, Ontario.

Galili, I. & Hazan, A.: 2000a, 'The Influence of an Historically Oriented Course on Students' Content Knowledge in Optics Evaluated by Means of Facets-Schemes Analysis', *American Journal of Physics* **68**(7), S3–S15.

Galili, I. & Hazan, A.: 2000b, 'Learners' Knowledge in Optics: Interpretation, Structure, and Analysis,' *International Journal of Science Education* **22**(1), 57–88.

Glasersfeld, R.: 1995, *Radical Constructivism: A Way of Knowing and Learning*, Falmer Press, London.

Griffiths, A.K. & Barman, C.R.: 1995, 'High School Students Views About the Nature of Science: Results from Three Countries', *Schools Science and Mathematics* **95**(5), 248–255.

Halloun, I. & Hestenes, D.: 1998, 'Interpreting VASS Dimensions and Profiles for Physics Students', *Science and Education* **7**, 553– 577.

Huygens, C.: 1690/1978, *Treatise on Light*, Encyclopedia Britannica, Chicago.

Jaffe, B.: 1938, 'The History of Chemistry and Its Place in the Teaching of Chemistry', *Journal of Chemical Education* **15**, 383–389.

Johnson, S. & Stewart, J.: 1991, 'Using Philosophy of Science in Curriculum Development: An Example from High School Genetics', in M. Matthews (ed.), *History, Philosophy and Science Teaching: Selected Readings*, OISE Teachers College Press, New York.

Kipnis, N.: 1993, 'Rediscovering Optics' , BENA Press, Minneapolis.

Klopfer, L.: 1969, 'The Teaching of Science and the History of Science' *International Journal of Science Education* **6**, 87–95.

Klopfer, L. & Cooly, W.: 1963, 'The History of Science Cases for High Schools in the Development of Students' Understanding of Science and Scientists', *Journal of Research in Science Teaching* **1**(1), 33–47.

Koulaidis, V. & Ogborn, J.: 1989, 'Philosophy of Science: An Empirical Study of Teachers' Views', *International Journal of Science Education* **11**(2), 173–184.

Koulaidis, V. & Ogborn, J.: 1995, 'Science Teachers Philosophical Assumptions: How Well Do We Understand Them? *International Journal of Science Education* **17**(3), 273–283.

Lederman G.N. & O'Malley M.: 1990, 'Students' Perception of Tentativeness in Science: Development, Use, and Sources of Change', *Science Education* **74**(2), 225–239.

Lederman, G.N., McComas, F.W. & Matthews, M.R.: 1998, 'The Nature of Science and Science Education – Editorial', *Science & Education* **7**(6), 507–509.

Lindberg, D.C.: 1976, *Theories of Vision from Al-Kindi to Kepler*, The University of Chicago Press, Chicago.

Lipson, H.: 1968, *The Experiments in Physics*, Oliver and Boyd, Edinburgh.

Lochhead, J. & Dufresne, R.: 1989; 'Helping Students Understand Difficult Science Concepts Through the Use of Dialogues with History', in D.E. Herget (ed.), *The History and Philosophy of Science in Science Education*, Tallahassee, Florida, pp. 221–229.

Loving, C.: 1991, 'The Scientific Theory Profile: A Philosophy of Science Model for Science Teachers', *International Journal of Science Education* **28**(9), 823–838.

Mach, E.: 1913/1926, *The Principles of Physical Optics. An Historical and Philosophical Treatment*, Dover, New York.

Matthews, M.R.: 1994, *Science Teaching: The Role of History and Philosophy of Science*, Routledge, New York.

Matthews, M.R.: 1997, 'The Nature of Science and Science Education – Editorial', *Science & Education* **6**(4) 323–329.

McComas, F.W., Clough, M. & Almazroa, H.: 1998, 'The Role and Character of the Nature of Science in Science Education', *Science & Education* **7**(6), 511–532.

Meichtry, Y.: 1993, 'The Impact of Science Curricula on Student Views About the Nature of Science', *Journal of Research in Science Teaching* **30**(5), 429–443.

Mellado, V.: 1997, 'Preservice Teachers' Classroom Practices and Their Conceptions of the Nature of Science', *Science & Education* **6**(4), 331–354.

Milgrom, I.: 1989, 'Designing Outlines of a Program for the Teaching of Social Aspects of Science and Technology', Ph.D. Thesis, The Hebrew University of Jerusalem (unpublished).

Monk, M. & Osborne, J.: 1997, 'Placing the History and Philosophy of Science on the Curriculum: A Model of Development of Pedagogy', *Science Education* **81**, 405–424.

NAS (National Academy of Science): 1996, *National Science Education Standards*, NAS, Washington, DC.

Newton, I.: 1704/1952, *Opticks*, Dover, New York.

Niedderer, H.: 1992, 'Science Philosophy, Science History and the Teaching of Physics', in S. Hills (ed.), *History and Philosophy of Science in Science Education*, Vol. II, Ontario Queen's University, Kingston, pp. 201–214.

Nuffield Advanced Science: 1988, *Teacher Guide 1*, Longman, Harlow, Essex.

Priestley, J.: 1772, *The History and Present State of Discoveries Relating to Vision, Light, and Colors*, London.

Ronchi, V.: 1970, *The Nature of Light*, Harvard University Press, Cambridge.

Rowell, J. & Cawthron E.R.: 1982, 'Images of Science: An empirical Study', *European Journal of Science Education* **4**(1), 79–94.

Schwab, J.: 1963, *Biology Teacher's Handbook*, Wiley, New York

Schwab, J.: 1964, 'The Teaching of Science as Enquiry', in *Teaching of Science*, Harvard University Press, Cambridge, MA.

Solomon, J., Duveen, J. & Scot, L.: 1992, 'Teaching About the Nature of Science through History: Action Research in the Classroom', *International Journal of Science Education* **29**(4), 409–421.

Tamir, P.: 1994, 'Israeli Students' Conceptions of Science and Views About the Scientific Enterprise', *Research in Science and Technological Education* **12**(2), 99–116.

Thomsen, P.: 1998, 'The Historical Philosophical dimension in Physics Teaching: Danish Experience', *Science & Education* 7, 493–503.

Tobin, K. & McRobbie, C.: 1997, 'Believes About the Nature of Science and the Enacted Curriculum', *Science & Education* 6(4), 355–371.

Wandersee, J.H., Mintzes, J.J. & Novak, J.D.: 1994, 'Research in Alternative Conceptions in Science', in D.L. Gabel (ed.), *Handbook of Research on Science Teaching and Learning*, MacMillan, New York.

Yoshida, A.: 1989, 'Results and Implications of Children's Views of Science Across the Six Countries', Paper Presented at the Annual Meeting of the National Association for Research in Science Teaching, San Diego, CA.

Scientific Controversies in Teaching Science: The Case of Volta

NAHUM KIPNIS
3200 Virginia Ave, South, #304, Minneapolis, MN 55426, USA; E-mail: nahumk@uswest.net

Abstract. This paper discusses a way of introducing a scientific controversy, which emphasizes objective aspects of such issues as multiple theoretical interpretation of phenomena, choosing a theory, insistence on the chosen theory, and others. The goal is to give students a better insight into the workings of science and provide guidelines for building theories in their own research.

1. Introduction

An in-depth discussion of scientific controversies in the classroom is one of the best ways to utilize the limited time teachers can spare for using the history of science in teaching science. Following a scientific debate can improve students' understanding of the inner workings of science, in particular, an introduction of a new scientific theory and its relation to experiment. Showing scientific results as debatable issues makes science more similar to other human activities that are easier to comprehend, such as a political debate or a court proceedings, which may sparkle an interest in science in some students. Finally, there is a pragmatic aspect in it as well: looking from different perspectives at a scientific concept can facilitate its understanding.

In this paper, I will focus on some aspects of the relationship between theory and experiment that have not yet attracted much attention. While discussing in the classroom why one theory replaces another, teachers usually emphasize that the new theory explains phenomena (experiments) unexplained in the old theory. The presumption behind this is that certain experiments naturally support one theory and contradict others. For instance, a teacher states that phenomena of interference and diffraction *prove* the wave nature of light and contradict the corpuscular theory. This statement, however, ignores the fact that throughout the eighteenth century these phenomena have been considered a strong argument *against* the wave theory of light (Kipnis 1992, pp. 193, 216). Apparently, the presumption mentioned above is nothing but a myth, and the true relationship between a theory and its experimental foundation is more complicated.

The best way to counter this myth is by showing how the same experiment gives rise to different theories. An excellent opportunity for such a study can be

F. Bevilacqua et al. (eds.), Science Education and Culture, 255–271.
© 2001 *Kluwer Academic Publishers. Printed in the Netherlands.*

found in two controversies associated with Alessandro Volta (1749–1827), Physics Professor at the University of Pavia: the debate on the "animal" electricity and the debate on the nature of the voltaic pile. First, I will present the relevant historical materials. Then, I will analyze them looking for such features of an interaction between experiment and theory that are common to both cases. Finally, I will suggest several experiments for a reproduction in the classroom. These experiments can enhance students' interest in the debate also show the limitations of an experiment as an argument in a scientific controversy.

2. The Story

2.1. THE "ANIMAL" ELECTRICITY

This controversy began in 1792 with the publication of the discovery by Luigi Galvani (1737–1798), Professor of Anatomy and Obstetrics at the University of Bologna. Like some other physiologists, Galvani believed that the "nervous fluid" responsible for animal movements was of an electric nature. This hypothesis was based on the existence of electric fish and a possibility of an electric stimulation of animals. After testing how a frog's muscle-nerve preparation reacted to static and atmospheric electricity, Galvani once noticed that a frog's leg contracted every time the muscle and the nerve were connected by a metal arc consisting of two different metals (Galvani 1791). To explain the new phenomena, later labelled "galvanic" or "galvanism," Galvani argued that the contractions occur when electricity flows between a nerve and a corresponding muscle through an external conductor, and that this electricity originates inside the frog. According to Galvani, this experiment (and some others) *proved* that electricity exists in every animal body, rather than being limited to electric fish. He had no hypothesis about its origin but offered one about its preservation: a nerve and a surrounding muscle make a sort of a Leyden jar that retains the charge until the nerve and the muscle are electrically connected. Galvani's arguments were based on experiments (Figure 1). To prove that the effect was due to an electricity of a new kind, he had to exclude other possible causes. First, he eliminated a possibility of a mechanical stimulation by laying down the nerve and the muscle on two metal plates and bringing the arc in touch with these plates rather than with the tissues: the contractions continued to occur. Also, he immersed the frog's feet in one glass of water and its crural nerve into another glass. When the arc touched the surface of water in both glasses, the legs contracted. Second, he proved the involvement of electricity by showing that if the connecting arc included a piece of a non-conducting material, such as glass or resin, there was no twitching. Third, he eliminated an involvement of static electricity coming from external bodies, by providing the arc with a glass handle and making the support for the frog from a conducting material. Finally, he excluded the role of atmospheric electricity by demonstrating that contractions continued when the frog's body was submerged in water. If the electricity involved, Galvani said, was neither static nor

Figure 1. Galvani's experiments.

atmospheric, it had to be a new kind of electricity, and probably it was the "animal electricity."

2.2. THE "CONTACT" ELECTRICITY

There were some experimental results that Galvani could not account for, in particular, that convulsions were much stronger when the arc consisted of two different metals rather than a single one. This circumstance appeared crucial to Alessandro Volta (1745–1827), Professor of Physics at the University of Pavia. Volta agreed that the phenomenon was electrical but he assumed the main source of electricity to be outside, in the contact of different metals, and the frog merely being a conductor.

Being unable to explain the convulsions produced with a single metal, Volta maintained for a while the "animal" electricity together with his "contact" electricity (Volta 1793). Then in 1794 he discovered experimentally that a difference in temperature or polish at the ends of a wire was sufficient to excite contractions. Thus, he concluded, the contact electricity was sufficient to explain all phenomena, since if a single wire was heterogeneous, it could be considered as two different metals. Giovanni Aldini (1762–1834), Professor of Physics at the University of Bologna and Galvani's nephew, countered this argument with a new experiment. He showed that mercury free of the heterogeneity described by Volta did produce contractions, and so did charcoal (Aldini 1794). Another strong blow to Volta's theory came from Galvani's experiment in which contractions occurred when a nerve directly touched the muscle without any intermediaries.

First, Volta tried to find a flaw in the opponents' experiments, such as mechanical pressure or a chemical difference at the ends of the connecting arc, but eventually he decided to modify his own theory. Prior to that he maintained, on the basis of his experiments, that the simplest circuits to create the contact electricity consisted of two different conductors of the first class (metals and some other solids) and one conductor of the second class (liquid or a humid body), or of two different conductors of the second class and one of the first class. Now, he adds to these two a third variety: a circuit made of three different conductors of the second class (Volta 1797a). This modification was specifically designed to explain the "all-animal circuit", in which the conductors in question were the nerve, the muscle, and the animal fluids. Thus, the key of the new theory was that a contact of *any* two different substances is a mover of electricity.

Although Volta's new theory (let us call it the "universal contact") explained all galvanic phenomena, this generality was a sign of weakness rather than strength. Indeed, his statement that any three conductors of the second class created galvanic electricity could not be independently verified, for the only experiment supporting it – the "all-animal circuit" – was the one that the hypothesis was created to account for. Besides, Volta has a difficulty in combining this generality with the fact that the contractions produced by two different metals and a liquid were much stronger than the rest. Actually, he opened a path for returning to his original theory that weakest contractions may result from internal animal electricity while the stronger ones from the external electricity. At the time, no one utilized this opportunity, but fifty years later this theory began to gain strength.

By 1795, Volta realized that he could not fully establish the existence of the "contact" electricity without eliminating the "animal" electricity. The main difficulty with this was that the frog's preparation was the only available sensitive detector of galvanic electricity: one could always say that electricity responsible for the contractions came from the frog itself rather than from the external part of the circuit. For this reason both theories had about the same standing among scientists. The only way to prove that a contact of different substances creates electricity was to replace a frog by a non-animal electric sensor, and Volta decided to try Nicholson's doubler of electricity.

The doubler consists of a sensitive electrometer and three polished discs of the same diameter made of brass. One disk is set on top of the electrometer, its upper surface being covered with a thin layer of an insulating warnish (Figure 2). The second disk is placed on top of the first, it is warnished on both sides and provided with an insulating handle attached to its edge. The third disk, warnished on its bottom side sits on top of the second, it also has an insulating handle perpendicular to its plane. In the beginning, only two lower disks are in use so that the instrument works as a condenser electrometer: the source of electricity touches the underside of the first disk and the upper side of the second disk. Upon removing the source of electricity the second disk is lifted by the insulating handle. If the deviation of the electrometer is still small, the third disk held by its insulating handle is placed on the elevated second disk. After this a finger briefly touches the upper side of the third disk and the two are separated. Next, the third plate touches the underside of the first while the second sits on top of it. Now, upon finger touching the second plate with a finger and removing it after first removing the third plate, the amount of

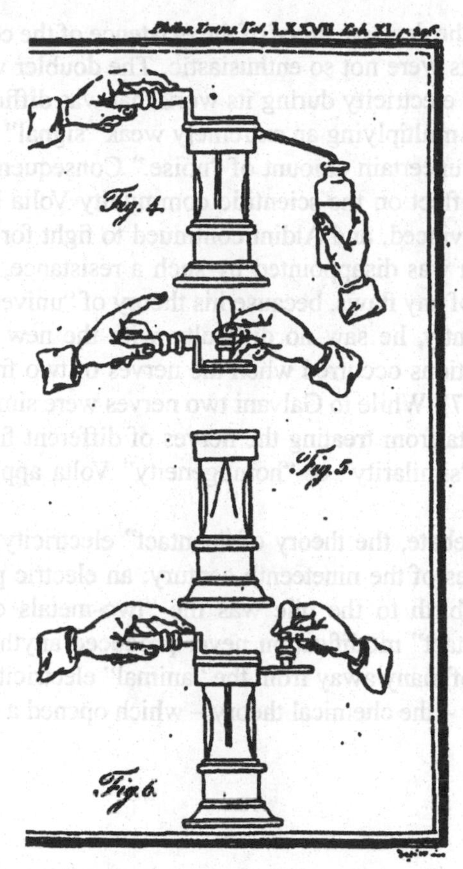

Figure 2. Bennet's doubler of electricity.

electricity on the electrometer is doubled. By repeating this procedure many times the amount of electricity can be considerably increased, which allows to measure very small charges.

Volta used a modification of the doubler invented by Cavallo, in which only the middle disk was mobile while others were fixed (Volta 1797b, c, d). He placed silver and tin rods on a wet cardboard, brought them in contact with two brass disks of the doubler and started the 'machine": after 20–30 turns the leaves of the electrometer diverged by 6 to 10 degrees. When he replaced the brass mobile disk with the tin disk and connected it a to brass rod while a brass disk touched a tin rod, the doubler showed a noticeable quantity of positive electricity. However, when he reversed the connections making each bar touching the disk of the same metal, there was no sign of electricity. Volta interpreted this result as a proof that electricity is created at the junction of different metals rather than that of a metal and a liquid (Volta 1797).

Volta was convinced that he found the decisive proof of the existence of the contact electricity, however other scientists were not so enthusiastic. The doubler was known for producing a "spontaneous" electricity during its work that was difficult to get rid of, which implied that while multiplying an extremely weak "signal" the instrument might have added to it an uncertain amount of "noise." Consequently, this experiment did not produce the effect on the scientific community Volta had expected. Galvani died in 1798 unconvinced, and Aldini continued to fight for his theory for many years to come. Volta was disappointed by such a resistance, for he considered his case clear and free of any flaws, because his theory of "universal contact" covered everything. Apparently, he saw no difficulty with the new experiment of Galvani in which contractions occurred when the nerves of two frogs touched one another (Pera 1992, p. 147). While to Galvani two nerves were similar substances, no one could prevent Volta from treating the nerves of different frogs as dissimilar. With no definition of "similarity" or "homogeneity" Volta applied this concept any way he wanted.

Regardless of its success in the debate, the theory of "contact" electricity led Volta to one of the greatest discoveries of the nineteenth century: an electric pile. More exactly, the theory that gave birth to the pile was the "two-metals contact" theory, while its "universal contact" modification never produced anything. While the pile diverted the attention of many away from the "animal" electricity, it brought to the fore another challenger – the chemical theory – which opened a new controversy.

2.3. THE "CHEMICAL" THEORY

This theory was initiated by Giovanni Fabbroni in 1792, but it became known only in 1797. According to Fabbroni, the primary cause of galvanic phenomena was chemical reactions rather than electricity. The basis for this theory was provided by an inability of electric theories to explain certain phenomena, in particular, why contractions occur even when the connecting circuit is open, or why the sensation of taste stimulated by a bimetal lasts after the bimetal is removed. Fabbroni experimented with different metals immersed in water and found that one of them oxidized, but only if the metals touched one another. This observation led him to the suggestion that the galvanic phenomena are due to an oxidation. Fabbroni did not try to eliminate electricity from galvanic phenomena altogether: he insisted that chemistry must have some role in them, in particular, in producing the sensations of taste or light. The "chemical" theory had a number of followers, including Alexander von Humboldt, however, it became really important only after the discovery of the electric pile when it redefined the role of chemical reactions in galvanic experiments from excluding electricity to being its cause.

2.4. THE PILE

The discovery became known after the publication of Volta's paper "On the electricity excited by the mere contact of conducting substances of different kinds" submitted to the Royal Society of London in April 1800 and read before the Society on June 26. The paper begins with a promise to inform of "some striking results I have obtained in pursuing my experiments on electricity excited by the mere mutual contact of different kinds of metal, and even by that of other conductors, also different from each other, either liquid or containing some liquid, to which they are properly indebted for their conducting power" (Volta 1800, PM, p. 289). This sentence already contains a complete theory of the phenomena to be described. By offering a theory up-front Volta implies that the purpose of the paper is not so much to get an additional support for this theory as to describe some remarkable phenomena he observed by means of a new apparatus. This apparatus consisted of many similar components, each of which included two different metals ("couples"), such as silver and zinc or copper and zinc, and a piece of cardboard or cloth moistened with pure or salt water. In one form of this apparatus (a "pile") all these components ("couples") made up a column arranged from the bottom up, for instance, as follows: copper, zinc, cloth, copper, zinc, cloth, etc. Another version of this apparatus, called a "chain of cups," consisted of a number of non-metal cups filled with salt water, each having a zinc and a copper plates immersed into water. The cups were arranged so that zinc of one cup was connected to copper of another cup, and so on (Figure 3).

Volta observed that when the number of couples was sufficiently high, the pile produced a shock similar to that of a Leyden jar. In addition to a shock, the apparatus could affect an electrometer and produce an electric spark, although these actions were less pronounced than the shock. For these reasons, Volta compares his apparatus (it became known later as "voltaic pile") to a battery of Leyden jars, "weakly charged" but of an "immense capacity." However, he emphasizes two important differences between them: (1) the pile acts continuously, providing repeating shocks without being recharged by an external electricity; and (2) it consists solely of conductors of electricity. Volta drew from this difference two consequences. One was that he discovered the first "perpetual" source of electricity. Another one, less known, was that he found an explanation of the torpedo fish.

The recurrent references to electric fish may appear intriguing to the reader viewing this paper in light of the debate about the existence of "animal" electricity, because the relevant experiments were carried out on frogs rather than torpedo. However, as Volta understood it, he had already refuted the "frog's electricity" in 1797, while electric fish remained a mystery. In fact, Volta never doubted that the shock produced by the torpedo was electrical, nor did he question that the electricity involved, unlike the case with frogs, resides inside the animal. His main argument with physiologists was that if electricity had any role in animal life, it could be explained by physical factors alone without bringing in any mysterious

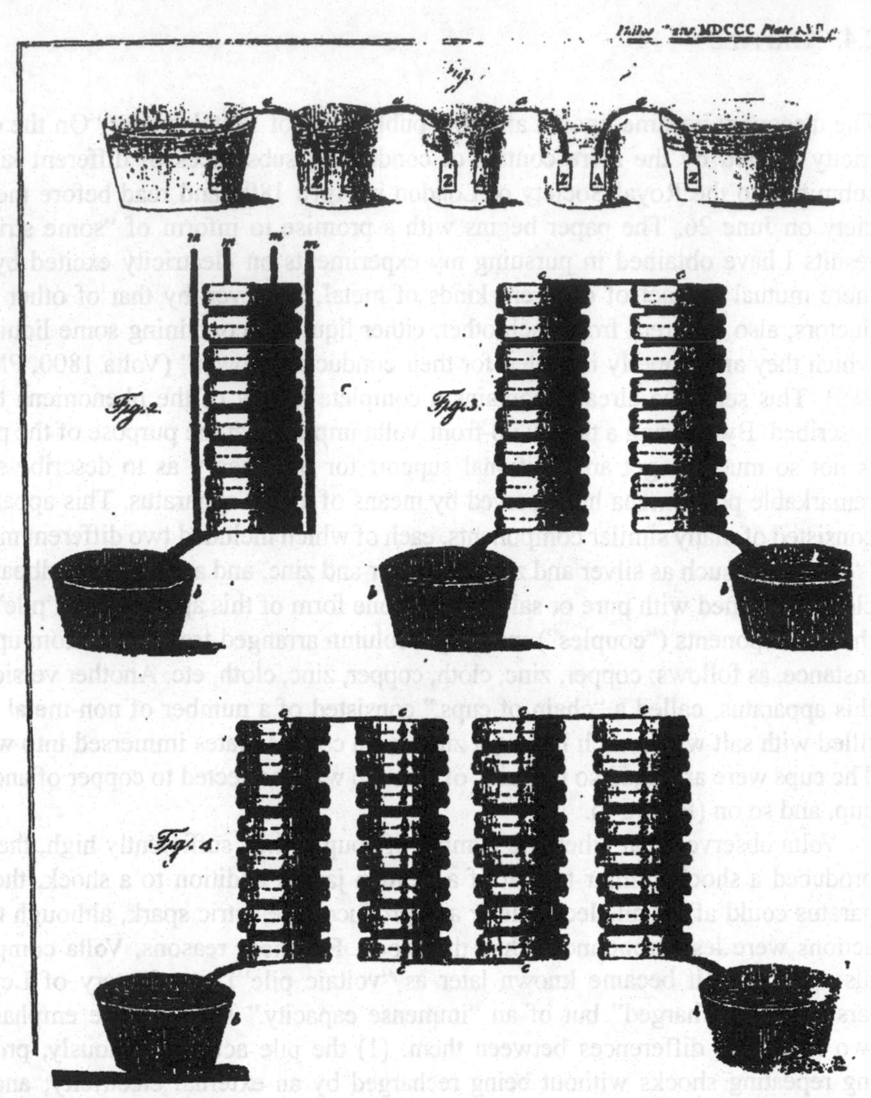

Figure 3. Volta's experiments.

"vital forces." As concerns frogs, he had already demonstrated – so he believed – that their contractions were due to an external "contact" electricity. In the case of the torpedo, however, his task was different: to conceive a physical model of its electric organ. Without fulfilling this task, Volta did not feel that his program of explaining life phenomena by physical processes was complete.

Volta begins with a critique of William Nicholson's theory of the electric organ of the torpedo, which compared it to a battery of Leyden jars. In Volta's view, since all membranes making up the columns of the electric organ are filled with fluids, they are comparatively good conductors. Since a Leyden jar cannot be made without an insulator, Volta concluded that electricity produced by the torpedo and some other fish cannot be static electricity. On the other hand, his pile consisted solely of conductors, and this, Volta supposed, could be the necessary model: the electricity of fish is galvanic, being produced by a contact of organic substances of different nature. To support this theory he indicates that the shocks produced by his apparatus are comparable in strength to those of a languid torpedo, and that it can give repeating shocks. He even calls his apparatus an "artificial electric organ." This name, as well as the initial shape of the apparatus – a column – show that his preoccupation with imitating the electric organ of the torpedo was an essential element of his research program.

To convince his readers, Volta wants them to succeed in repeating his experiments. Thus, he explains in detail how to build the apparatus and how to use it. In particular, he recommends to increase the number of couples and wet the fingers touching a pile, or, better still, by immersing a part of the hand in water that is connected to the pile. These tips are more than empirical findings, they are closely correlated with his theory. Volta maintains that only a junction of two metals is an "electric motor," while the liquid itself is merely a conductor. Since one of the two metals attracts electricity stronger than the other, each couple moves electricity in a certain direction, e.g. from zinc to copper. Thus, if several couples have the same orientation, their efforts combine, and electricity moves faster: the more couples, the better. As to wetting the hand connected to the pile, it is to reduce its resistance. Using the whole hand instead of fingers serves the same purpose by increasing the area of its contact with the liquid. This idea provides a fine opportunity for physics teachers to expand their teaching of resistance to non-metal conductors, especially because its meaning is not clear to some people even now (Mentens 1998, p. 309).[1]

Although Volta insisted that he does not need the pile to support his theory of contact electricity, it was the pile that made many scientists to turn to Volta's theory from that of Galvani. They reasoned as follows: (1) the actions of the pile are electrical; (2) since its effect is nothing else as a multiplied effect of a single couple, thus a contact of two different metals creates electricity; and (3) the electricity created by a bimetal is the same whether it is detected by an electrometer or by a frog, thus Galvani's experiments were due to the "contact electricity" rather than "animal electricity." In fact, the last conclusion was not logical, since a circuit with a frog could have had both sources of electricity, but this detail went unnoticed. Likewise, few people noticed that they had begun using the term "galvanic" ("galvanic circuit,' "galvanic current," etc.) to refer to phenomena produced by a voltaic pile rather than to those involving frogs (Kipnis 1987, p. 135). Yet, while securing Volta's victory over Galvani and Aldini, the pile brought to life even more powerful objections to his theory than before.

2.5. THE CHICKEN-AND-EGG PROBLEM

The first objections came from England. Having seen the first part of Volta's paper before it was read in full at the Royal Society on June 26, 1800, several scientists constructed voltaic piles and began experimenting with them. In addition to the effects described by Volta, they found that a pile can produce various chemical phenomena. In particular, Anthony Carlisle and William Nicholson observed the release of oxygen and hydrogen, which they attributed to decomposition of water, and also an oxidation of metals, while William Cruickshank precipitated a number of metals. A few months later, on the basis of these and other experiments, William Hyde Wollaston (1766–1828) and Humphrey Davy (1778–1829) suggested that, contrary to Volta's opinion, liquids play an active role in galvanic phenomena, and chemical reactions may be the cause of electricity rather than its consequence (Wollaston 1800). To show the active role of the liquid Davy performed the following experiment (Davy 1800). In an iron-copper pile with water, iron is charged positively, but if water is replaced with sulfate of potassium it changes its electrization to negative. He also created a pile of a single metal but of two different liquids: metal, cloth wetted with nitric acid, cloth with water, cloth with sulfate of potassium, same metal, nitric acid, etc. The acid and the alkaline at the ends of the pile were connected by paper strips moistened with water. When the metal was replaced by charcoal the pile worked too.

In his first responses to this criticism, Volta insists that he had already proved the role of a contact of different metals as an "electric motor." He describes an experiment in which he held a zinc and a copper plates by insulating handles, touched them to one another, separated, and brought them in contact with the plates of a condenser electrometer. The leaves of the electrometer diverged by 1 to 2 degrees. Since no third wet object was involved, Volta concludes that the electricity is caused by a mere contact of two metals, without any chemical interaction.

Soon Volta changes his tactics claiming that the objections to his theory actually support it, including Davy's experiment with one metal and two liquids. In particular, he says, since a contact of any two different substances produces tension, and since the metal (or charcoal) is positive relative to one liquid and negative relative to the other, both tensions move electricity in the same direction. As to the role of chemical reactions, he observes that since adding salt to pure water does not change the tension or polarity, chemical reactions are of no consequence to producing electricity. Volta agrees however to give chemical reactions a role in improving the conductivity of the pile: when an acid, for instance, attacks a metal surface, it adheres closer to it than water and thus diminishes the resistance of this contact (Volta 1802).

The inconsistency of these two arguments is caused by Volta's usage of different detectors of electricity in the two cases. If the main criterion of the pile's power is tension as measured by an electrometer, then, indeed, different liquids produce about the same effect. However, when the power of a pile is measured by a shock or the rate of a chemical reaction, both of which are derivatives of current, changing the liquid does change the outcome. In fact, Volta himself confirms this by noting that adding salt to water made the shock much stronger.

His second argument would have been unassailable, if he could provide an independent evidence that the conductivity of a liquid depends on the substance dissolved, its concentration, the area of contact, etc. However, for static electricity, such measurements were limited to comparing pure water with sea water. As to galvanic electricity, no such data existed at all. In fact, eventually it became clear that the only way to compare the conductivity of two piles made of the same number of couples was by comparing an effect that depends on current, such as a shock received by the same person. Thus, Volta had no ground to *assume* that given the same number of couples, the pile made of copper and zinc immersed into a weak sulfuric acid acts stronger than the one made of silver and zinc immersed in saline water, because the former pile has a greater conductivity than the latter.

Initially, Volta's theory of the pile had a considerable support, especially from Parisian physicists, but with time, especially since the 1830s, the chemical theory gained the upper ground. Interestingly, after 1802, Volta himself no longer participated in the debate.

3. Lessons from History

3.1. DEBATING THE TRUTH IN SCIENCE

A teacher should expect students to be surprised by the fact that scientists' behavior does not suit the image they have of a "scientific discussion," where the participants are attentive to the views of one another, passionless, and pursue no other goal as finding the truth. What they learn from the story, however, is quite different.

First, they see a rigid, uncompromising attitude, where each side claims to have the whole truth and insists on it until death. As shown above, there was some evidence that both "animal" and "contact" electricity exist. However, neither Galvani and Aldini, nor Volta were interested in a compromise. One finds this even more surprising when one remembers that Volta's first theory contained both sorts of electricity. Actually, instead of eliminating "animal electricity" Volta's last theory of the "universal contact" created a new model for it: electricity inside an animal originates at the contact of different tissues. Bearing this in mind, it would have been fair for him to say either in 1797 or in 1800: "There is no way I can disprove the existence of electricity inside an animal shown in the "all-animal circuit." But I can offer a better model of this electricity: animal electricity is produced not by "vital forces" which exist exclusively in animals but by a universal physical cause such as a contact of different tissues." This would have been perfectly consistent with his physical model of the electric organ of the torpedo. However, Volta never said this, and no one raised this sort of objections to his theory, at least not publicly.

Second, neither side was willing to reveal the shortcomings of one's own theory until forced to do so. Galvani knew that both muscles and nerves are conductors, thus to preserve his Leyden-jar model he invented an oily substance that supposedly insulated the sciatic nerve from the muscle. Likewise, Volta could not have been unaware that his latest theory of the "universal contact" was unverifiable. Indeed, to test, for instance, that a contact of two different tissues creates the "contact" electricity one cannot use another frog's preparation as a detector, because the latter could generate the "animal" electricity. Nor can one employ an electrometer, since in this case the true source of electricity could be a contact of an animal tissue with a metal part of the instrument rather than a contact of two animal organs.

Third, while comparing a theory to experiment, its partisans emphasized only those aspects of an experiment that suited their theory. For instance, Volta preferred to measure the "power" (or "activity") of a voltaic pile by its tension, because tension does not depend on the chemical activity of the liquid. Davy, on the other hand, chose for that purpose the amount of gas released by the pile, because it depends on the liquid.

Fourth, when a logical connection between experiments and a theory was necessary but difficult to establish, a circular reasoning was called to help. For instance, to prove that the true "mover" of electricity is a contact of two different metals rather than that of a metal and a liquid, Volta carried out an experiment with dry zinc and copper plates connected to a condenser-electrometer. Whatever was the nature of the electricity he measured (probably it was static electricity), his conclusion that it was the same electricity he had obtained with the two metals touching animal tissues was unfounded. Such conclusion would be logical only if one *presumes* that the cause of the phenomena is solely in the metals, which is what one has to prove.

Finally, unlike the case of Galvani, in his debate with Davy, Volta does not offer any new experiments that could have shed new light on the matter. Apparently, he believes that his theory does not need an additional support, thus he focuses on counter-charges which could weaken the claims of the chemical theory. For instance, he questions the evidence that certain chemical reactions, such as oxidation, create electricity. Or, he attacks the usage of the shock as a gauge of a voltaic pile. In his view, "the electrometer is the best judge of the electric force, that is, it provides us with a more reliable and more exact measurement of this force than the commotion" (Volta 1802, p. 343). This change of his original position might have resulted from Volta's inability to explain why a shock produced by a pile does not depend on the size of its plates.

Without denying that the "human factor" affected somewhat the rhetoric of these controversies, there are details which do not fit the picture of a scientific debate as a purely social activity. Let us check the possibility of objective factors at work, by asking questions about the "deviations" from such a social discourse.

3.2. OBJECTIVE GROUNDS

The first question is: why did scientists divide for a long time between two theories instead of agreeing to embrace one of them? We see that in both cases each of the two competing theories was internally consistent, supported by experiment, and initially enjoyed about the same standing among scientists. If two theories appear to be equally legitimate, choosing one of them should be a matter of a personal choice, subject to various factors. For instance, we see an influence of professional interests, since physiologists primarily supported Galvani, physicists sided with Volta, and chemists preferred the "chemical theory."

The fact that the experimental results cited by each side did not contradict those of the opponents means that there was enough room for two theories. This was possible because the two theories focused on different aspects of the same complex phenomenon combining physiological and physical components. It was possible for Galvani to see the origin of electricity inside an animal and for Volta, outside it, because in most of these experiments both electricities were present: bio-potentials always exist in animal tissues, and employing metal conductors to connect them introduces electrochemical potentials. This was proven only around 1850, when the invention of depolarizing electrodes permitted scientists to separate the two electricities. While such technology did not exist at Galvani's time, still his experiment with the "all-animal circuit" should have warned against dismissing the idea of "animal" electricity too easily.

In the case of the pile, one could attribute the origin of electricity to a contact of two different metals (the "contact" electricity) with the liquid being a conductor, or to a contact of these metals with a liquid (the "chemical" electricity). As long as one considered the effect of the pile to be a multiplication of the effect of a single pair, it did not matter how the pair functioned. To account for a rise in a pile's ability to produce gas when a diluted acid replaced water, or with an increase of the size of its plates, one could say either that a greater chemical activity released more electricity at the contact, or that it improved electric conductivity: the difference was semantic, because the two concepts could not have been measured independently.

The second question is: why did not contenders try to reach a compromise? This appears to be quite possible in the case of Galvani and Volta because initially Volta accepted both "animal" and "contact" electricities. The main reason for him to eventually eliminate the former was adhering to the "Occam's razor", the principle of reducing the number of possible causes of phenomena to a minimum. As soon as he found that he could explain all galvanic phenomena by the "contact" electricity, he pronounced the concept of "animal" electricity unnecessary. Scientists had used "Occam's razor" frequently but not always successfully, and Volta was out of lack.

The third question is: why did each contender stubbornly adhere to a chosen theory ignoring its deficiencies? Since some defects of each theory were obvious to its defenders from its inception, apparently they decided to support it because of the theory's *positive* contribution, with the hope that future research will resolve its difficulties and prove *their* theory to be the winner.

Thus, the behavior of Volta and his opponents in a debate was in part determined by impersonal factors that depended either on the phenomena under investigation or on common practices of scientists. It is quite clear, though, even from the materials presented above that personal factors were also involved, for some arguments of some participants suggest that winning an argument was no less important to them than finding the truth. However, this subject is beyond the scope of this paper.

4. The Experiments

These experiments are to be conducted in conjunction with the corresponding part of the story. To a large extent they may be open-ended. If a teacher wants to use these experiments to teach students the art of an investigation (Kipnis 1992), it is desirable to do them when students know only some of the historical results: in this way students will have an opportunity to compare their own results with those of Galvani or Volta. The experiments, especially those of Volta, allow many easily achievable modifications, and students are to be encouraged to be creative and devise new experiments to resolve the issue between Volta and Davy.

To give students a better feel of the original experiments, the emphasis here is on using the apparatus similar to the historical one and employing the original procedure. In the case of Volta, an electrometer had to be replaced with a voltmeter. Measuring current with a multimeter is a modern addition; however, lighting a bulb is a modification of a historical experiment of fusing a thin wire.

4.1. GALVANI'S EXPERIMENTS

If possible, do these experiments in a laboratory setting. Otherwise, the teacher should do them with students' help as a demonstration, making them visible to the whole class by means of a camcorder and a television set.

Materials: A frog's preparation (prepared not in front of students): it consists of hind legs and a part of the vertebra, from which sciatic nerves should be uncovered. Needle, electrostatic generator, a wire insulated at one end, strips of various metals (zinc, copper, aluminum, steel, etc.), insulated wires.

Procedure: First, demonstrate stimulation means known before Galvani: mechanical, static electricity, and chemical. Prick the nerve with a needle until a leg twitches. Take a metal rod with an insulating handle. Charge it from any source of static electricity and bring the metal in touch with the nerve or the muscle: you should see a brief contraction. Put a grain of salt on the nerve: the leg will start twitching.

Then show Galvani's experiments: (1) Touch the nerve with a zinc strip then touch the muscle with a copper strip: you should not have any reaction. Then bring the other ends of the metals into a contact: a contraction occurs. Try other pairs of metals, then try two strips of the same metal; (2) place a piece of aluminum foil under the nerve and another one under the muscle and touch them instead of the tissues: you should see contractions; (3) cover the top parts of zinc and copper strips with an insulating tape and repeat the procedure: you should see no twitching at all.

4.2. VOLTA'S EXPERIMENTS

4.2.1. *The Pile*

Materials: Zinc and copper squares with a side of 4 cm or larger, paper towel, diluted sulfuric acid or lemon juice, multimeter, a small 25 mA bulb, wire leads with alligator clips, a plastic cup.

Procedure: Cut pieces of a paper towel slightly smaller than the metal squares, moisten it with water squeezing the extra liquid out, make a pile of 30 couples: Zn, paper, Cu, Zn, paper, Cu, etc. Attach the leads to the ends of the pile and grasp them firmly with fingers of both hands previously well moistened. If the shock is not felt, immerse the leads into two cups with water and put the fingers into the cups.

Touch the leads to the tip and the back of your tongue: you should feel a sour taste. Touch the leads to the tongue and the upper gum: you shall see a flash of light (close the eyes). While doing physiological experiments, you can start with connecting one wire to an intermediate plate rather than the end plate and gradually increase the number of couples in the circuit (5, 10, etc.)

Measure the voltage across the pile and the current. Calculate the resistance. Try to light a small bulb.

Reassemble the pile, replace the paper squares with others moistened with a weak acid. Repeat all experiments and compare the results. Does the choice of a liquid affect the power of a pile?

4.2.2. *The Chain of Cups*

Materials: 30 plastic cups which are large enough to host zinc and copper electrodes, wire leads with alligator clips, multimeter, a small bulb. If possible, make electrodes of the size of 2 cm × 10 cm; otherwise, use the squares from the previous part.

Procedure: Place a zinc and a copper electrode in each cup, taking care they do not touch one another (use a separator made out of an insulator), fill the cups with water, and connect the electrodes so that zinc of one cup were connected to copper of the next cup, and so on.

Repeat all the experiments described in the previous part and compare the results. Replace water with the weak acid and repeat the experiments. Does the choice of a liquid affect the power of the battery? Compare to the case of the pile.

4.3. STUDENTS LEARN FROM SCIENTISTS

If a teacher is keen on the investigative experimentation, the story supported by some historical experiments may have a practical importance in teaching students how to go about creating a theory explaining their own investigative experiments. In particular, they learn that as long as a theory is consistent and explain several experiments, it has a right to exist even if it cannot account for other experiments. Here the concept of the "partial truth" is very useful: within the given range of phenomena studied and time spent the "partial truth" is *the* truth, and if future research modifies it, the original conclusion still preserves its validity within the original range. Although both Galvani and Volta believed to have discovered the whole truth, since their theories did not cover all known phenomena, they come under the above stated definition of a "partial truth". And if something was good (in the modern view) for Galvani and Volta, it is good for students too. This means that students should not fear of inventing a false theory in their investigations, provided they take care to make their conclusions sufficiently consistent and based on a sufficient number of experiments.

Naturally, students will find out soon the ambiguity of the word "sufficient": one cannot know in advance how many times to repeat each experiment and how many times to modify it in order to arrive at a "partial truth" that is closer to the correct result rather than a false one. Having learned this from their own experience students will become more critical to the certainty of the results of historical experiments and their validity as arguments in a theoretical debate.

5. Conclusion

Learning about a historical controversy may improve students' understanding of how scientists defend a new theory. While the conclusions drawn here are consistent with other cases not discussed in this paper, teacher is not advised to present them as general: it is better to discuss other cases (at least one) and let students do the generalization. Our analysis is not complete, because the role of "human factors" in a scientific debate is left out. This subject certainly deserves a separate study, however, its absence should not preclude teachers from discussing the impersonal factors, especially because the latter are more relevant to improving students' skills in conducting their own investigative experiments.

Note

[1] The author treats the immersion of the whole hand into water as nothing more than a showmanship.

References

Aldini, G.: 1794, *De Animali Electricitate. Dissertationes Due*, Bologna, **I**, 5–8.

Davy, H.: 1800, 'Notice of Some Observations on the Causes of the Galvanic Phenomena', *Nicholson's Journal of Natural Philosophy* **4**, 337–342.

Galvani, L.: 1953, *Commentary on the Effects of Electricity in Muscular Motion* [1791], transl. by M.F. Foley, Norwalk, CT.

Galvani, L.: 1794, *Dell'uso et dell'attivita dell'arco Conduttore nelle Contrazione dei Muscoli. Supplemento*, pp. 4–6, Bologna. See an English translation of the description of this experiment in Dibner, B.: 1952, *Galvani-Volta*, Norwalk, CT, pp. 50–51.

Kipnis, N.: 1987, 'Luigi Galvani and the Debate on Animal Electricity, 1791–1800', *Annals of Science* **44**, 107–142.

Kipnis, N.: 1992, *Rediscovering Optics*, BENA Press, Minneapolis.

Mertens, J.: 1998, 'Shocks and Sparks: The Voltaic Pile as a Demonstration Device', *Isis* **89**, 300–311.

Nicholson, W.: 1800, 'Account of the New Electrical Apparatus of Sig. Alex. Volta ...', *Nicholson's Journal of Natural Philosophy* **4**, 179–187.

Pera, M.: 1992, *The Ambiguous Frog*, transl. by J. Mandelbaum, Princeton University Press, Princeton.

Volta, A.: 1793, 'Account of Some Discoveries Made by Mr. Galvani .... In a letter to Mr. Tiberius Cavallo', *Phil. Trans. Roy. Soc. Lond.* **83**, 10–44; also *Le Opere di Alessandro Volta*, 7 vols (Milano, 1918; reprint: New York, 1968), **I**, 173–208.

Volta, A.: 1797a, 'Lettera Prima al Prof. Gren di Halla, 1 August 1796', *Neues Journal der Physik*, **3**, 479–481; also *Opere* **I**, 395–413.

Volta, A.: 1797b, 'Lettera Seconda al Prof. Gren [August 1796]', *Neues Journal der Physik* **4**, 107–135; also *Opere* **I**, 417–431.

Volta, A.: 1797c, 'Lettera Terza al Prof. Gren [March 1797]', *Brugnatelli's Annali di chimica* **14**, 40ff; also *Opere* **I**, 435–447.

Volta, A.: 1797d, 'Mémoire sur l'Électricité Excitée par le Contact Mutuel des Conducteurs Même les Plus Parfaits ... en une Suite de Lettres au Dr. Van Marum', unpublished, *Opere* **I** 493–516.

Volta, A.: 1800, 'On the Electricity Excited by the Mere Contact of Conducting Substances of Different Kinds' (in French), *Phil. Trans. Roy. Soc. Lond.* **90**(2), 403–431; see also the English translation in: 1800, *Philosophical Magazine* 7, 289–311.

Volta, A.: 1802, 'Électricité Voltaïque', *Bibliotheque Britannique* **19**, 270–289, 339–350.

Wollaston, W.H.: 1801, 'Experiments on the Chemical Production and Agency of Electricity', *Phil. Trans. Roy. Soc. Lond.* **90**, 427–434.

Newton and Colour: the Complex Interplay of Theory and Experiment

ROBERTO DE ANDRADE MARTINS and
CIBELLE CELESTINO SILVA

Group of History and Theory of Science, DRCC, Instituto de Física 'Gleb Wataghin', P.O. Box 6165, UNICAMP, 13083-970 Campinas, SP, Brazil; E-mail: rmartins@ifi.unicamp.br

ABSTRACT. The general aim of this paper is to elucidate some aspects of Newton's theory of light and colours, specially as presented in his first optical paper of 1672. This study analyzes Newton's main experiments intended to show that light is a mixture of rays with different refrangibilities. Although this theory is nowadays accepted and taught without discussion it is not as simple as it seems and many questions may arise in a critical study. Newton's theory of light and colour can be used as an example of the great care that must be taken when History of Science is applied to science teaching. An inadequate use of History of Science in education may convey to the students a wrong conception of scientific method and a mythical idea of science.

1. INTRODUCTION

It is nowadays assumed that the use of the history of science may improve the teaching of science. Accordingly, there has been an increasing use of history of science by teachers – both at high-school and university levels. In the specific case of physics, the development and status of uses of history of physics in education has been recently reviewed (Bevilacqua and Giannetto 1996).

There are, however, some pitfalls on the way to this approach. The history of science can be misused (as anything else) and lead to a mistaken view of science. The general aim of this paper is to elucidate some of those dangers and to show that, given suitable precautions, the history of science may indeed help science teaching. Great care, however, must be taken to ensure adequate use of historical resources in education.

Instead of discussing those dangers in an abstract way, this paper will focus upon one recent attempt to apply History of Physics to education: Dudley Towne's use of Newton's colour theory (Towne, 1993). Towne used Newton's 1672 original presentation of his theory, together with experiments and other aids, in teaching beginning, nonscientist students. He claimed that Newton's work is clear, easy (and even 'delightful') to read and understand. He stated that the original paper is a model for the presentation of the scientific method. He also emphasized how easy it is to draw the correct inferences from Newton's experiments.

Both the analysis of Newton's work and its educational use as presented by Towne are highly problematical. Newton's arguments are not as straightforward as they seem. Besides, the interpretation of scientific

F. Bevilacqua et al. (eds.), Science Education and Culture, 273–291.
© 2001 *Kluwer Academic Publishers. Printed in the Netherlands.*

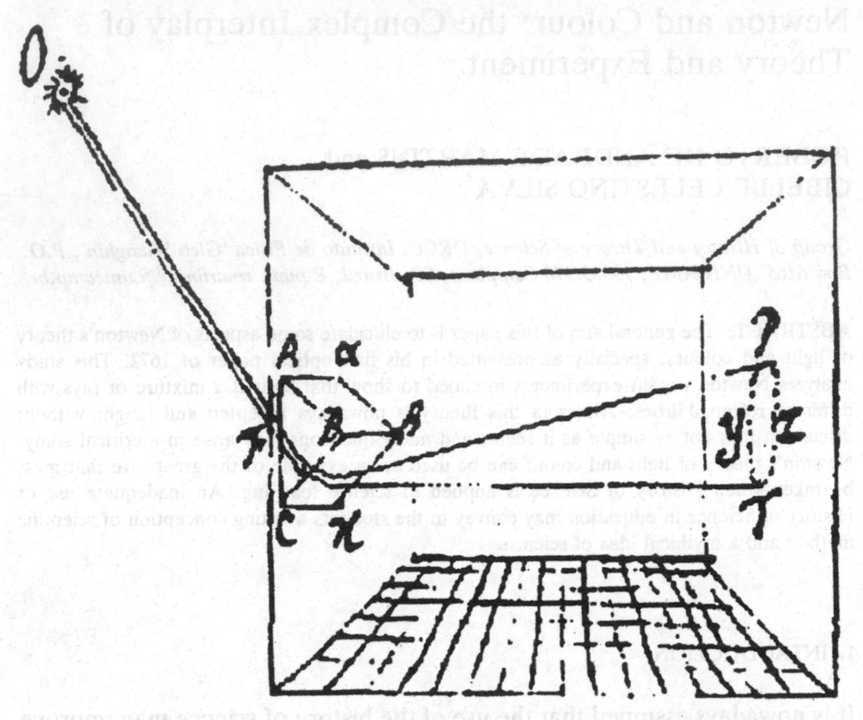

Figure 1. Newton's scheme for the first experiment in his 1672 paper.

method implicit in his paper is at variance with current historical and philosophical knowledge.

The general aim of this paper is to elucidate Newton's work and to show how it may improve science teaching.

2. THE DEFLECTION OF LIGHT BY A PRISM

In a paper published in 1672 (Newton 1672a), Newton presented his concept that light is a 'heterogeneous mixture of differently refrangible rays' – each colour corresponding to a different refrangibility. He presented several experiments to corroborate this theory. In the first one (Figure 1),[1] a beam of sun light passed through a prism and formed a spot[2] on the wall of his chamber. He noticed that the spot was not circular as the disk of the sun – it was oblong (Kuhn 1978, p. 35; Lohne 1968, p. 172). To explain this effect he assumed that the white light of the sun was composed of many different rays. Each kind of ray is refracted in a different direction and is associated with a different colour: 'the least refrangible rays are disposed to exhibit red colour, and (. . .) the most

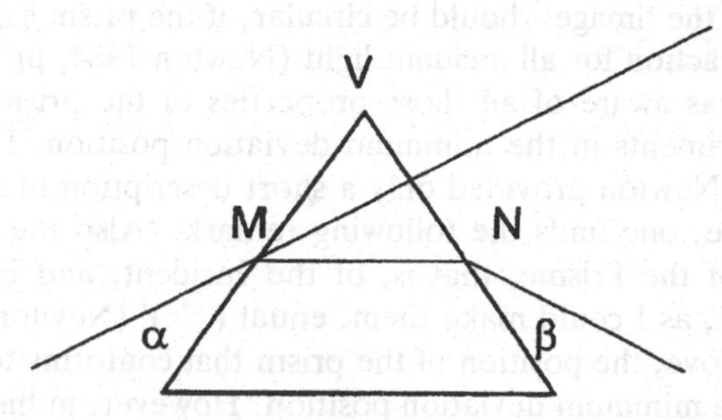

Figure 2. A prism in the minimum deviation position.

refrangible rays are all disposed to exhibit a deep violet colour' (Newton 1672a, p. 53).

One important fact in favor of Newton's theory was his *Experimentum Crucis*. In this experiment, light passed through two prisms. The first one produced a coloured spectrum and the second was used to study the deviation of each colour. The experiment showed that each colour of the spectrum suffered no further division at the second prism, and that each colour was deflected at a different angle.

In modern textbooks either the first or both those experiments are usually introduced as sufficient evidence for Newton's theory of composition of white light.

3. THE MINIMUM DEVIATION POSITION OF THE PRISM

When Newton described the single prism experiment, he remarked that the spot projected on the wall should be circular and not oblong, according to the 'received laws of refraction'.

Why did Newton state that he expected that the 'image' should be circular? Ask this question to undergraduate physics students, and you will notice the difficulty of that point. To understand what Newton meant, it is necessary to take into account the details of his experiment and some *implicit* considerations concerning the exact position of the prism.

There is one single position of the prism that would produce a circular 'image', according to the Cartesian law of refraction. It is the so-called 'minimum deviation' position. If the prism is slowly rotated around the axis that passes through the centre of the triangular faces, one observes that the direction of the deflected beam changes. There is one special position where the angle between the initial direction of the beam and its direction after passing through the prism is a minimum. In this position, the incident and refracted beams make equal angles with the sides of the prism (Alonso and Finn 1972; Figure 2). It is possible to prove that at this position the 'image' should be circular, if the prism exhibited a single index of refraction for all incident light (Newton 1984, pp. 53–54).

Newton was aware of all those properties of the prism and made his colour experiments in the minimum deviation position. However, in his 1672 paper, Newton provided only a short description of the first experiment.[3] There, one finds the following remark: 'Also the Refractions on both sides of the Prisme, that is, of the Incident, and Emergent Rays, were as near, as I could make them, equal (. . .)' (Newton 1672a, p. 49). As stated above, the position of the prism that conforms to this condition is exactly the minimum deviation position. However, in his paper Newton neither stressed the importance of this position, nor did he state that only in this position one would expect a circular 'image', according to the 'received laws of refraction'.

Did Newton know all that in 1672? In his published articles of that time he presented neither a proof similar to the one provided in the *Lectiones opticae* nor even the simpler one published in the *Opticks* (Newton 1704, p. 49). However, he clearly stated in the 1672 paper that he *computed* the angle between the rays coming from the Sun after they passed through the prism 'and found, that the emergent Rays should have comprehended an angle of about 31′, as they did, before they were incident' (Newton 1672a, p. 49). However, the measured divergence of the deflected beam was 2° 49′ instead of 31′. The discrepancy between the predicted and observed angle required an explanation, and Newton's theory provided it.

All this shows that the minimum deviation position is a necessary condition of Newton's first experiment.[4] On the other hand, if one reads critically the 1672 paper, it becomes evident that Newton's article is far from being clear and didactic, since Newton did not make it clear that the minimum deviation position of the prism was important. He also did not tell how to find this position (Sabra 1981, p. 237).

4. MISUNDERSTANDING OF NEWTON'S PAPER

When Newton published his first paper, many people were unable to understand that the whole argument depended on the choice of the minimum deviation position. The first critic of Newton's theory was the French priest Ignace Pardies.

Pardies stated that two rays that arrive at the prism would suffer no change in their relative angles in planes parallel to the axis of the prism. However, in a plane perpendicular to the axis, the angle after passing through the prism might be different from the initial angle. To substantiate his claim, Pardies presented the detailed computation corresponding to a special position of the prism. He concluded that two rays arriving at the first surface of the prism encompassing an angle of 30′ might leave the prism forming an angle of more than 3°, depending on the angle of incidence (Pardies 1672a, p. 87)[5].

In his answer to Pardies, Newton accepted the method and computations of the priest. However, he remarked that in his own experiment and calculations, he had assumed that the incident and emergent rays had equal inclinations relative to the sides of the prism, whereas in Pardies' calculation the angles were widely different:

> But the Rev. Father is under a mistake. For he has made the refractions by the different parts of the prism to be as unequal as possible, whereas in the experiments, and in the calculation from them, I employed equal refractions (Newton 1672b, p. 90).

Newton then presented a general (geometric) proof that when his experimental conditions are satisfied, the angle of the deflected rays should be equal to that of the incident rays.

Once Pardies understood the required conditions of the experiment, he agreed with Newton that the 'image' should be round, according to the usual optical theory. Pardies' behavior shows that he did not understand from Newton's first paper that the minimum deviation position was a crucial condition. It also shows that Pardies' criticism was not as silly as it seems at first sight.

Let us now consider Towne's account of Newton's first experiment. Nowhere in his article does he refer to the relevance of the minimum deviation position. On the contrary: in his footnote 7 he says that 'although it is not essential to do so for any of the experiments described in Newton's paper, to preserve a sense of reproducibility it is advisable to turn the prism so that some colour is at minimum deviation'. That is wrong. If the prism were not in this position in the first experiment, nothing could be concluded from it – as shown by Pardies.

5. ELIMINATION OF DIFFERENT HYPOTHESES

After one understands the theory behind Newton's first experiment, it is possible to grasp his first conclusion: the facts are in disagreement with the accepted theory of refraction. What else could be concluded from this experiment?

Towne stated that this experiment alone is sufficient to conclude that the light of the Sun was heterogeneous:

> (. . .) the oblong shape of the spectrum can be measured with a ruler, and is sufficient evidence for the declaration that light consists of 'difform rays, some of which are more refrangible than others'. (Towne 1993, p. 115)

It was *not* possible to conclude that, since other explanations were possible. Indeed, both Newton and his contemporaries (Pardies, Hooke, Huygens, etc.) suggested *several* explanations for this effect. In the 1672 paper, Newton explored many conjectures that occurred to him. He tested whether the oblong shape of the spot could be due to the different thickness of the prism, or to the size of the hole, or to the position of the prism (inside or outside the dark room). In all those variations of the first experiment, the spot remained oblong. Newton then devised a second experiment:

> Then I suspected, whether by any *uneveness* in the glass, or other contingent irregularity, these colours might be thus dilated. And to try this, I took another Prisme like the former, and so placed it, that the light, passing through both, might be refracted contrary ways, and so by the latter returned into that course, from which the former had diverted it. For, by this means I thought, the *regular* effects of the first Prisme would be destroyed by the second Prisme, but the *irregular* ones more augmented, by the multiplicity of refractions (Newton 1672a).[6]

The test showed that the spot was now circular. So, the irregularities of the glass were not the cause of the oblong shape.

Another interesting conjecture of Newton's was that light might travel in *curved* lines after passing through the prism. If light travels in a straight line, and if the hole is of negligible size, the dimensions of the spot will be proportional to the distance between the hole and the screen. If one takes into account the dimensions of the hole, then it is the *difference* between the dimensions of the spot and the diameter of the hole that should be proportional to the distance – as Newton indeed observed. So, light travels in straight lines after passing through the prism (Newton 1672a, p. 50).

In Newton's *Opticks* there is a much clearer presentation of the evidence. The second proposition of part 1, book 1, states that 'The light of the Sun consists of rays differently refrangible'. In the proof of this proposition, Newton presented his experiment of the oblong spot, but afterwards remarked:

> So, then, by these two experiments it appears that in equal incidences there is a considerable inequality of refractions. But whence this inequality arises, whether it be that some of the incident rays are refracted more, and others less, constantly, or by chance, or that one and the same ray is by refraction disturbed, shattered, dilated, and as it were split and spread into many diverging rays, as Grimaldi supposes, does not yet appear by these experiments, but will appear by those that follow (Newton 1704, p. 34).[7]

After this remark, Newton presented experiments #5 to #10 (Newton 1704, pp. 34–61), together with many variations and commentaries, before he concluded the proof of the proposition. Therefore, Newton himself clearly perceived that the first experiment was not sufficient to prove that the light of the Sun contains 'rays differently refrangible'.

After eliminating several alternative explanations, Newton presented a new important experiment. He called it the *Experimentum Crucis* – an obvious reference to Francis Bacon – and he probably intended it to be decisive.

6. THE 'EXPERIMENTUM CRUCIS'

A planned experiment is always undertaken after theoretical analysis. The naive belief that one must go to the laboratory with an 'empty mind' or that 'experiments talk by themselves' (as Towne's paper puts it) is an old scientific myth – and here 'myth' means 'outside reality'. When Newton undertook his study of colours, he was deeply concerned with a few theories about light. He was trying to find out which one was correct. Theory guided experiments – not the converse.

As Newton tells us, he was not the first one to observe the colours produced by a prism. Indeed, he stated that 'I procured me a Triangular glass-Prisme, to try therewith the celebrated *Phænomena of Colours*' (Newton 1672a, pp. 47–48). It was well known that prisms produced an effect similar to a rainbow – the phenomenon was described by Robert Boyle, René Descartes, Robert Hooke (Boyle 1664; Descartes 1637; Hooke 1665) and several different authors of that time. Several explanations had already occurred to many people.

In Newton's first experiment, the oblong shape of the spot was produced by different colours. Each colour emerged from the prism in a different direction. Nowadays, we interpret this as a *separation* of colours that are already present in white light. However, that was not the only (or even the most 'intuitive') interpretation.

The first idea that occurred to everybody – including Newton himself – was that the prism *produced* colours – that is, white light was *transformed* into a set of colours. Indeed, white light always seemed to be the simplest kind of light. When light passes through a transparent or translucent coloured body it acquires colour – and this seemed a *transformation* of light. In the same way, it was believed that the prism *created* the colours – it was not just a *separation* of colours.

When Newton published his studies of light and colour, Hooke's *Micrographia* (published in 1665) was an influential work. Hooke had presented in that book a very obscure theory about the transformation of white light when it is obliquely refracted.

In his 1672 paper, Newton had already arrived at the 'correct' conclusion: each spectral colour has fixed, unchangeable properties; and each colour has a specific refrangibility.

In Newton's theory, the least refrangible rays correspond to red and the most refrangible correspond to violet. This is a delicate point of Newton's theory. The relative refrangibilities of different colours vary in different substances. It is possible to find transparent bodies that deflect blue and violet light less than red light – contrary to Newton's belief.

Newton's idea that white light is not simple but a mixture of all colours is not intuitive. It did not arise at once in his mind, but evolved slowly from his intensive work. The main point was to find out whether colours can be transformed and created or not. This is the central aim of Newton's *Experimentum Crucis* (Lohne 1968).

In that experiment, a beam of solar light passes through a first prism and strikes a board with a small hole in it, so that only a small portion of the spectrum (a single colour) passes through it. This secondary beam reaches a second prism. Newton observed that the second prism did not change the colour of the secondary beam. He also noticed that different colours presented different deflections in the second prism: the red light suffered again the least deviation, and violet the greatest (Newton 1672a, pp. 50–51).

Newton compared this experiment to what happened in the case of white light in a single prism: different colours appear and each colour is deflected in a different direction. His explanation was that white light consists of a mixture of all colours that appear in the spectrum, each colour being separated from the others – but not created – by the prism, because of their different refrangibilities. This hypothesis also explained the oblong form of the spot in the first experiment:

> (. . .) the true cause of the length of the image was to be no other, then the *Light* consists of *Rays differently refrangible* which without any respect to a difference in their incidence, were, according to their degrees of refrangibility, transmitted towards divers parts of the wall (Newton 1672a, p. 51).

The relation between colour and refrangibility stated by Newton did not cause great controversy. The problematic question was the composition of white light. The statement that white light is a 'Heterogeneous mixture of differently refrangible Rays' led to a strong controversy between Newton and Hooke, Huygens and Pardies (Sabra 1962).

For Hooke, white light was a simple kind of vibration and coloured light was a modification of white light. He supposed that light was some kind of non-periodic wave that would acquire different properties near the edge of the light beam. Hooke believed that the wave front would become inclined relative to the direction of propagation when light was obliquely refracted – as in a prism. The extremity of the wave front that came first would become red and the end extremity would become blue. Near the prism we do not observe all spectral colours. We see exactly what Hooke describes: a white beam with small blue and red fringes on opposite sides.

If one observed the beam very far from the prism, the red and blue regions would expand and overlap. Hooke believed that all colours were produced by the blending of blue and red. So, it was possible to explain the colours produced by the prism.

In answer to Hooke's letter (Hooke 1672) Newton presented many experiments to show that white light is a mixture of different rays (Newton 1672c). In the 1672 paper, he had already combined the colours produced by the prism with the aid of a converging lens and produced white light.

> (. . .) all Colours of the Prisme being made to converge, and thereby to be again mixed as they were in the light before it was Incident upon the Prisme, reproduced light, intirely and perfectly white, and not at all sensibly differing from a *direct* Light of the Sun, unless

when the glass, I used, were not sufficiently clear; for then they would a little incline it to *their* colour. (Newton 1672a, p. 55)

The plan of Newton's experiment may be found in Newton's original drawing published in the 1672 paper. Towne referred to this experiment, but the conception of his drawing is unintelligible (Towne 1993, p. 115). It presents a parallel beam that becomes divergent without apparent cause. The beam leaving the prism is divergent, not parallel as he represents it. It will be difficult for any student to understand that drawing. It would be better to use Newton's original scheme, as it is much clearer than Towne's.

The composed white light produced by Newton was visibly equal to solar light. Nevertheless, neither this experiment nor the *Experimentum Crucis* proved that this resulting light was really equal to solar light. It could happen – as Hooke believed – that the white light of the Sun was simple, and that the different modifications of white light (the several colours) could be combined to produce another kind of white, by mutual compensation of their differences.

In all of Newton's experiments, light is refracted at least once. It could happen that the refracting medium acted upon light by changing it, in such a way that this modification remained unchangeable in subsequent refractions.

The choice between Newton's theory and the 'modification theory' could not be decided by experiment alone. Indeed, it was impossible to perceive the existence of all colours in white light, before it was refracted. Hence, it was always possible to maintain that, before any transformation, white light is simple and not composite.

Newton at last perceived that the distinction should be grounded upon methodological arguments. In his answer to Hooke, he said:

I see no reason to suspect, that the same *Phænomena* should have other causes in the Open Air. (Newton 1672c, p. 134)

This means that he saw no reason to introduce a distinction between two kinds of white light, if they exhibited the same properties in all experiments. One should not multiply entities if this is not necessary: one should choose the simplest theory, according to the methodological rule known as *Occam's razor*.[8]

Returning to Towne's paper, one sees that it does not discuss those questions.

He states that

(...) the simplicity of the experiments and the order in which Newton presents them allow the theory to form in the reader's mind before Newton makes a formal statement of the hypotheses. (Towne 1993, p. 113)

According to Towne, students will be led to the same theory as Newton and will conclude that white light is a mixture of rays. However, it was shown above that this conclusion is not straightforward and that there are other possible interpretations of Newton's experiments.

7. CONSTANCY AND COMPOSITION OF COLOURS

An important part of Newton's argument is contained in his experiments intended to show that spectral colours cannot be transformed into different colours. In his first paper Newton already stated the immutability of colours. He made several experiments intended to modify them and never observed any change. In his experiments he

> (...) refracted it with Prismes, and reflected it with Bodies, which in Day-light were of other colours; I have intercepted it with the coloured film of Air interceding two compressed plates of glass; transmitted it through coloured Mediums, and through mediums irradiated with other sorts of Rays, and diversly terminated it; and yet could never produce any new colour of it. (Newton 1672a, p. 54)

The *Experimentum Crucis* showed that a second refraction did not decompose the colours that came from the first prism. It was also necessary to show that when pure light (e.g. a spectral red light) is diffused by a coloured body (e.g. a blue paper) or passes through a transparent coloured glass its colour does not change – it only suffers an intensity change. Additional experiments devised by Newton to show that spectral colours do not change in those conditions were highly relevant to support his claim of the constancy of pure colours.

According to Newton's theory, coloured bodies do not transform the colour of the light they receive: they act as filters, allowing some colours to be reflected and absorbing other colours. Newton stated that the colours of natural bodies

> (...) have no other origin than this, that they are variously qualified to reflect one sort of light in greater plenty than the other. (...) that means any body may be made to appear of any colour. They have there no appropriate colour, but ever appear of the colour of the light cast upon them, but yet with this difference, that they are most brisk and vivid in the light of their own day-light-colour (Newton 1672a, p. 56).

This is another very important point of Newton's theory that Towne was unable to grasp. To illustrate this theory, Towne suggested an experiment that contradicts Newton's concept. He stated that two strips of blue and red paper illuminated by the spectrum will appear black and then turn into white depending on the part of the spectrum that shines upon them.

According to Newton's theory a paper will appear white if it reflects light of all colours of the spectrum, in a proportion similar to that of the Sun's light. This can never occur in the suggested experiment and therefore the strips of paper would never appear white. Besides, a paper will appear black if it absorbs most or all incident light. This would not occur with common blue or red paper under red or blue light – as stated in the suggestion. They must look dark but will reflect a small part of the incident light.

In the 1672 paper, Newton described experiments with red and blue pigments. When he threw different colours of the spectrum upon those pigments, he observed that they appeared of the same colour used to illuminate them, although they appeared more bright when their natural colour was cast upon them.

8. PRIMARY AND COMPOUND COLOURS

To understand Newton's argument, it is also necessary to stress his concept of simple (or primary) colour. Our common sense accepts that colours can be changed in several circumstances, such as in the case of mingling pigments or beams of light. If we regard colour as the qualitative property of light perceived by our senses, colour can indeed be changed. It is possible to produce orange colour from yellow and red paint. So, according to common sense, colours are not immutable as Newton asserted.

To develop his theory, Newton created a new concept of colour. He distinguished between our sensation and the properties of light itself. He carefully stated that different rays of light have different 'disposition to exhibit this or that particular colour'. The same kind of light always produces the same sensation, but the same sensation is sometimes due to different kinds of light.

Newton introduced a theoretical distinction between simple (or primary) colour and compound colour. The first one (primary colour) corresponds to a homogeneous light, one that cannot be decomposed into different components. The second one (compound colour) corresponds to a heterogeneous light, one that can be decomposed into different components. Our eyes cannot distinguish primary from compound colours: they may look exactly alike.[9] However, the two kinds can be distinguished by experiment: compound light can be decomposed in two or more components by a prism. Primary light cannot be so decomposed.

It follows from this *definition* that white light is not simple or primary. It is compound, since it may be decomposed into several different colours by a prism.

Now, it might seem as though Newton was merely playing with words: if he *defined* in this way simple and compound colour, it follows *from the definition* that white light is not simple. So, the whole question is reduced to a choice of definition. It seems that Newton did not need much to attain his objective.

This, of course, is an oversimplification of the problem, but that is the way it is understood by most students and – unfortunately – by teachers. If one accepts Newton's definition, then *one single experiment* – the 'decomposition' of white light by a prism – is sufficient to prove that white light is compound.[10]

One must remark, however, that definitions and distinctions are not arbitrary. Newton proposed a dichotomy between primary and compound colour (or light). This dichotomy is philosophically adequate if any colour (or light) can be exclusively classified *either* as primary *or* as compound, but *never* as both or neither. His concept will be *useful* if both sets are not empty. Only experience can show whether it is adequate or not.

The *Experimentum Crucis* is instrumental in showing that there are, indeed, pure colours. If one separates from the coloured spectrum a narrow beam of light, its colour will not be changed by a second prism. Besides, it is also necessary to show that this colour cannot be decomposed or altered by other means (for instance: by passing it through a coloured glass).

It is also necessary to test whether the concept of compound colour holds water. Suppose one joins two pure beams of light (for instance, red and yellow), producing orange light. According to the *concept* of compound colour, this orange cannot be pure or primary. However, only *experience* can show whether this orange light will be decomposed by a prism. It could happen (in principle) that the combination of two different primary colours would, in some cases, yield another different colour that could not be decomposed by a prism.[11] For this reason, Newton had to test this, too. So he did, and he observed that the simple colours used to form a compound colour could be always retrieved again by passing the compound light through a prism.

Several other points of Newton's work could deserve discussion. Let us, however, discuss the moral of this history.

9. HISTORY OF SCIENCE AND EDUCATION

There are several ways of using history of science as an aid in teaching. The choice depends on the educational aim and on the kind of students in view. The public may include science students, future teachers, non-scientists, etc. The aim may include learning scientific theories and concepts, the nature of science and its method, the relation of science to its social context, and so on.

The use of history of science has been particularly popular among people who address non-scientists (Gross 1980; Hetherington 1982). This is the specific case of Towne's use of Newton's work on colour. It seems that his aims in using Newton's paper were:

- to exhibit a particular concept of (inductive) scientific method;
- to show that scientific works can be clear and interesting even when read by non-scientists;
- to teach some physics (the classic theory of colours).

Let us discuss each of these points in turn.

9.1. *Scientific Method*

Physics teachers (even at university level) sometimes do not understand the nature of science. There is still a widespread belief in an inductivist model of scientific inquiry, of the worst positivist kind (Abimbola 1983; Hodson 1985). Teachers who do not have interest and competence in history and philosophy of science will usually transmit a distorted view of the scientific enterprise to their students (Matthews 1988). They may try to show how one gets a theory from observation and experiment or how one can *prove* a theory – notwithstanding the philosophical impossibility of both attempts. Sometimes they are not aware of their lack of understanding and even try to use history of science to improve their teaching. However, the kind of history of science they use is distorted and oversimplified – the kind of thing historians of science call 'Whig history' (Brush 1974; Siegel 1979).

The careful study of history of science can teach a lot about the nature of science. Pumfrey (1991), for instance, lists a few important components of the contemporary view of scientific endeavor:

1. Meaningful observation is not possible without a pre-existing expectation.
2. Nature does not yield evidence simple enough to allow one unambiguous interpretation.
3. Scientific theories are not inductions, but hypotheses which go imaginatively and necessarily beyond observations.
4. Scientific theories cannot be proved.
5. Scientific knowledge is not static and convergent, but changing and open-ended.
6. Shared training is an essential component of scientific agreement.
7. Scientific reasoning is not itself compelling without appeal to social, moral, spiritual and cultural resources.
8. Scientists do not draw incontestable deductions, but make complex expert judgments.
9. Disagreement is always possible.

It is easy to perceive that the analysis of Newton's 1672 paper presented in this paper provides an example of most of those components of the nature of science. However, this cannot be achieved by the mere reading of Newton's 1672 paper. It is necessary to *discuss* it and to *read it in the light of its context*.

It is very misleading to study a detached piece of scientific work, without a knowledge of its context. For this reason, a teacher who is not fully conversant with the context had better use 'case studies' produced by professional historians of science – such as Conant's (1966) *Harvard Case Histories in Experimental Science* – rather than attempting to use a detached piece of primary source. A fine scientific appreciation of Newton's 1672 paper requires some knowledge of Newton's other works on optics, and also some knowledge of previous and contemporary optical studies by other researchers. Depending on the aim, it will be necessary also to study the philosophical, technological and social contexts behind Newton's work. Only in this way can a nice picture of the scientific practice emerge.

9.2. *Science for Non-scientists*

Many science teachers are eager to show that science is not an esoteric discipline: anyone may understand and enjoy science. There is some truth in this statement: anyone may understand and enjoy *some part or aspect* of science. However, science itself **is** an esoteric discipline – exactly as music, for instance, is. Most people can enjoy music, but only a few persons are able to understand its structure, to play it well or to compose good music. To be a competent piano player, any person must undergo a technical training that may last for many years. To become a good composer, the training will be even more difficult and sometimes painful. The same kind of thing occurs in science. One should not present scientists as demigods (it is always nice to remember that scientists are human and fail). On the other hand, the difficulties of scientific training should not be underestimated.

When teaching physics to non-scientists, there is always the danger of presenting some kind of 'watered-down science', which avoids difficult aspects – such as measurement, equations, complex arguments, and so on. There are, indeed, many interesting things about science that can be learned without entering into technical details. It seems, however, that history of science is not the best way to present the simple aspects of science. Of course, one can use the 'external' history of science to discuss issues such as the relation between scientific and technical development without the analysis of 'difficult' aspects. However, if one intends to teach science itself through the history of science, it will be impossible to avoid technical details. Indeed, it may be easier to present or to learn a textbook version of any scientific subject than to present or to learn its conceptual history.

9.3. *Scientific Knowledge*

There is an important distinction between scientific *knowledge* and scientific *belief*. A person has scientific knowledge about some subject if he knows the scientific results, accepts this knowledge, and *has the right to accept it*, because he knows how this knowledge was justified and grounded.[12] Scientific belief, on the other side, corresponds to the knowledge of the scientific results, together with its acceptance as true, when this acceptance is due to mere belief in the authority of the teacher or of 'the scientists'. Scientific belief is just a modern kind of superstition. However, it is much easier to acquire than scientific knowledge.

One possible way to acquire scientific knowledge, in the above sense, it is to study the history of science[13] – but not 'Whig history'. It is necessary to study the scientific context, the experimental basis, the several alternatives of the time, and the dynamic process of discovery (or invention), justification, discussion, and diffusion. In this way can one learn how a theory was justified and why it was accepted. At the same time, one will learn a lot about the very nature of science.

10. CONCLUSION

Newton's first paper presented an experiment where a beam of solar white light passed through a prism set at minimum deviation position and perpendicularly reached a wall. According to common refraction laws (that is, the Snell-Descartes law), the spot at the wall should be circular – but only a complex theoretical computation can prove it.

Newton found that the spot was oblong. The explanation provided by Newton for this new phenomenon was that white light is a mixture of rays, of different colours, which differ in refrangibility. Newton justified this statement by a smart combination of experiment and theoretical argument.

Newton studied the relation between colour and refrangibility in the *Experimentum Crucis*. He stated that to each colour corresponds a well-defined refrangibility, and conversely. This property only applies to pure or primary colours – those that cannot be decomposed by a prism. This new concept introduced by Newton was central to his argument.

By a set of experiments, he showed that pure coloured light is immutable in several circumstances where composed colour changes. Since pure colours are immutable and since each colour is related to a given refrangibility, this last also must be immutable.

In no experiment with pure or compound colours did Newton observe the change or creation of new colours, or the change of their refrangibility.

Since the refrangibility of the rays is immutable they must be the same before any refraction, that is, prisms do not modify this characteristic of the rays. Hence the coloured rays are already present in white light before it passes through a prism.

To confirm his theory, Newton presented another experiment: the coloured rays emerging from a prism passed through a convergent lens and at its focus white light was produced, with the same characteristics as those of the Sun. Since entities should not be multiplied without necessity, these two white lights – the solar one and the produced by the convergence of the coloured rays – must be accepted to be equal.

Newton's complex argument does not correspond to a mere 'induction' from experiments. If one wants to teach Newton's theory of light, it is necessary to present it as it is: a fine but difficult piece of scientific work that exhibits the complex interplay of theory and experiment.

A correct understanding of the structure and dynamics of science is essential to education. Without such an understanding, many mistakes may easily occur – as happened with Towne.

Towne's paper does not exhibit the structure of Newton's argument. Many of the misunderstandings pointed above may be attributed to the fact that Newton's argument is not as simple and direct as it was supposed to be. Indeed, below the apparent simplicity of Newton's theory there is a deep and complex work. The detailed discussion of Newton's argument seems a nice example of how the history of science may be used in teaching to discuss the complexity of actual scientific work.

ACKNOWLEDGEMENTS

The authors acknowledge support received from the Brazilian National Council for Scientific and Technological Development (CNPq), the São Paulo State Research Foundation (FAPESP) and the Ministry of Education (CAPES/MEC). We are also grateful for the useful comments of one of the referees.

NOTES

[1] The description of this experiment by Newton in his 1672 paper is accompanied by no drawing. The draft reproduced here is from Newton's manuscript *Lectiones opticae* (*circa* 1672): MS. Add. 4002, fol. 3 of the Cambridge University Library, reprinted in Whiteside (1973).

[2] It is not really correct to say that the prism projects an *image* on the wall, although Newton himself uses this expression. When an optical device produces an image of an object, each point of the object corresponds (ideally) to one single point in the image. When a system of lenses produces a real image of a small light source, the light 'rays' converge after passing through the lens and concentrate to form the image. When an image of the Sun is formed with the aid of a converging lens, for instance, it is possible (with suitable magnification) to see in the image the sunspots that may happen to be visible in the Sun's disk. If we use a divergent lens, it will be possible to project upon a surface a round 'image' of the Sun, but it will be impossible to see sunspots. A prism will produce only a *virtual image* of real objects. This virtual image can be seen if one looks towards the object through the prism. Only if we use both a prism and a converging lens, then it will be possible to produce a real image on the wall. In Newton's first experiment, however, we can only talk about the light *spot* – not the *image* – on the wall. By the way: the distinction between 'objective' and 'subjective' experiments stressed by Towne (1993, p. 117) is nothing but the difference between observing the virtual image and the spot projected on the wall. One way is no more 'objective' than the other: in both cases, light is seen with the use of the observer's eyes.

[3] In this article we shall refer to the first experiment described by Newton in his 1672 paper as 'Newton's first experiment'. One should remember, however, that this was not the very first optical experiment made by Newton. It is possible to find a description of his first observations in the notebooks he kept during the period 1664–1665. See McGuirre and Tammy (1983).

[4] Let us remark how difficult it is to obtain the required angular conditions for Newton's experiment. It is necessary that the prism be put in the minimum deviation position and, *at the same time*, the deflected beam must be perpendicular to the wall of the room where the experiment is being done. If the axis of the prism is horizontal and parallel to the wall (as shown in all drawings), the experiment can be performed only on two precise days each

year. Some difficulties of Newton's experiments are discussed by Lohne (1964, pp. 125–139).

[5] Pardies' computation is wrong, although his method seems correct. Re-doing his calculations, one finds that instead of the divergence of 2° 23′ that Pardies obtained for incidences of 30° and 29° 30′, the correct divergence is 1° 40′. For the incidences of 29° 30′ and 29°, the correct divergence is 1° 57′. For the incidences of 29° and 28° 30′ the divergence would be 2° 29′ and for incidence of 28° 30′ and 28′ the divergence would be 4° 17′. So, *in principle* Pardies is correct: it is possible to explain the length of Newton's oblong spot supposing that all rays have the same refractivity. It is remarkable that Newton did not point out Pardies' calculation mistake.

[6] About Newton's modifications of his experiment, see Mamiani (1976, p. 115).

[7] For more information about the optical theory of Grimaldi see Hall (1987).

[8] Newton made constant use of this kind of simplicity arguments in his work. In his *Philosophiae naturalis principia mathematica* one finds a set of philosophical rules (*Regulae philosophandi*). Two of them, that were already found in the first edition of this book, read: 'Rule 1: We are to admit no more causes of natural things than such as are both true and sufficient to explain their appearances. Rule 2: Therefore to the same natural effects we must, as far as possible, assign the same causes'. This is a clear presentation of the methodological rule he had already used in his optical work. (See: Koyré 1972, Vol. 2, pp. 550–6).

[9] Notice that this is a strange property of light. In the case of sound, human sensation is able to distinguish pure tones (those corresponding to a single frequency) from compound tones. There is a subjective quality (pitch) that allows us to distinguish between notes of the same main frequency produced by different instruments. When two different notes are played together, they do not produce a single intermediary sound: they are heard separately and can be harmonious or otherwise. There is nothing of this kind in light and colour – but there is no *a priori* reason why light and sound should lead to different sense structures.

[10] Newton's first experiment is usually called 'the experiment of decomposition of white light'. The name itself implies the conclusion.

[11] To understand this possibility, one may compare the phenomena of light to those that occur in chemistry. In some cases, when we join two pure substances it is possible to separate them again by physical procedures (distillation, or another process). However, in other cases, the union of two pure substances produces a third pure substance that cannot be decomposed by physical procedures. It could happen, *in principle*, that something similar occurred to light: in some cases we could have a mere *mixture* of colours, in other cases a *combination* of colours. It could also happen that the result of the combination of colours could not be decomposed by a prism.

[12] This distinction has been pointed out by Rogers (1982), although in a slightly different way – he assumed that scientific knowledge is *true*. Of course, scientific knowledge may be useful, well grounded and acceptable, but it is temporary and not *true*, in a philosophical sense.

[13] Another way is, of course, the practice of scientific research. However, in an educational context, it seems that the only way of acquiring scientific knowledge about 'established' science is the historical one.

REFERENCES

Abimbola, I. O.: 1983, 'The Relevance of the "New" Philosophy of Science for the Science Curriculum', *School Science and Mathematics* **83**, 181–193.

Alonso, M. & Finn, E.: 1972, *Physics*, Addison-Wesley Publishing Company, Massachusetts.

Bevilacqua, F. & Giannetto, E.: 1996, 'The History of Physics and European Physics Education', *Science & Education* **5**, 235–246.

Boyle, R.: 1664, *Experiments and Considerations Touching Colours*, H. Herringman, London. Reprinted: Johnson Reprint, New York, 1964.

Brush, S. G.: 1974, 'Should the History of Science be Rated X?', *Science* **183**, 1164–1172.

Conant, J. B. (ed.): 1966, *Harvard Case Histories in Experimental Science*, 2 vols., Harvard University Press, Cambridge, MA.

Descartes, R.: 1937, *Discours de la Méthode pour Bien Conduire sa Raison, et Chercher la Verité dans les Sciences. Plus la Dioptrique. Les Meteores. Et la Geometrie. Qui Sont des Essais de cette Methode.* I Maire, Leyde. Reprinted in: Charles Adam & Paul Tannery (eds.), *Oeuvres de Descartes*, Vrin, Paris, 1964–74, Vol. 6.

Gross, W. E.: 1980, 'The History of Science in the Two-Year College Curriculum', *Journal of College Science Teaching* **10**, 19–21.

Hall, A. R.: 1987, 'Beyond the Fringe: Diffraction as Seen by Grimaldi, Fabri, Hooke and Newton', *Notes and Records of the Royal Society* **41**, 111–143.

Hetherington, N. S.: 1982, 'The History of Science and the Teaching of Science Literacy', *Journal of Teaching* **17**, 53–66.

Hodson, D.: 1985, 'Philosophy of Science, Science and Science Education', *Studies in Science Education* **12**, 25–57.

Hooke, R.: 1665, *Micrographia or Some Physiological Descriptions of Minute Bodies Made by Magnifing Glasses. With Observations and Inquires Thereupon*, J. Martyn and J. Allestry, London. Reprinted: Dover, New York, 1961.

Koyré, A., Cohen, I. B. & Whitman, A.: 1972, *Isaac Newton's Philosophiae Naturalis Principia Mathematica, Third Edition (1726) with Variant Readings*, 2 vols, Cambridge University Press, Cambridge.

Kuhn, T.: 1978, 'Newton's Optical Papers', in I. B. Cohen & R. E. Schofield (eds.), *Isaac Newton's Papers & Letters on Natural Philosophy*, Harvard University Press, Cambridge, MA, pp. 27–45.

Lohne, J. A.: 1964, 'Isaac Newton: The Rise of a Scientist, 1661–1671', *Notes and Records of the Royal Society of London* **20**, 125–139.

Lohne, J. A.: 1968, 'Experimentum Crucis', *Notes and Records of the Royal Society* **23**, 169–199.

Mamiani, M.: 1976, *Isaac Newton Filosofo della Natura: le Lezioni Giovanili di Ottica e la Genesi del Metodo Newtoniano*, La Nuova Italia Editrice, Firenze.

Matthews, M. R.: 1988, 'A Role for History and Philosophy of Science Teaching', *Educational Philosophy and Theory* **20**, 67–81.

McGuirre, J. E. & Tammy, M.: 1983, *Certain Philosophical Questions: Newton's Trinity Notebook*, Cambridge University Press, Cambridge.

Newton, I.: 1672a, 'A Letter of Mr. Isaac Newton, Professor of the Mathematicks in the University of Cambridge; Containing His New Theory about Light and Colours; Sent by the Author to the Publisher from Cambridge, Febr. 6. 1671/72; in Order to be Communicated to the R. Society', *Philosophical Transactions of the Royal Society* **6**(80), 3075–3087. Reprinted in I. Bernard Cohen & R. E. Schofield (eds.): 1978, *Isaac Newton's Papers & Letters on Natural Philosophy*, Harvard University Press, Cambridge, MA, pp. 47–59. Also reprinted in: Newton, I.: 1993, 'A New Theory about Light and Colors', *American Journal of Physics* **61**, 108–112.

Newton, I.: 1672b, 'Mr. Newton's Letter of April 13, 1672, O. S. Written to the Editor Being an Answer to the Foregoing Letter of F. Pardies. Translated from the Latin', *Philosophical Transactions of the Royal Society* **7**, 730–732. Reprinted in I. Bernard Cohen & R. E. Schofield (eds.): 1978. *Isaac Newton's Papers & Letters on Natural Philosophy*, Harvard University Press, Cambridge, MA, pp. 90–92.

Newton, I.: 1672c, 'Mr. Isaac Newton's Answer to Some Considerations upon his Doctrine of Light and Colours; Which Doctrine was Printed in Numb. 80 of these Tracts', *Philosophical Transactions of the Royal Society* **7**, 5084–5103. Reprinted in I. Bernard Cohen & R. E. Schofield (eds.).: 1978, *Isaac Newton's Papers & Letters on Natural Philosophy*, Harvard University Press, Cambridge, MA, pp. 116–135.

Newton, I.: 1704, *Opticks Or, A Treatise of the Reflections, Refractions, Inflexions and Colours of Light*, London; Reprint: 1952, Dover, New York.

Newton, I.: 1984, *The Optical Papers of Isaac Newton, vol 1: The Optical Lectures (1670–1672)*, A. E. Shapiro (ed.), Cambridge University Press, Cambridge.

Pardies, I. G.: 1672, 'Some Animadversions on the Theory of Light of Mr. Isaac Newton, Prof. of Mathematics in the University of Cambridge, Printed in No. 80. In a Letter of April 9, 1672. N. S. from Ignatius Gaston Pardies, Prof. of Mathematics in the Parisian College of Clermont. Translated from Latin', *Philosophical Transactions of the Royal Society* **7**, 726–729, 1672. Reprinted in I. Bernard Cohen & R. E. Schofield (eds.), 1978, *Isaac Newton's Papers & Letters on Natural Philosophy*, Harvard University Press, Cambridge, MA, pp. 86–89.

Pumfrey, S.: 1991, 'History of Science in the National Science Curriculum: A Critical Review of Resources and Aims', *British Journal for the History of Science* **24**, 61–78.

Rogers, P. J.: 1982, 'Epistemology and History in the Teaching of School Science', *European Journal of Science Education* **4**, 1–27.

Russell, C.: 1984, 'Whigs and Professionals', *Nature* **308**, 777–778.

Sabra, A. I.: 1981, *Theories of Light from Descartes to Newton*, Cambridge University Press, London.

Siegel, H.: 1979, 'On the Distortion of the History of Science in Science Education', *Science Education* **63**, 111–118.

Towne, D. H.: 1993, 'Teaching Newton's Color Theory Firsthand', *American Journal of Physics* **61**, 113–116.

Whiteside, D. T.: 1973, *The Unpublished First Version of Isaac Newton's Cambridge Lectures on Optics, 1670–1672*, Cambridge University Library, Cambridge.

Methodology and Politics in Science: The Fate of Huygens' 1673 Proposal of the Seconds Pendulum as an International Standard of Length and Some Educational Suggestions *

MICHAEL R. MATTHEWS
School of Education, University of New South Wales, Sydney 2052, Australia; E-mail: m.matthews@unsw.edu.au

Abstract. This paper is part of a larger work on the history, philosophy and utilisation of pendulum motion studies (Matthews 2000). The paper deals with the fate of Christiaan Huygens 1673 proposal to use the length of a seconds pendulum (effectively one metre) as a universal, natural and objective standard of length. This is something which, if it had been adopted, would have been of inestimable scientific, commercial and cultural benefit. Why it was not originally adopted in the late seventeenth century, and why it was again rejected in the late eighteenth century (1795) when the Revolutionary Assembly in France adopted the metric system with the metre being defined as one ten-millionth of the quarter meridan distance – raise interesting questions about the methodology and politics of science. Given that pendulum motion is a standard component of all science courses throughout the world, and given that most science education reforms, including the US National Science Education Standards and recent Australian state reforms, require that something of the 'big picture' of science be conveyed to students (the relationship of science to culture, commerce, history and philosophy) – it is suggested that these educational goals can be advanced by teaching about the fate of Huygens' proposal.

In the past two decades, science curricula in many parts of the world have endorsed a liberal or contextural science programme where students are expected to learn something of the philosophical, social, historical and ethical dimensions of science, as well as acquire scientific knowledge and process skills (Matthews 1994, chap. 3; McComas and Olsen 1998). These curricula expect students to learn *about* science as well as learning the *content* of science. This is sometimes stated as learning about 'the nature of science' (McComas et al. 1998), and sometimes it is expressed as developing a deeper, more informed or 'critical' scientific literacy (Bybee et al. 1992, chap. 2).

The US *National Science Education Standards* (NRC 1996) , for instance, ask that students should learn how:

science contributes to culture (NRC 1996, p. 21);
Technology and science are closely related. A single problem has both sci-

* Parts of this paper have appeared in M.R. Matthews *Time for Science Education* (Plenum, 2000)

F. Bevilacqua et al. (eds.), Science Education and Culture, 293–309.
© 2001 *Kluwer Academic Publishers. Printed in the Netherlands.*

entific and technological aspects (NRC 1996, p. 24);
curriculum will often integrate topics from different subject-matter areas ... and from different school subjects – such as science and mathematics, science and language arts, or science and history (NRC 1996, p. 23);
scientific literacy also includes understanding the nature of science, the scientific enterprise, and the role of science in society and personal life (NRC 1996, p. 21);
progress in science and technology can be affected by social issues and challenges (NRC 1996, p. 199);

Comparable statements can be found in Canadian, Japanese, British and Australian (specifically the state of New South Wales) curricular documents.

Unfortunately teacher education programmes have lagged behind these curricular developments. Most programmes do not adequately prepare teachers to cope with the broader cultural dimensions of the new curricula. The joint BSCS-SSEC document recognised this when, after developing its curriculum framework for a more historical and philosophical approach to the teaching of natural and social science, it said that the first barrier to implementing such a curriculum was that 'the preparation of teachers is inadequate' (Bybee et al. 1992, p. xiii).

One clear need is to provide teachers with accessible case studies that can be utilised in classrooms to illustrate the historical, philosophical and cultural dimensions of science. The fate of Huygens' 1673 proposal of the seconds pendulum as a universal standard of length is one such case.[1]

1. Christiaan Huygens

Christiaan Huygens (1629–1695) was one of the preeminent minds of the seventeenth century. He was born in The Hague in 1629. By the age of thirteen he had built himself a lathe, by seventeen he had independently discovered Galileo's time-squared law of fall and Galileo's parabolic trajectory of a projectile,[2] by twenty he had completed and published a study of hydrostatics, by twenty-three he formulated the laws of elastic collision, by twenty-five he was an optical lens grinder of national renown, by twenty-six, and using one of his own telescopes, he had discovered the ring of Saturn. When he was in his mid-thirties many thought of him as the greatest mathematician in Europe – no slight claim given that his contemporaries included Pascal, Mersenne, Fermat, Descartes, Leibniz and Newton. In 1663 he was made a member of England's newly founded (1662) Royal Society. In 1666, at age thirty-seven, he was invited by Louis XIV to be founding president of the *Académie Royal des Sciences*, an invitation he accepted, and a position he held till 1681, even throughout the war between France and Holland.[3]

Huygens made fundamental contributions to mathematics (theory of evolutes and probability), mechanics (theory of impact), optics (both practical with his own ground lenses, and theoretical with his wave theory of light), astronomy (discovery of Saturn's rings and determination of the period of Mars), and to philosophy (his elaboration of the mechanical world view, and proposals for a form of hypothetico-deductive methodology in science). In 1695 he died where he was born.[4] Upon his death, Leibniz wrote: 'The loss of the illustrious Monsieur Huygens is inestimable; few people knew him as well as I; in my opinion he equaled the reputation of Galileo and Descartes and aided theirs because he surpassed the discoveries that they made; in a word, he was one of the premier ornaments of our time' (Yoder 1991, p. 1).

Huygens refined Galileo's theory of the pendulum.[5] His first book on the subject was the *Horologium* (1658),[6] his second and major work was the *Horologium Oscillatorium* ('The Pendulum Clock' (1673). He recognized the problem identified by Descartes – 'the simple pendulum does not naturally provide an accurate and equal measure of time since its wider motions are observed to be slower than its narrower motions' (Huygens 1673/1986, p. 11). Huygens changed two central features of Galileo's theory, namely the claims that period varied with length, and that the circle was the tautochronous curve (the curve on which bodies falling freely under the influence of gravity reach its nadir at the same time regardless of where they were released). In contrast Huygens showed mathematically that period varied with the square root ('subduplicate' as Huygens refers to it) of length, and that the cycloid was the tautochronous curve.[7]

Huygens thought so highly of the second result that he described it in a 1666 letter to Ismaël Boulliau as 'the principal fruit that one could have hoped for from the science of accelerated motion, which Galileo had the honor of being the first to treat' (Blay 1998, p. 19).[8] Huygens realised that such an isochronic motion could be the regulator for a new, and accurate clock, the pendulum clock which he proceeded to make and with which he hoped to win the various longitude prizes offered by governments and monachs.[9]

2. The Seconds Pendulum as a Universal Length Standard

In the process of elaborating his theory of pendulum motion and clockwork design Huygens argued that the seconds pendulum could provide a new international standard of length. Undoubtedly this would have been a major contribution to simplifying the chaotic state of measurement existing in science and everyday life. Within France, as in other countries, the unit of length varied from city to city. A not insignificant problem for commerce, trade, construction and technology. To say nothing of science. Many attempts had been made to simplify and unify the chaotic French system. The emperor Charlemagne in 789 was among the earliest to issue edicts calling for a uniform system of weights and measures in France. Henry II repeated these calls, issuing a decree in 1557 stating that:

> Weights and measures shall be reduced to clearly defined forms and shall bear the appellation of royal weights and measures. Since in all duchies, marquisates, counties, viscountcies, baronies, castellanies, cities and lands observing the laws of our kingdom, weights and measures are of diverse

names and dimensions, wherefore many of them do not correspond to their designations; and often, indeed, there coexist two weights and two measures of different sizes, the smaller one being used in selling and the other in buying, whence innumerable dishonest deals arise ... All shall be required to regulate their measures according to ours. (Kula 1986, pp. 168–169)

This had the same effect as King Canute's admonition to the rising tide. One traveller to France in 1789 was infuriated by a country 'where infinite perplexity of the measures exceeds all comprehension. They differ not only in every province, but in every district and almost in every town' (Alder 1995, p. 43). There were 700 or 800 different units or measures, with different towns and localities having their own version of the measure or unit. As Heilbron notes, 'In Paris a pint held a little less than a liter; in Saint-Denis, a liter and a half; in Seine-en-Montagne, two liters; and in Précy-sous-Thil, three and a third' (Heilbron 1989, p. 989). One estimate is that in France alone there were 250,000 different, local, measures of length, weight and volume (Alder 1995, p. 43).

The situation was little better in the German states. Although the 'Common German Mile' was widely used, and taken to be 1/15th of a degree at the equator, in Vienna it was sub-divided into 23,524 'work shoes', whilst in Innsbruch it was sub-divided into 32,000 'work shoes'. Other cities had their own idiocyncratic divisions. Some Italian States used the 'Italian Mile' which was 1/60th of an equatorial degree, and contained 5,881 Vienna work-shoes. The multiplicity of measures facilitated widespread fraud, or just smart business practice: merchants routinely bought according to 'long' measures and sold according to 'short' measures.

The English situation was at least more uniform, if more 'unnatural'. In 1305 Edward I had established the standard yard as the length of the Iron Ulna kept in the Royal Palace, and one thirty-sixth part of it was to be an inch. Away from the Palace, in in places without access to a copy of the Royal Ulna it was ordained that 'three grains of barley, dry or round, make an inch; twelve inches make a foot; three feet make an ulna; $5\frac{1}{2}$ ulna make a perch'. A slightly shorter yard standard was decreed by Henry VII in 1497, and a slightly shorter one again by Elizabeth I in 1588.

These English standards were arbitrary and artificial; the yard was not a natural unit. The Royal Society at its inception was asked to investigate the reform of the length standards, and Christopher Wren proposed, as did Huygens, the length of a pendulum beating seconds as an English, and also international, length standard.[10] Both Huygens and Wren assumed that a seconds pendulum would beat seconds no matter where in the world it was taken. The technical problems were firstly, how to get a pendulum to beat seconds and next, how to measure its effective length.

Huygens, in the early pages of his *Horologium*, after giving the length of his seconds pendulum, says:

When I say 'three feet', I am not speaking of 'feet' as this term is used in various European countries, but rather in the sense of that exact and eternal measure of a foot as taken from the very length of this pendulum. In what

> follows I will call this an 'hour foot', and the measurement of all other feet must be referred to it if we wish to treat matters exactly in what follows. (Huygens 1673/1986, p. 17)

This topic is taken up later in the book when he discusses the centre of oscillation of bodies, there he remarks:

> Another result, which I think will be helpful to many, is that by this means I can offer a most accurate definition of length which is certain and which will last for all ages. (Huygens 1673/1986, p. 106)

Elaborating this 'useful result', he says:

> A certain and permanent measure of magnitudes, which is not subject to chance modifications and which cannot be abolished, corrupted, or damaged by the passage of time, is a most useful thing which many have sought for a long time. If this had been found in ancient times, we would not now be so perplexed by disputes over the measurement of the old Roman and Hebrew foot. However, this measure is easily established by means of our clock, without which this either could not be done or else could be done only with great difficulty. (Huygens 1673/1986, p. 167)

In brief, Huygens says that first a seconds clock is built and tested against the rotation of the fixed stars (as described on pp. 23–25 of his 1673 book), then a pendulum is to be set swinging with a small amplitude and its length adjusted until it swings in time with the seconds clock, then:

> ... measure the distance from the point of suspension to the center of the simple pendulum. For the case in which each oscillation marks off one second, divide this distance into three parts. Each of these parts is the length of an hour foot ... By doing all this, the hour foot can be established not only in all nations, but can also be reestablished for all ages to come. Also, all other measurements of a foot can be expressed once and for all by their proportion to the hour foot, and can thus be known with certainty for posterity. (Huygens 1673/1986, p. 168)

Thus his basic unit of length was to be three horological feet (0.9935 m, $39\frac{1}{8}$ English inches), less than a millimetre short of the original metre adopted a century later. The astronomer Picard concurred in this recommendation, saying that a 'universal foot' should be one-third the length of a seconds pendulum.

3. An Astonishing Discovery: Jean Richer's Cayenne Voyage of 1672–1673

Huygens thought that his universal length standard, the seconds pendulum, was dependent only upon the force of gravity, which he took to be constant all over the earth, and thus the length standard would not change with change of location. The standard was to be portable over space and time. He did recognise that the centrifugal force exerted by the rotating earth (the force that tends to throw bodies off the earth's surface) varied from the equator (largest) to the pole (smallest), but he did not think that its effect on the pendulum would be measurable.[11] In this he was wrong.

As soon as the *Académie Royale des Sciences* observatory was established by Jean-Dominique Cassini in 1669, the *Académie*, according to its secretary 'began to discuss sending observers under the patronage of our most munificent King into the different parts of the world to observe the longitudes of localities for the perfection of geography and navigation' (Olmsted 1942, p. 120). The *Académie* instituted the modern tradition of scientific voyages of discovery, and Jean Richer's voyage to Cayenne in 1672–1673 was the second such purely scientific voyage undertaken[12] – the first being Jean Picard's voyage to Uraniborg in Denmark in 1671. Cayenne was in French Guiana, at latitude approximately 5° N. It was chosen as a site for astronomical observations because equatorial observations were minimally affected by refraction of light passing through the earth's atmosphere – the observer, the sun and the planets were all in the same elliptic.

The primary purpose of Richer's voyage was to ascertain the value of solar parallax and to correct the tables of refraction used by navigators and astronomers. A secondary consideration was checking the reliability of marine pendulum clocks which were being carried for the purpose of establishing Cayenne's exact longitude.

The voyage was spectactularly successful in its primary purposes: the obliquity of the ecliptic was determined, the timing of solstices and equinoxes was refined and, most importantly, a new and far more accurate value for the parallax of the sun was ascertained – 9.5" of arc. But it was the unexpected consequences of Richer's voyage which destroyed Huygens' vision of the seconds pendulum as a universal standard of length 'for all nations' and 'all ages'.

Richer found that a pendulum set to swing in seconds at Paris, had to be shortened in order to swing in seconds at Cayenne. Not much – 2.8 mm (0.28%) – but nevertheless shortened. Although, with good reason, many doubted his experimental ability,[13] Richer found that a Paris seconds-clock lost $2\frac{1}{2}$ minutes daily at Cayenne. Richer was adamant that Huygens' clocks slowed, that they had to be shortened. This was tantamount to saying that the force of gravity, and hence the weight of bodies, diminished from Paris to the equator – an astonishing conclusion.[14]

Richer's demonstration raised the problem of an independent measure of time. He did not have a second timepiece (a digital watch, for instance) against which to measure the speeding up or slowing down of his pendulum clock. The only independent clock he had was the clock of the heavens. He probably measured the number of pendulum swings against the number of seconds in a solar day (noon to noon) or a sidereal day, or measured portions thereof. This was a difficult enough technical exercise, and it was compounded by the fact that the solar day actually varies in length by plus or minus 15 minutes through the year. But the yearly variation, the Equation of Time, was known, and the technical problems of timing the sun's transit were overcome.

4. Some Methodological Matters

Richer's claim that the pendulum clock slows in equatorial regions nicely illustrates some key methodological matters about science, and about theory testing. The logic of the accommodation of theory to evidence constitutes the *methodology* of science. *Methodology* is usefully distinguished from the *method* of science which involves how one conducts experiments, gathers data, seeks information, and selects appropriate tools, instruments and means of analysis for the conduct of an investigation. Methodology involves what one does with the data, and how one relates it to hypotheses and theories: How many white swans does one have to see in order to conclude that 'all swans are white'? How does finite evidence bear on the truth of scientific hypotheses that are universal in their scope? And so on. These methodological questions have been, since Aristotle, the core subject matter of philosophy of science.

Huygens himself provided perhaps the first statement of the hypothetico-deductive account of the methodology of science. In the Preface to his *Treatise on Light* (Huygens 1690/1945) he wrote:

> One finds in this subject a kind of demonstration which does not carry with it so high a degree of certainty as that employed in geometry; and which differs distinctly from the method employed by geometers in that they prove their propositions by well established and incontrovertible principles, while here principles are tested by the inferences which are derivable from them. The nature of the subject permits of no other treatment. It is possible, however, in this way to establish a probability which is little short of certainty. This is the case when the consequences of the assumed principles are in perfect accord with the observed phenomena, and especially when these verifications are numerous; but above all when one employs the hypothesis to predict new phenomena and finds his expectations realized.

The entrenched orthodoxy since at least the second century BC was that the earth was spherical (theory T). On the assumption that gravity alone affects the period of a constant length pendulum, the observational implication was that period at Paris and the period at Cayenne of Huygens' seconds-pendulum would be the same (O). Thus:

$$T \rightarrow O$$

But Richer seemingly found that the period at Cayenne was longer (~O). Thus, on simple falsificationist views of theory testing:

$$T \rightarrow O$$
$$\sim O$$
$$\therefore \sim T$$

But theory testing is never so simple. Many upholders of T just denied the second premise, ~O. The astronomer Jean Picard, for instance, did not accept Richer's findings. Rather than accept the message of varying gravitation, he doubted the messenger. Similarly Huygens was not favourably disposed towards Richer. In 1670, on one longitude testing voyage to the West Indies and Canada, Richer had behaved irresponsibly with regard to Huygens' clocks – he did not immediately restart them when they stopped in a storm, and finally he allowed them to crash to the deck (Mahoney 1980, p. 253). Huygens did not require much convincing that it was Richer's ability, not gravity, that was lacking at Cayenne.

Others saw that theories did not confront evidence on their own, there was always an 'other things being equal' assumption made in theory test; there were *ceretis paribus* clauses (C) that accompanied the theory into the experiment. Thus:

$$T + C \rightarrow O$$
$$\sim O$$
$$\therefore \sim T \text{ or } \sim C$$

These people maintained belief in T, and said that the assumption that other things were equal was mistaken – humidity had interferred with the swings, heat had lengthened the pendulum, friction increased in the tropics, and so on. These, in principle, were legitimate concerns. But, one by one, these explanations were ruled out. And others, including Sir Edmund Halley, confirmed Richer's finding that the pendulum slowed in equatorial regions.[15] Thus ~O became established as a scientific fact, and upholders of T, the spherical earth hypothesis, had to adjust to it. This was not easy.

Newton, for instance, acknowledged the veracity of Richer's claims, writing in his *Waste Book* of 1682, that:

> Monsr. Richer sent by ye French King to make observations in the Isle of Cayenne (North Lat 5^{gr}) having before he went thither set his clocke exactly at Paris found there in Cayenne that it went too slow as every day to loose two minutes and a half for many days together and after his clock had stood & went again it lost $2\frac{1}{2}$ minutes as before. Whence Mr Halley concludes that y^{e} pendulum was to be shortened in proportion of – to – to make y^{e} clock true at Cayenne. In Gorea y^{e} observation was less exact. (Cook 1998, p. 116)

In his *Principia* (Bk. III, Prop. XX, Prob. IV), Newton utilised Richer's, and Halley's comparable observations from St. Helena, to develop his oblate account of the Earth's shape.

In 1738 Voltaire, a champion of Newtonian science, wrote on the Richer episode, drawing attention to the problems of adjustment that scientists experienced:

> At last in 1672, Mr Richer, in a Voyage to Cayenna, near the Line, undertaken by Order of Lewis XIV under the protection of Colbert, the Father of all Arts; Richer, I say, among many Observations, found that the Pendulum of his Clock no longer made its Vibrations so frequently as in the Latitude of

Paris, and that it was absolutely necessary to shorten it by a Line, that is, eleventh Part of our Inch, and about a Quarter more.

Natural Philosophy and Geometry were not then, by far, so much cultivated as at present. Who could have believed that from this Remark, so trifling in Appearance, that from the Difference of the eleventh of our Inch, or thereabouts, could have sprung the greatest of physical Truths? It was found, at first, that Gravity must needs be less under the Equator, than in the Latitude of France, since Gravity alone occasions the Vibration of a Pendulum.

In Consequence of this it was discovered, that, whereas the Gravity of Bodies is by so much the less powerful, as these Bodies are farther removed from the Centre of the Earth, the Region of the Equator must absolutely be much more elevated than that of France; and so must be farther removed from the Centre; and therefore, that the Earth could not be a Sphere. Many Philosophers, on occasion of these Discoveries, did what Men usually do, in Points concerning which it is requisite to change their Opinion; they opposed the new-discovered Truth. (Fauvel and Gray 1987, p. 420)

5. From the Seconds Pendulum to the Original Metre: Some Political Matters

Richer's findings did, in an Age of precision, rule out the seconds pendulum as an invariant universal standard of length, verifiable for all nations. But once a location, or latitude, was specified, then the length of the seconds pendulum would be invariant, and it could still be a universal standard. Moreover such a standard would be natural, not arbitrary. It would of course be a matter of some national pride as to what location was chosen: Paris, London, Madrid, Berlin or Rome, for instance. La Condamine and others, in an expedition to Peru in 1735, ascertained the length of the equatorial seconds pendulum to be 439.15 lines. In 1739 the length of a Paris seconds pendulum was determined to be 440.5597 lines. In 1745 in the Academy of Sciences, La Condamine proposed, perhaps as a way of rising above nationalist rivalries in Europe, that there be a universal and invariable measure based upon the length of the second pendulum at the equator (all major European powers having equatorial colonies where they could do their own measurements).

The French revolutionaries of 1789 not only proposed to do away with the arcane feudal system of the *ancien régime*, but also with the cacophony of measures that people saw as robbing them in every transaction, and as requiring elitist knowledge to judge and manipulate. The people wanted a simple, rational, democratic and universal system; one where, as Sir John Riggs Miller the English champion of measure reform would say, 'the meanest intellect is on a par with the most dextereous' (Heilbron 1989, p. 990). And exactly one thousand years after Charlemagne's call for unified French measures, the anti-Royalists of the French Revolution resolved to do something effective about the matter.

One of the first decisions of the Estates General was to direct the *Académie* to establish a Committee on Weights and Measures to recommend reform of French measurement. This committee was duly established and included Lavoisier, Coulombe, Delambre, Lagrange and Laplace. Talleyrand, the Bishop of Autun, was not too worried about nationalist bias and, in 1790, suggested to the new, post-Revolution, National Assembly that the unit of length be the seconds pendulum at the 45° latitude, a latitude conveniently running through France. Talleyrand wrote to Riggs Miller in the English parliament urging him to use his influence to have a common system adopted between France and England, saying 'Too long have Great Britain and France been at variance with each other, for empty honour or for guilty interests. It is time that two free Nations should unite their exertions for the promotion of a discovery that must be useful to mankind' (Berriman 1953, p. 141). Alas this entreaty, even when followed by Louis XVI being asked by the Assembly to write to George III seeking a unified standard, fell upon deaf ears.

The committee's second report, in 1791, rejected the pendulum measure and instead revised a version of Abbe Gabriel Mouton's 1670 suggestion for using the length of a geodetic minute of arc decimally divided. Cassini had also advocated this geodetic foot measure in 1720, wanting it to be 1/6,000 of a terrestrial minute of arc. The committee recommended that the length standard be one ten-millionth of the distance of the quadrant of the arc of meridian from the north pole to the equator passing through Dunkirk, Paris and Barcelona. Dunkirk and Barcelona 'anchored' the segment of meridian at sea level, and the Paris meridian was the obvious choice for the French standard. This was painstakingly measured by the astronomer-geodesists Delambre and Méchain during the years 1792–1799 using classical surveying techniques and the Toise (approximately 2 m) as their unit of length (Chapin 1994, p. 1094).

The committee's third report, 1793, named the new geodesic unit a *metre* (from the Greek word *metron*, meaning 'measure'). The Convention, which had replaced the Assembly, accepted this, and the new standard was enacted in the law of the 18th Germinal, Year III (7 April 1795), 5th Article. These new metric measures were officially termed 'republican', indicating their political dimension. As the Minister of Finance in year II of the Republic said, 'the introduction of new weights and measures was extremely important on account of its association with the Revolution, and for the enlightenment and in the interest of the people' (Kula 1986, p. 239). This association also meant that Lavoisier, Laplace and Coulombe were dismissed from the Agency of Weights and Measures for un-Republican associations. The brass standard metre was put in place in 1799.

It is problematic why the Academy's committee opposed Talleyrand's proposal of the seconds pendulum, which was Huygens' suggestion of one hundred years earlier. The length of the seconds pendulum at a given latitude was constant, public, recoverable, natural and portable. It seemingly fitted all the criteria for a good standard. Ostensively the committee's reason was that it introduced temporal considerations into a length standard. Some have suggested that the real reason is that the Academy wanted to preserve its intellectual and technical territory, and boost its funding. The determination of *its* standard was a highly complex and elitist matter, that required not only the most sophisticated technology, but also agreement being reached on the degree of oblateness of the earth. On August 8th 1791, it received 100,000 livres, about twice its normal annual budget, as a down payment for the geodesic survey of the meridian sector. Estimates of the total cost of determining the length of the 10 millionith part of the quarter meridian through Paris vary from 300,000 livres to millions of livres (Heilbron 1989, p. 991). There was, in the Committee's recommendation, a certain element of venal self interest hiding behind noble academic ideals. As one commentator at the time, the surveyor Jean Baptiste Biot, wrote: 'If the reasons that the Academy presented to the Assembly [to obtain support for the project] were not altogether the true ones, that is because the sciences also have their politics' (Heilbron 1989, p. 992). It is salutory to recognise that politics, even fairly venal politics, was involved in the determination of the length standard upon which all measurement in modern science is predicated.

There was more than just impersonal scientific interest involved in adopting a unified system of measurement. As one early nineteenth century French commentator remarked:

> The conquerors of our days, peoples or princes, want their empire to possess a unified surface over which the superb eye of power can wander without encountering any inequality which hurts or limits its view. The same code of law, the same measures, the same rules, and if we could gradually get there, the same language; that is what is proclaimed as the perfection of the social organization ... The great slogan of the day is uniformity. (Alder 1995, p. 62).

Huygens' pendulum standard did survive its rejection by the *Académie*. After years of patient measurement of the meridian sector, and the expenditure of a great deal of state money, the *Académie* choose a fraction of the meridian distance that coincided with Huygens 'three horological feet', and accepted the seconds pendulum as a *secondary* reference for its new length standard. Their metre differed from that of Huygens by only 0.3 of a millimetre, or 0.0003 m. This is a remarkable cosmic coincidence. The number of seconds in a day – 86,400 – is purely conventional, yet the length of a pendulum beating seconds turns out to be exactly one 40th million part of the circumfrence of the earth.

The *Académie's* argument about not mixing temporal and length considerations failed at the time to convince sceptics; two hundred years later in 1983, it did not convince delegates to the General Conference on Weights and Measures meeting in Paris who defined the standard universal metre as 'the length of path travelled by light in vacuum during a time interval of 1/299,792,458 of a second'. Thus time and space become inextricably linked. Again, pleasingly, this seemingly arbitrary figure, that is among the first things to confront students opening modern textbooks, is within a millimetre of Huygen's original and entirely natural length standard.[16]

Educational Suggestions

The fate of Huygen's 1673 proposal for a universal standard of length well illustrates a number of central methodological features of science, as well as the interrelations of science, technology, culture and politics. Given that the *US National Science Education Standards* explicitly advocate teaching about such interrelationships, and given that the document devotes two pages to the pendulum, it is a great pity that absolutely none of this story is mentioned.[17] Its neglect indicates the degree of rapprochement still needed between science education and the history and philosophy of science. The pendulum figures in just about every science curriculum in the world – and is standardly voted the most boring topic. There is an opportunity to embelish, with rich and informative history and philosophy, the dry practicals, formulae and exercises. Whether the opportunity is realised will depend upon the knowledge and enthusiasm that teachers have for the history and philosophy of the subject they teach.

This paper is part of a book that indicates a little of the cultural, historical, technological, religious and philosophical dimensions of the analysis of pendulum motion and of timekeeping (Matthews 2000). It is unrealistic to think that all these aspects of pendulum motion can be covered in a science course. But it is not unrealistic to hope that some coordination between school subjects can be achieved, and thus for teachers in related fields to work together on the 'big picture' of pendulum motion and timekeeping. This is precisely what, n the United States, the BSCS/SSEC project, and Project 2061 recommend.

The school curriculum appears to both students and teachers as being completely compartmentalised: subjects are isolated from each other, and topics within subjects are frequently unrelated. In school, knowledge seems to be truly fragmented.

However, well chosen themes that are heuristically rich, can organise a curriculum to maximise the degree to which the interdependence of knowledge becomes more transparent. It may be, minimally, a matter of looking at existing independently generated curricula and simply pulling the related parts together and arranging for some coordination and cross-referencing. But it can be more than this.

The following diagram illustrates how cross-disciplinary approaches to pendulum motion can bring some degree of cohesion and interrelatedness to the school curriculum.

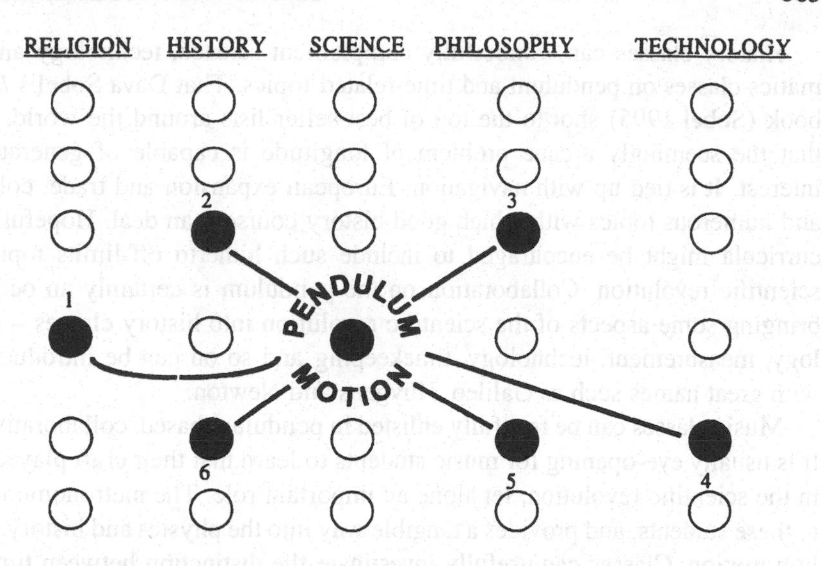

1. The Design Argument
2. European Voyages of Discovery
3. Aristotelian Physics and Methodology
4. Pendulum Clock
5. Idealisation and Theory Testing
6. Industrial Revolution

Timekeeping, navigation, and the longitude problem, for instance, provide rich materials for multicultural science teaching and comparative anthropology. We know that the polynesians made great sea journeys. Although it needs be admitted that we know little about the failures of their journeys, nevertheless comparing their navigation methods with the marine chronometer methods of Europe is instructive.

Technology classes are one clear field for cooperation. The making of a simple pendulum clock is a wonderful opportunity to appreciate the combination of brain and hand that is so essential to science. Likewise the making of water clocks, candle clocks, sundials and so on. The making of a sundial is the occasion for the world of astronomy to be opened up for students – the path of the earth, solstices, seasons, the names of the months, and so on. Questions such as whether or not the school flag pole might function as a sundial can be investigated. These technical activities can be combined with some basic historical study of ancient societies, and simple anthropology of contemporary societies that are not awash with clocks and watches. More sophisticated questions about timekeeping, wage labour, time-related cultural mores such as punctuality, sporting records, and so on, can also be raised.

Mathematics is another obvious area for collaboration. Geometrical depiction of the pendulum, and the progressive working out of its 'geometrical' properties, can give an oft-missing, applied focus, to basic geometry. How geometrical depiction allows pendulum motion to be related to freefall, and to inclined-plane motion, can be investigated.

History classes can wonderfully complement science, technology and mathematics classes on pendulum and time-related topics. That Dava Sobel's *Longitude* book (Sobel 1995) shot to the top of best-seller lists around the world, indicates that the seemingly arcane problem of longitude is capable of generating great interest. It is tied up with navigation, European expansion and trade, colonisation and numerous topics with which good history courses can deal. Hopefully history curricula might be encouraged to include such hitherto off-limits topics as the scientific revolution. Collaboration on the pendulum is certainly an occasion for bringing some aspects of the scientific revolution into history classes – methodology, measurement, technology, timekeeping and so on can be introduced, along with great names such as Galileo, Huygens and Newton.

Music classes can be fruitfully enlisted in pendulum based, collaborative efforts. It is usually eye-opening for music students to learn that their craft played any role in the scientific revolution, let alone an important role. The metronome is familiar to these students, and provides a tangible way into the physics and history of pendulum motion. Classes can usefully investigate the distinction between timekeeping and time measurement, and practical work on the reliability of rhythmic beat as a measure of time can be undertaken.

And much more is possible. Religion classes can deal with the chequered history of the design argument and its utilisation of the clock as a metaphor for the world. Literature classes can study the rich genre of prose and poetry occasioned by the scientific revolution, and of course write their own verse on time-related themes. Philosophy classes can take up methodological matters raised by the study of pendulum motion, they can deal with the relationship of theory to evidence, and of experiment to both theory and evidence. Drama classes could conduct a dramatic reenactment of the deliberations of the French Academy's metrication committee and its decision to opt for the expensive and elitist geodesic unit of length rather than the cheap and democratic chronological measure of Huygens. Such a reenactment would involve historical, philosophical and scientific studies, as well as the full range of dramatic competencies.

Notes

[1] This case study is part of a larger investigation of the history of the pendulum, its utilisation in timekeeping, and its philosophical and cultural ramifications (Matthews 2000).

[2] He sent these proofs to his father for possible publication, but almost immediately he obtained and read a copy of Galileo's *Discourse* wherein was Galileo's original proofs. He wrote to his father retracting the request for publication, saying 'I did not wish to write the *Iliad* after Homer had' (Yoder 1988, p. 9).

[3] This was an early manifestation of the view that science should be above, or at least beyond, politics. During the Napoleonic Wars the English scientist Davy would be allowed to travel freely in France conducting his research; during the First World War, the French scientist and historian, Pierre Duhem, defended the achievements of German science (Duhem 1916/1991); during the Second World War, the German scientist Max Planck appealed to such a separation to justify his remaining as director of the Kaier Wilhelm Institute in Nazi Germany (Heilbron 1986).

[4] The life and achievements of Huygens are written up in Bell (1947), Bos (1980a) and in the entry in *The Dictionary of Scientific Biography*.

[5] For Galileo's influence on Huygens horological investigations, see Dobson (1985).

[6] A translation of this work is appended to Edwardes (1977).

[7] The cycloid curve had been studied by Galileo and his pupil Torricelli. Huygens recognised that, for large amplitudes, the effective length of the pendulum had to be shortened with respect to the amplitude; a cycloid enabled this to happen.

[8] No small praise for something so seemingly abstruse and arcane; and a reminder, given its utilisation in producing truly isochronic pendulum motion in clocks, of the utility of 'theoretical' research.

[9] Dava Sobel provides a very readable, best-seller, account of the British efforts at solving the problem of longitude – ultimately done in the 1770s by John Harrison (Sobel 1995). The problem of finding longitude at sea is dealt with in many places, but see especially Gould (1923, pp. 1–17), Williams (1992, chap. 6), Landes (1998), Matthews (2000, chaps. 2, 7) and Stimson (1998). See also other contributions to Andrewes (1998), these being the papers presented at the 1993 Harvard Longitude Symposium on which Sobel's book was based.

[10] Nothing came of Wren's proposal, and the English persisted with their artificial yard standard. In 1758 it was found that even the two standard yards preserved at the Exchequer were not the same length (Heilbron 1989, p. 990)! An Act of 1824 did specify that if the Imperial Standard Yard were to be destroyed, then it should be replaced by the natural unit of a seconds pendulum. Such a catastrophe did happen in 1834 when a fire in the House of Parliament destroyed the standard yard. However by then it was realised that the length of a seconds pendulum was not as invariant as originally thought; and so the English, and their colonies, persisted with the artificial yard standard (Scarr 1967, pp. 2–5).

[11] He had calculated that the radius of the earth would have to be hundreds of times larger than it is for the centrifugal force to equal a body's weight, and hence for the body to be thrown off the rotating earth (Matthews 2000, pp. 150–152). Galileo had made this point in his arguments with those who claimed a rotating earth would hurl bodies, and buildings, into space.

[12] Full details of Richer's voyage, its aims and accomplishments, can be found in Olmsted (1942).

[13] Richer had previously not excelled as a careful experimentalist (Mahoney 1980, p. 253).

[14] The fact that the weight of a body changed from place to place, as was manifest in the variation of the pendulum's period, sowed the seed for the conceptual distinction between weight and mass. The intuition was that although weight changed with change in gravity, nevertheless something about the 'massiness' of the body remained the same. Jean Bernoulli first introduced the distinction between mass and weight, and Newton, as will shortly been seen, clarified it by introducing the idea of inertial mass.

[15] In 1677 Edmund Halley took a London seconds-pendulum to St. Helena (16° Lat. South) and found that it also had to be shortened (Cook 1998, p. 87). However he attributed the effect to the height above sea level at which he took his readings on the island. That is, the greater distance from the earth's center and hence the less gravitational attraction. In 1682 Robert Hooke informed the Royal Society that a London clock beat faster, that is, its period was shorter, when taken to Tangiers. On these and other examples, see Ariotti (1972, p. 406–407).

[16] Accounts of the development of the standard metre can be found in Alder (1995), Thompson (1967), Kline (1988, chap. 9), Kula (1986), Heilbron (1989), Petry (1993, pp. 302–303) and Berriman (1953, chap. XI). Additionally see *L'Epopé du Metre Historie Systeme Decimel*, Ministry of Weights and Measures, Bicentennary of the French Revolution Documentation, Paris, September 1989.

[17] For elaboration of this matter, see Matthews (1998).

References

Alder, K.: 1995, 'A Revolution to Measure: The Political Economy of the Metric System in France', in M.N. Wise (ed.), *The Values of Precision*, Princeton University Press, Princeton, NJ, pp. 39–71.

Andrewes, W.J.H. (ed.): 1998, *The Quest for Longitude: The Proceedings of the Longitude Symposium, Harvard University, Cambridge, Massachusetts, November 4–6, 1993*, Collection of Historical Scientific Instruments, Harvard University, Cambridge, MA.

Ariotti, P.E.: 1972, 'Aspects of the Conception and Development of the Pendulum in the 17th Century', *Archive for History of the Exact Sciences* **8**, 329–410.

Bell, A.E.: 1947, *Christiaan Huygens and the Development of Science in the Seventeenth Century*, Edward Arnold, London.

Berriman, A.E.: 1953, *Historical Metrology: A New Analysis of the Archaelogical and Historical Evidence Relating to Weights and Measures*, J.M. Dent & Sons, London.

Blay, M.: 1998, *Reasoning with the Infinite: From the Closed World to the Mathematical Universe*, University of Chicago Press, Chicago.

Bos, H.J.M.: 1980a, 'Christiaan Huygens – A Biographical Sketch', in H.J.M. Bos et al. (eds), *Studies on Christiaan Huygens*, Swets & Zeitlinger, Lisse, pp. 7–18.

Bybee, R.W., Ellis, J.D, Giese, J.R. & Parisi, L.: 1992, *Teaching About the History and Nature of Science and Technology: A Curriculum Framework*, BSCS/SSEC, Colorado Springs.

Chapin, S.L.: 1994, 'Geodesy', in I. Grattan-Guinness (ed.), *Companion Encyclopedia of the History and Philosophy of the Mathematical Sciences*, Routledge, London, pp. 1089–1100.

Cook, A.: 1998, *Edmond Halley: Charting the Heavens and the Seas*, Clarendon Press, Oxford.

Dobson, R.D.: 1985, 'Galileo Galilei and Christiaan Huygens', *Antiquarian Horology* **15**, 261–270.

Duhem, P.: 1916/1991, *German Science*, trans. J. Lyon and S.L. Jaki, Open Court Publishers, La Salle IL.

Edwardes, E.L.: 1977, *The Story of the Pendulum Clock*, J. Sherratt, Altrincham.

Fauvel, J. & Gray, J. (eds): 1987, *The History of Mathematics: A Reader*, Macmillan, London.

Gould, R. T.: 1923, *The Marine Chronometer, Its History and Development*, J.D. Potter, London. Reprinted by The Holland Press, London, 1978.

Heilbron, J.L.: 1986, *The Dilemmas of an Upright Man: Max Planck as Spokesman for German Science*, University of California Press, Berkeley.

Heilbron, J.L.: 1989, 'The Politics of the Meter Stick', *American Journal of Physics* **57**, 988–992.

Huygens, C.: 1673/1986, *Horologium Oscillatorium. The Pendulum Clock or Geometrical Demonstrations Concerning the Motion of Fpendula as Applied to Clocks*, R.J. Blackwell, trans., Iowa State University Press, Ames.

Huygens, C.: 1690/1945, *Treatise on Light*, translated and edited by S.P. Thompson, University of Chicago Press, Chicago.

Kline, H.A.: 1988, *The Science of Measurement: A Historical Survey*, Dover Publications, New York.

Kula, W.: 1986, *Measures and Man*, Princeton University Press, Princeton NJ.

Landes, D.S.: 1998, 'Finding the Point at Sea', in W.J.H. Andrewes (ed.), *The Quest for Longitude*, The Collection of Historical Scientific Instruments, Harvard University, Cambridge, MA, 2nd Edition, pp. 20–30.

Mahoney, M.S.: 1980, 'Christiaan Huygens: The Measurement of Time and of Longitude at Sea', in H.J.M. Bos et al. (eds), *Studies on Christiaan Huygens*, Swets & Zeitlinger, Lisse, pp. 234–270.

Matthews, M.R.: 1994, *Science Teaching: The Role of History and Philosophy of Science*, Routledge, New York.

Matthews, M.R.: 1998, 'How History and Philosophy in the US Science Education Standards Could Have Promoted Multidisciplinary Teaching', *School Science and Mathematics* **98**(6), 285–293.

Matthews, M.R.: 2000, *Time for Science Education: How Teaching the History and Philosophy of Pendulum Motion can Contribute to Science Literacy*, Plenum Publishers, New York.

McComas, W.F. & Olsen, J.K.: 1998, 'The Nature of Science in International Science Education Standards Documents', in W.F. McComas (ed.), *The Nature of Science in Science Education:Rationales and Strategies*, Kluwer Academic Publishers, Dordrecht, pp. 41–52.

McComas, W.F., Clough, M.P. & Almazroa, H.: 1998, 'The Role and Character of the Nature of Science in Science Education', *Science & Education* **7**(6), 511–532.

NRC (National Research Council): 1996, *National Science Education Standards*, National Academy Press, Washington.

Olmsted, J.W.: 1942, 'The Scientific Expedition of Jean Richer to Cayenne (1672–1673)', *Isis* **34**, 117–128.

Petry, M.J.: 1993, 'Classifying the Motion: Hegel on the Pendulum', in M.J. Petry (ed.), *Hegel and Newtonism*, Kluwer Academic Publishers, Dordrecht, pp. 291–316.

Scarr, A.J.T.: 1967, *Metrology and Precision Engineering*, McGraw-Hill, London.

Sobel, D.: 1995, *Longitude: The True Story of a Lone Genius Who Solved the Greatest Scientific Problem of His Time*, Walker & Co., New York.

Stimson, A.: 1998, 'The Longitude Problem: The Navigator's Story', in W.J.H. Andrewes (ed.), *The Quest for Longitude*, The Collection of Historical Scientific Instruments, 2nd ed., Harvard University, Cambridge, MA, pp. 72–84.

Thompson, E.P.: 1967, 'Time, Work-Discipline and Industrial Capitalism', *Past and Present* **38**, 56–97.

Williams, J.E.D.: 1992, *From Sails to Satellites: The Origins and Development of Navigational Science*, Oxford University Press, Oxford.

Yoder, J.G.: 1988, *Unrolling Time: Christiaan Huygens and the Mathematization of Nature*, Cambridge University Press, Cambridge.

Reconstructing the Basic Concepts of General Relativity from an Educational and Cultural Point of View

OLIVIA LEVRINI
Physics Department, University of Bologna, Italy
E-mail: levrini@df.unibo.it

Abstract. Research in the history and philosophy of physics has shown that the formalism of General Relativity can be interpreted in several different ways and, consequently, its teaching is very problematic. The present contribution is an example of a reconstruction of the debate concerning the foundations of General Relativity on the basis of cultural and educational criteria. In particular, the debate will be presented as guided by the concept of space, and by the different perspectives from which such a concept can be viewed. It will be pointed out that the various ways of looking at space give powerful criteria not only to create an order among the interpretations, but also to exploit the educational and cultural value of the debate.

1. Introduction

Current teaching of General Relativity (GR) is based on two strong assumptions:

1. The theory should only be taught in highly specialised university courses, as advanced mathematics is necessary. Hence GR is seldom mastered; neither its general outlines nor its basic ideas are widely known, at least in Italy.
2. The educational and cultural aspects to be dealt with are the 'emblematic' story of non-euclidean geometry and the 'emblematic' success of a theory constructed on formal criteria and able both to explain 'ancient' observations (the precession of the perihelion of Mercury) and to predict new and surprising experimental results (the deflection of light by gravitational fields).

GR is usually introduced, when taught, in much the same way Einstein introduced it:

- recognition that Lorentz frames, in Special Relativity (SR), have maintained a residual privilege, in other words an absolute character;
- formulation of the problem of making motion completely relative, and expression of Mach's Principle (MP);
- introduction of the Equivalence Principle (EP);
- discussion of the lift thought experiment;
- generalisation of the Principle of Relativity (RP) through the Principle of Equivalence and introduction of the concept of general covariance (GC);

F. Bevilacqua et al. (eds.), Science Education and Culture, 311–326.
© 2001 *Kluwer Academic Publishers. Printed in the Netherlands.*

- construction of the equation of geodesics and recognition of Newton's theory as the limiting case of particles moving slowly in a weak stationary gravitational field;
- development of differential geometry and construction of Einstein's field equation.

Research in the history and philosophy of physics have shown how problematic the links among the various points can be. Each step has also given rise to questions which leave the debate lively and open. For example:

- What is the real nature of the privilege maintained by Lorentz frames in SR? Epistemological (a characteristic of our description) or ontological (a feature of spacetime)?
- What is the real physical meaning of the EP? Is gravitation 'apparent', like the Coriolis force? Or do apparent forces have a gravitational nature?
- What is the cognitive role played by the lift thought experiment? Does it bring out the inertial nature of gravitation or the gravitational nature of inertial effects?
- Does GR satisfy a Principle of General Relativity? If so, why do neither Misner, Thorne and Wheeler nor Hawking and Ellis mention such a principle in their well-known textbooks? What is the meaning of GC? Is it a physical principle or only a prescription regarding the form to be given to equations?

The presentation and interpretation of the formalism of GR depends significantly on how the basic principles are interpreted, as a glance at popular university textbooks on GR and cosmology shows. Misner, Thorne and Wheeler, for instance, base their interpretation on the synthetic power of geometry to suggest imagery and design spacetime curved by matter (Misner et al., 1970), whereas Weinberg uses tensors and the other geometrical objects as pure formal instruments characterised by particular transformation properties under arbitrary changes of coordinates (Weinberg, 1972).

The following considerations have motivated the present study:

- the choices made by the authors of textbooks are not always altogether explicit in their cultural and educational presuppositions and students can find it difficult to compare different textbooks;
- the current debate on the history and philosophy of GR is interesting and rich from a cultural and educational point of view, but is not exploited enough for improving teaching, even at university level;
- though the basic ideas of modern physics (at least those of quantum physics and relativistic theories) have introduced radical modifications in human thought, they have not influenced the general outlook of educated citizens. One reason could be that research in science education has not developed strategies for making the basic ideas of modern physics accessible to secondary school students; in other words for reaching the physical and cultural core of modern theories without getting trapped in their formalisms.

Specific aims of the study are:

- analysing and reshaping the current debate on the meanings of GR in order to design teaching materials that can support and guide the reading of the most widely used university textbooks (Misner, Thorne and Wheeler; Shutz, Weinberg, Hawking and Ellis, for example);
- singling out those concepts and ideas that should be known by educated citizens, and about which materials for preservice teacher education can be produced.

2. The Concept of Space as an Educational Criterion for Reconstructing the Current Debate on the Foundations of GR

From an educational perspective a reading of the actual debate on the foundations of GR can be based on looking at it as the most recent expression of the traditional dispute between two particular images of space.

In fact the crucial role assumed by geometry in GR has re-opened the question concerning the *real nature* of space: Is it a physical object endowed with substantiality (the 'space-container' or 'substantival space') or is it no more than a set of formal relations among objects or possible positions of objects (if we refer to space), or among events or possible events (if we refer to spacetime), constructed by human reason to organise or comprehend the factual world ('relational space')? Such a question is at the basis of the historical debate between Newton and Leibniz about the meaning of classical mechanics (Friedman, 1983) and of the debate about the different interpretations of SR given by Minkowski, Einstein and Poincaré (Levrini, 1999). Nowadays space is seen as substantival in the so-called geometrodynamical interpretation of GR given by Wheeler, whereas an explicit denial of such a space is at the basis of the interpretations given by Weinberg, and by Einstein's followers, such as Sciama.

The concept of space can be seen as an *educational* criterion for reorganising the discussion about the meaning of GR for several reasons. The main ones are that space, being an intuitive and primitive concept, is a powerful tool for:

- anchoring the interpretation of the meaning of GR to a primitive question ('What is space?') that students should recognise as a real cultural problem. Hence GR can be presented as a real cultural product and not simply as an empty, abstract algorithm of raising and lowering indexes, of which one should only consider the predictive power;
- guiding the comparison of the different interpretations of the same formalism expressed in different textbooks, bringing out the fact that behind each interpretation there is a specific epistemological and ontological dimension;
- providing students with conceptual tools for taking a personal position on the meaning of the theory and, consequently, for feeling themselves more and more engaged in the learning process.

In the next section I shall present the main results of the analysis we have carried out in order to reshape the debate on the foundations, history and philosophy of GR from an educational perspective. The results are meant to be a guide for university teachers who want to provide students with tools for reading and interpreting different textbooks. Moreover they represent our starting point for selecting the basic ideas of GR which, in our opinion, should be part of the general knowledge of each secondary school teacher. Indeed, on the basis of these results, conceptual paths relating to 'spacetime physics', to be used at preservice teachers education level, have been designed (Grimellini et al., 1999; Levrini, 2000a, 2000b).

3. The Different Conceptions of Space and Their Implications for the Different Interpretations of General Relativity

"*Space acts on matter, telling it how to move. In turn, matter reacts back on space, telling it how to curve*" (Misner et al., 1970, p. 5): This statement describes the basic image of the geometrodynamical interpretation of the theory. Such an interpretation cannot be understood unless one assumes – as Misner, Thorne and Wheeler do – a substantival space, conceived of as a real entity whose existence is independent of the contained matter, even if the two interact.

Despite the wide acceptance among physicists of the geometrodynamical interpretation and despite its educational effectiveness,[1] such a space is not unanimously accepted: It is criticised above all on grounds of its unobservability and, hence, of its absoluteness. To this objection the 'substantivalist' usually answers that since GR expresses a physical-causal interaction between matter and space, the latter, even if not directly observable, is not causally inert. Its existence manifests itself in the fact that, if postulated, it can convincingly explain observable gravitational effects. This kind of argument is known as "inference to the best explanation" (Boniolo and Dorato, 1997): Existence is ascribed to theoretical entities, even if these are not directly observable, because their postulation allows the formulation of the explanation providing the most persuasive interpretation of the empirical data. It is easy to understand that such an argument cannot solve the question because, for the 'relationist', a substantival space is unacceptable in any case, even if presented as a hypothesis at the basis of an 'effective' interpretation.

The best argument against substantivalism is, however, the "principle of the identity of indescernibles" used by Leibniz against Newton's absolute space (Alexander, 1956): If everything in the world were reflected East to West (or better, translated a few meters East), retaining all the relations between bodies, would we have a different world? Unlike the relationistis, the "Newtonian" substantivalist must answer "yes" since the bodies in the world are now in different spatial positions, even though the relations between them are unchanged.

Such an argument can be translated into the context of GR, where the diffeomorphisms take the place of Leibniz' displacements of all bodies in space in such a way that their relative relations are preserved. In other words, substantivalists must nowadays bear in mind that their position leads them to deny the so-called "Leibniz equivalence": diffeomorphic models represent the same physical world. But by denying Leibniz equivalence, substantivalists must also accept a radical form of indeterminism, because a corollary of Leibniz equivalence states that diffeomorphic models can differ in properties that remain undetermined (the "hole argument"). Therefore substantivalists must not only accept that diffeomorphic models represent different worlds, but also that some of these worlds have undetermined properties.[2]

Nevertheless, it cannot be denied that the assumption of a substantival space introduces a strong criterion for interpreting the basic principles of GR, and this may be the reason for its widespread acceptance, even if it is not always made explicit. If a substantival spacetime continuum is accepted, every principle is interpreted as a statement about the continuum. The substantivalist aims at constructing a 'geometrical theory of the physical world', which means, from the substantivalist perspective, explaining physical processes and events in terms of their relations with spacetime, conceived of as a primitive entity. The substantivalist project is illustrated by Friedman as follows:

> [...] According to the present point of view, then, the basic or primitive elements of our theories are of two kinds: space-time and its geometrical structure; and matter fields – distributions of mass, charge, and so on – which represent the physical processes and events occurring within space-time. Our theories seek to explain and predict the properties of material processes and events by relating them to the geometrical structure within which they are "contained". [...] In the present treatment we explicitly take the more abstract geometrical entities as primitive and define the more observational entities in terms of them. (Friedman, 1983, p. 32)

The substantivalist project of geometrizing gravitation arises from the interpretation of the Equivalence Principle (EP) due to Pauli:

> *For every infinitely small region (i.e., a world region which is so small that the space- and time-variation of gravity can be neglected in it) there always exists a coordinate system K_0 (X_1, X_2, X_3, X_4) in which gravitation has no influence either on the motion of particles or any other physical processes.* In short, in an infinitely small world region every gravitational field can be transformed away.
>
> [...] The special theory of relativity should be valid in K_0. All its theorems have thus to be retained, except that we have put the system K_0, defined for an infinitely small region, in place of the Galilean coordinate system. (Pauli, 1921, p. 145)

Such a definition led Pauli to a very radical interpretation of gravitation as an *apparent force*, in the sense that changing the frame of reference can locally eliminate it:

> Gravitation in Einstein's theory is just as much of *an apparent force* as the Coriolis and centrifugal forces are in Newtonian theory. (We would be equally

> justified in taking the view that neither of the two forces should be called as apparent force in Einstein's theory.) It does not affect the argument that the gravitational force cannot, in general, be transformed away in finite regions, whereas the other forces can. The gravitational force can always be transformed away in infinitely small regions, and this fact alone is decisive. (Pauli, 1921; p. 148)

The interpretation of gravitation as an apparent force is no longer accepted because all the relativists agree that only the uniform component of gravitation can be considered a fictitious term and that a coordinate transformation or change in the state of motion of the observer have no effects on the presence or absence of a gravitational field. The presence of a 'true' gravitational field is determined by an invariant criterion, the curvature of the metric.

Pauli's formulation has, however, opened the possibility of interpreting gravitation, if not as an apparent force, as a geometrical one: a force which manifests itself as a modification of spacetime identified with the differentiable manifold. Such a possibility did not figure in Einstein's original intentions. As we shall see, Einstein saw the EP as the key to interpreting the inertial effects as being of gravitational origin and, at least when he was trying to construct GR as a 'Machian theory of inertia', he was not at all interested in eliminating gravitation (Norton, 1989).

Correspondingly, the lift thought experiment is used by substantivalists as an argument supporting the idea that gravitation can be locally transformed away: The lift is imagined as falling freely in a gravitational field, so as to suggest an image of the world whose real essence can be revealed by eliminating gravitational effects. The imagery adopted is that of spacemen hovering in their spaceships, because the new privileged class of frames of reference to which GR would have led is the class of Local Inertial Frames (free-falling, nonrotating frames).

From the substantivalist perspective GC and the principle of GR cannot be considered equivalent. Their meanings, identified by Einstein, must be separated.

GC is said to have lost its physical meaning because each spacetime theory (including Newtonian mechanics) can be given a generally covariant formulation.[3] Consequently, this property does not provide any information about the characteristics of the manifold which means, for the substantivalist, that it cannot have any physical meaning. It only tells us the form to be given to physical equations, for the geometrical essence of the world to be made more explicit.

On the other hand, a RP is related to a concept of *invariance*, i.e., to spacetime symmetries. Indeed both the Galilean and Lorentz groups of transformations can be viewed as the symmetry groups of classical and Minkowski spacetime, respectively, as Poincaré firstly pointed out in 1905. But if one looks at the principles of relativity in this way, the conclusion is that the principle cannot be extended to GR, because its curved continuum has no other symmetries but the trivial ones:

> [...] although there are no inertial frames in general relativity, there are inertial (geodesic) *trajectories*, and these trajectories give rise to an absolute distinction between inertial and noninertial (accelerated or rotating) motion,

> just as in previous theories. It is not true, therefore, that all frames of reference are 'equivalent' or 'indistinguishable'. We still have a privileged subclass of frames, the local inertial frames [...]. The existence of such a privileged subclass of frames clearly shows that the general theory does not institute a thoroughgoing relativity of motion. (Friedman, 1983; p. 27)[4]

The interpretation of the RP as related to the concept of invariance with respect to a group of transformations cannot be considered peculiar to the sustantivalist interpretation, and is widely accepted from a relationist point of view as well. The substantivalist can, nevertheless, be more inclined to accept the idea that the RP cannot be extended because his choice of assuming a substantival space already expresses his preference for absolute over relative concepts. This interpretation can be seen, therefore, as lending support to his point of view. The relationist, instead, argues that such an interpretation attributes to the RP a weaker role than it had in SR and makes it marginal in the entire structure of the theory (Norton, 1989). This can account for the omission of the RP from certain textbooks.

To complete the substantivalist framework, it must be pointed out that Mach's Principle (MP), in its more famous formulation that traces the origin of inertia back to gravitation,[5] plays no role in the theory structure. It would in fact undermine the real meaning of GR, which, for the substantivalist, expresses the geometrical nature of gravitation and not the gravitational origin of inertial effects.

Einstein's position and the relationist position, represented nowadays by Sciama, for instance, are basically different from the substantivalist one. They arise from the intention of eliminating from physics every absolute concept, like the 'unjustified' privileged role attached to inertial states of motion in classical mechanics and in SR. The main goal of GR is therefore to address such an "inherent epistemological defect" (Einstein, 1916a, 112), i.e., to find the physical nature of this 'mysterious' privilege. Einstein's refusal of a substantival space is expressed as follows:

> space as opposed to 'what fills space' has no separate existence [...]. There is no such thing as an empty space, i.e., a space without field. Spacetime does not claim existence on its own, but only a structural quality of the field. (Einstein, 1952, p. 375)

The quotation indicates that the primitive entity is the field, a physical object, while space is a derived entity: Space is an interpretative category constructed by humans in order to "comprehend" the natural world, i.e., to create:

> an order among sense impressions, this order being produced by the creation of general concepts, relations between the concepts and by definite relations of same kind between the concepts and sense experience. (Einstein, 1954, p. 292)

At the basis of such a relationist view is the Machian conviction, in Einstein's interpretation, that inertial effects have a gravitational origin. The substantivalist interpretation of the EP has upset the original interpretation given by Einstein: the EP was the indication of the gravitational origin of inertia, and not of the geometrical nature of gravitation. In other words, while gravitational effects must, for a substantivalist, be interpreted in terms of spacetime geometry, for the relationist *a la* Mach inertial effects derive from an interaction among all the masses of the universe. The gravitational nature of the inertial effects can be understood in the particular setting of the uniformly accelerated lift. The thought experiment is used by Einstein and Sciama to *create* a gravitational field in vacuum and not to *eliminate* gravitation in a gravitational field. So gravitation is present everywhere in Einstein's universe, because every relative accelerated motion produces a gravitational field. His universe is as full as the Cartesian universe, as Einstein himself recognises:

> [...] Descartes was not so far from the truth when he believed he must exclude the existence of an empty space. The notion indeed appears absurd, as long as physical reality is seen exclusively in ponderable bodies. It requires the idea of the field as the representative of reality, in combination with the general principle of relativity, to show the true kernel of Descartes' idea: there exists no space 'empty of field'. (Einstein, 1952, pp. 375–376)

The discussion about the role and the *status* of the so-called MP within the structure of GR is still open. It is nevertheless accepted that the implementation of the MP in GR is not a logical necessity, since non Machian theories, like de Sitter's, also exist. On the other hand Machian models, such as Sciama's, have so far presented neither logical nor physical contradictions. In other words, the choice among possible models has its roots in meta-physical reasons. The role of meta-physical beliefs is clear in Einstein's position, when he refuses space as a primitive entity – which he considers antimachian – and gives to the basic principles of GR the following meanings:

- *The EP:* besides the manifestation of the gravitational origin of inertia, the equivalence between inertia and gravitation allows us to interpret the distinction between inertial and non-inertial frames not as an inherent property of the frames, but as a distinction induced by the structure of the gravitational field.

  > The structure responsible for inertial and gravitational effects is the metric tensor. The space-time manifold itself has no properties that would enable us to designate the motion associated with any given world line as privileged, that is as 'inertial' or 'unaccelerated'. This designation depends entirely on the metric and the affine structure for space-time that it determines. (Norton, 1989, p. 41)

  Correspondingly, the determination of Lorentz frames depends on the Minkowski tensor which is interpreted, by Einstein, as a particular gravitational field: Minkowski spacetime, indeed, represents not space in the absence of gravitation, but an *example* of the four-dimensional generalization of a special gravitational field:

> If we imagine the gravitational field, i.e., the functions g_{ik}, to be removed, there does not remain a space of the type, but absolutely *nothing*. [...] [Minkowski's space-time] is not a space without field, but a special case of g_{ik} field, for which – for the coordinates system used, which in itself has no objective significance – the functions g_{ik} have values that do not depend on the coordinates.[6] (Einstein, 1952, p. 375)

- *Principle of GC:* From the relationist point of view the principle still conserves a real physical meaning, since it plays the role of a 'principle of reality': It allows us to distinguish mathematical properties from the physical world. From the relationist point of view there is no substantival space literally identified with nature to ensure that the theory is referring the factual world. A criterion of reality must be sought within the equations, and "Leibniz equivalence" is the criterion chosen: physical properties are those invariant under the general group of diffeomorfisms, because properties of the physical world must not depend on our ways of describing them.[7] The application of this principle of reality implies that only topological space is real, because the other geometrical structures (i.e., the metrical, affine and conformal properties) depend on the coordinate system. Einstein writes:

> *Reality* is physically nothing other than the totality of space-time point coincidences. If, for example, all physical occurrences were constructed from the motion of material points alone, then the meetings of the points, i.e., the intersections of their world lines, would be the only reality, i.e., that which is in principle observable. (Einstein, 1916, in Speziali, 1972, p. 64)

In the famous article on the foundations of GR Einstein wrote:

> That this requirement of general co-variance [...] is a natural one, will be seen from the following reflection. All our space-time verifications invariably amount to a determination of space-time coincidences. If, for example, events consisted merely in the motion of material points, then ultimately nothing would be observable but the meanings of two or more of these points. Moreover, the results of our measuring are nothing but verifications of such meetings of the material points of our measuring instruments with other material points, coincidences between the hands of a clock and points on the clock dial, and observed point-events happening at the same place at the same time. (Einstein, 1916a, p. 117)

Einstein is so convinced that a principle of GR must hold in physics:

> The laws of physics must be of such a nature that they apply to systems of reference in any kind of motion. (1916a)

that he reformulates the RP as follows:

> Principle of relativity: The laws of nature are only assertions about time-space coincidences; therefore, they find their only natural expression in general covariant equations. (1916a)

> [The] exact formulation of the general principle of relativity [...]. *All Gaussian co-ordinate systems are essentially equivalent for the formulation of the general laws of nature.* (1916b, p. 108)

It is clear that this formulation of the principle of GR has nothing to do with the standard formulation, according to which special relativistic physics would hold in arbitrary frames of reference. Einstein's formulation of the principle of GR expresses the need to reduce the physical content of natural laws to the catalogue of spacetime coincidences and to consider every further structure as a construction, because such a catalogue is the same from any frame of reference in any kind of motion.

Nowadays relationists also prefer to detach the principle of GR and the PGC, for their coincidence leads to a rather weak interpretation of the former and of the physical content of theories. Sciama for example, aiming at the construction of a Machian theory of inertia in analogy to the electromagnetic theory, was able to implement a general principle of relativity. His laws of gravitation, containing a static and a dynamical part of inertial/gravitational interaction, are the same not only for observers moving at any constant speed but to all observers, however their laboratories may accelerate or rotate. The price of the requirement was, however, certain modifications to SR (Sciama, 1953, 1959).

Beside Mach's form of relationism another interpretation exists which refuses to attribute a geometrical character to gravitational force or, in other words, to identify the differentiable manifold with spacetime. Its main exponent is Weinberg, and the position is expressed in his text *Gravitation and Cosmology* (Weinberg, 1972). An interesting point is the explicit analogical role ascribed to geometry. In the paragraph titled "The Geometric Analogy" Weinberg writes:

> We have seen in this chapter that the nonvanishing of the tensor $R_{\lambda\mu\nu\kappa}$ is the true expression of the presence of a gravitational field. We also saw in Chapter 1 that Gauss was led to introduce the Gaussian curvature $K = -R/2$ as the true measure of the departure of a two-dimensional geometry from that of Euclid, and that Riemann subsequently introduced the curvature tensor $R_{\lambda\mu\nu\kappa}$ to generalize the concept of curvature to three or more dimensions. It is therefore not surprising that Einstein and his successors have regarded the effects of a gravitational field as producing a change in the geometry of space and time. At one time it was even hoped that the rest of physics could be brought into geometric formulation, but this hope has met with disappointment, and the geometric interpretation of the theory of gravitation has dwindled to a mere analogy, which lingers in our language in terms like "metric", "affine connection", and "curvature", but is not otherwise very useful. The important thing is to be able to make predictions about images on the astronomers' photographic plates, frequencies of spectral lines, and so on, and it simply doesn't matter whether we ascribe these predictions to the physical effect of gravitational fields on the motion of planets and photons or to a curvature of space and time. (The reader should be warned that these views are heterodox

> and would meet with objections from many general relativists.) (Weinberg, 1972, p. 147)

The analogy in Weinberg shows all the characteristics of a "theory-constitutive metaphor" (Boyd, 1979). A metaphor, in other words, whose main feature is that of orienting a "programmatic research". Its primary system (*explanandum*) is an only partially understood physical phenomenon; whilst the secondary system (model) is a strongly structured system which has not yet been completely explored, i.e., which is characterised by "conceptual open-endedness". A theory-constitutive metaphor is introduced:

> when there is (or seems to be) good reason to believe that there are theoretically important respects of similarities or analogy between the literal subjects of the metaphors and their secondary subjects. The function of such metaphors is to put us on the track of these respects of similarity or analogy. (Boyd, 1979, p. 363)

The metaphor represents a theoretical device, because the dynamics of its construction allows the discovery of new features both of the physical phenomenon under investigation and of the formal language used to describe it. In other words, a theory-constitutive metaphor is:

> one of many devices available to the scientific community to accomplish the task of *accommodation of language to the causal structure of the world*. By this I mean the task of introducing terminology, and modifying usage of existing terminology, so that linguistic categories are available which describe the causally and explanatorily significant features of the world. (Boyd, 1979, p. 358)

In Weinberg's analogy the natural phenomena to be studied are the effects of gravitation (primary system) and the structured linguistic system is the Gauss–Riemann theory of curved surfaces (secondary system). The two systems have already been viewed by Einstein as the terms of the analogy constructed on the identification of spacetime and the manifold. Such an analogy has been exploited to such an extent by the geometrodynamical interpretation that the two systems have finally coincided and the metaphor was dead. In Weinberg's opinion, the main reason to separate the two terms of the metaphor is that the geometrical nature of gravitation introduces a difference between this interaction and the other fundamental ones. And this difference is unacceptable for a scientist who intends to unify natural interactions. The project itself would suggest that the core identification be moved toward the EP, considered by Weinberg the expression of a principle of symmetry (the equivalence among systems of coordinates) and therefore closer to the basic principles of quantum field theories. The EP can be seen as the analogical term of Gauss' postulate because:

- the former states that at any point in space-time we may erect a locally inertial coordinate system in which *matter* satisfies the *laws of special relativity*;

- the latter states that at any point on a curved surface we may erect a locally Cartesian coordinate system in which *distances* satisfy *Pythagoras' theorem.*

The laws of gravitation must, as a consequence, be very similar to those of Riemannian geometry. The modified core of the analogy and the attention paid to possible links with the other theories introduce further elements to be explored, new relations to be constructed: a new dynamism typical of a theory-constitutive metaphor.

The central role attached by Weinberg to the EP is emphasised by the meanings ascribed to the other basic principles of GR. The principle of GC is read by Weinberg as the operative reformulation of the EP.

> [...] we shall follow a different method, one that is of precisely the same physical content [of the EP], but is much more elegant in appearance and convenient in execution. This method is based on an alternative version of the Principle of equivalence, known as *the Principle of General Covariance.* (Weinberg, 1972, p. 41)

The Principle of GC gives explicit expression to the character of dynamical symmetry inherent in EP, and by this its analogy with the local gauge symmetries.

The EP is also seen as a possible explanation of the origin of inertia: in this sense it is considered an alternative to the MP (Weinberg, 1972).

In conclusion I would like to point out the characteristics of the EP that, in my opinion, could have led Weinberg to emphasise it. The EP:

- allows the construction of a bridge toward differential geometry and its powerful formal system;
- constrains the theory to a well corroborated experimental result (the equivalence between inertial mass and gravitational mass) and to accepted theories, like SR and the Newtonian theory of gravitation.

In other words, focusing the interpretation of GR on the EP exploits the power of the geometrical formalism, while having at the same time the minimum number of ontological constrains. These aspects are not trivial for a physicist engaged in research in particle physics, where a highly elaborate formalism coexists with ontological constraints loose enough to allow new interpretations. This particular perspective allows us to understand why Weinberg sees the equivalence between inertia and gravitation as a coincidence. It derives from experiments and this is enough to place it at the basis of a theory, but not enough to lead us to interpret gravitation in terms of inertia or inertia in terms of gravitation. Weinberg's opinion is that the current state of the research prevents us from constraining the theory in excessively tight ontological interpretations. GR is a theory still very internal to its language and its formal rules. Only the unified theory will give us the tools to anchor the theory to the natural world and to interpret correctly its terms and the relationships among them.

4. Cultural and Educational Analysis of the Reconstruction of General Relativity's Basic Ideas

The concept of space as criterion to re-organise the discussion leads us to identify three possible interpretations of the formalism of GR. The interpretations are different because they evoke strongly different imagery, attach a different role to geometry, look at physical knowledge from a different perspective, and use different criteria to decide what is 'real'.

In the substantival world, as much empty of gravitation as it is full of geometry, the latter is the essence and the structure of the natural world. Physics reveals its real nature through geometry and therefore a spacetime theory must be written as a geometrical-axiomatic 'theory of principles'.[8] Reality is attached to those entities whose existence, if postulated, can simplify the theoretical explanation.

Einstein's preference for a 'theory of principles' does not imply that reality can be attached to non-observational entities. A distinction between reality and physical knowledge persists and the role of the latter is to make a "conceptual construction or model for the real world" (Einstein, 1954). In Einstein's world full of gravitational field, geometry is the structure of our knowledge, the formal relations built to create "an order among sensory impressions" and among observations to which reality is attached.

In Weinberg's interpretation, physical phenomena and geometry are the two terms of a theory-constitutive metaphor, terms that must be explored and investigated together. In this case the role of geometry as language is explicit and GR is presented as a 'constructive theory', in which some relationships are taken as the basis for the creation of new links with phenomena different form the initial ones. The interpretation of EP as a principle of symmetry, indeed, creates links with the other quantum theories of fields. If we look at GR from the overall perspective outlined by Weinberg, we are led to think that the weakness of the ontological commitments does not imply giving up realism but only a temporary judgement suspension while we search for a better ontological interpretation.

Thus, differences in interpretation are deep and belong to various levels (imaginative, epistemological, ontological, metaphysical ...). Their explication is important for exploiting cultural and cognitive potentialities of GR and for comparing university texts (Misner, Thorne and Wheeler, Shutz as example of the substantivalist perspective and Weinberg, Hawking and Ellis as relationalist presentations). This does not mean that a university course of GR must address in detail all the interpretations. On the contrary, I believe that the choice of one leading interpretation (and one leading text) is necessary, provided that it is explicit and presented as one possible interpretation of the theory. A teacher should situate the chosen conceptual path within the overall cultural context and give the students the main conceptual tools for reading several different texts and comparing different interpretations. "Teaching controversy" should help students to focus the peculiar aspects of each interpretation and to elaborate critical criteria both to express their own preferences and to give them a cultural collocation (Bevilacqua, 1999). Furthermore, a critical comparison between different interpretations of the same formalism can capture the interest of a larger number of students and remove the idea of physics as a sterile and static collection of truths.

As far as teaching the basic ideas of GR is concerned, the main problem is that only few concepts of GR can be taught avoiding the use of the tensorial formalism. Hence teaching at secondary school level does not usually go beyond the presentation of the EP, the gravitational red-shift and the lift thought experiment. My opinion is that these topics can be enough to give students ideas about the physical and cultural value of GR, provided that their discussion is anchored to the controversy about space and if the classical roots of the controversy are made explicit. In this case teaching should aim at guiding students to compare some popular expositions of GR such as, for instance, Einstein (1916b), Sciama (1959, 1969), Smolin (1992), Weinberg (1992), and Wheeler (1990).

5. Conclusion

The analysis allowed us to point out criteria for interpreting also the historical debate on the foundations of SR among Minkowski, Einstein and Poincaré and that on classical mechanics between Newton and Leibniz (Levrini, 1999, 2000a). Therefore we can state that it is possible to start from classical physics and reconstruct fundamental issues of physics on the basis of an intuitive or imaginable concept, such as that of space (Grimellini et al., 1999) and that is possible to lead students to recognise that at basis of modern physics there are primitive ideas, even if they have been transfigured by an increasingly difficult formal language.

I believe that this can be a powerful way to make the most important concepts of modern physics more easily accessible and, at the same time, classical physics more interesting as it becomes a field open to updated interpretations in the light of present physical research.

Acknowledgements

The present study is part of my PhD Dissertation. I wish to thank first of all my supervisors, Silvio Bergia and Nella Grimellini Tomasini, whose competence has made the work extremely stimulating and interesting. I want to express my gratitude to Carla Casadio for the long and lively discussions which stimulated and supported my research work. I also give my thanks to Alexander Afriat, Barbara Pecori and Rolando Rizzi for their competence and patience in improving editing for the English translation.

Notes

[1] Important university textbooks follow, indeed, such an interpretation. Besides the quoted Misner et al. (1970), see for example, Schutz (1985).

[2] For more details, see, for example, Earman and Norton (1987), Norton (1992).

[3] Kretschmann, in 1917, was the first to point out that each physical law can be given a general covariant formulation. A complete and comprehensible discussion of the issue can be found in Friedman (1983).

[4] More details can be found, for example, in Friedman (1983, pp. 46–61).

[5] The so-called MP has been formulated in many ways, as one can infer from the Barbour and Pfister, eds. (1995) volume.

[6] Italics added by author.

[7] More details can be found, for example, in Norton (1992).

[8] 'Theories of principles' and 'constructive theories' are here used in the sense indicated by Einstein. See, for example, Pais (1982).

References

Alexander, H.G. (ed.): 1956, *The Leibniz-Clark Correspondence*, Manchester University Press, Manchester.

Barbour, J. & Pfister, H. (eds.): 1995, *Mach's Principle. From Newton's Bucket to Quantum Gravity*, Einstein Studies, Vol. 6, Birkhäuser, Boston.

Bevilacqua, F.: 1999, 'Teaching Controversy in Teaching Science', plenary lecture at the Second Joint Conference – Science as Culture – Como-Pavia (www.cilea.it/volta99).

Boniolo, G. & Dorato, M.: 1997, 'Dalla relatività galileiana alla relatività generale', in G. Boniolo (ed.), *Filosofia della fisica*, Mondadori, Milano, pp. 5–167.

Boyd, R.: 1979, 'Metaphor and Theory Change: What is "Metaphor" a Metaphor for?', in A. Ortony (ed.), *Metaphor and Thought*, Cambridge University Press, Cambridge, pp. 356–408.

Earman, J. & Norton, J.: 1987, 'What Price Substantivalism? The Hole Story', *Brit. J.; Phil. Sci.* **38**, 515–525.

Einstein, A.: 1916a, 'Die Grundlage der allgemeinen Relativitätstheorie', *Annalen der Physik*, Vol. 49 ('The Foundation of the General Theory of Relativity', in H.A. Lorentz, A. Einstein, H. Minkowski & H. Weyl (1952), *The Principle of Relativity. A Collection of Original Memoirs on the Special and General Theory of Relativity* (with notes by A. Sommerfeld), Dover Publications, New York, pp. 109–164).

Einstein, A.: 1916b, *Über die spezielle und allgemeinen Relativitätstheorie (gemeinveständlich) (Relativity, The Special and the General Theory: A Popular Exposition*, 15th edn, R. W. Lawson, trans., Methuen, London, 1954).

Einstein, A.: 1952, *Relativity and the Problem of Space*, in A. Einstein (1954), *Ideas and Opinions*, trans. Sonja Bergmann, Crown Publisher, New York.

Einstein, A.: 1954, *Ideas and Opinions*, trans. Sonja Bergmann, Crown Publisher, New York.

Friedman, M.: 1983, *Foundations of Space-Time Theories*, Princeton University Press, Princeton.

Giannetto, E.: 1995, 'Henri Poincaré and the Rise of Special Relativity', *Hadronic Journal Supplement* **10**, 365–433.

Grimellini Tomasini, N., Levrini, O., Casadio, C., Clementi, M. & Medri Senni, S.: 1999, 'Insegnare fisica per nuclei fondanti: un esempio riferito al concetto di spazio', *La Fisica nella Scuola* **XXXII**(4), 202–213.

Hawking, S.W. & Ellis, G.F.R: 1973, *The Large-Scale Structure of Space-Time*, Cambridge University Press, London and New York.

Kretschmann, E.: 1917, 'Über den physikalischen Sinn der Relativitätspostulate, A. Einstein's neue und seine ursprüngliche Relativitätstheorie', *Annalen der Physik* **53**, 575–614.
Levrini, O.: 1999, 'Relatività ristretta e concezioni di spazio', *Giornale di fisica* **XL**(4), 205–220.
Levrini, O.: 2000a, *Analysing the Possible Interpretations of the Formalism of General Relativity: Implications for Teaching*, Ph.D. dissertation, Physics Department, University of Bologna, Italy.
Levrini, O.: 2000b, *Teaching Modern Physics by a Modern Teaching of Classical Physics. The Case of General Relativity*, pre-print.
Mach, E.: 1883, *Die Mechanik in ihrer Entwickelung, historisch-kritisch dargestellt (The Science of Mechanics: A Critical and Historical Account of its Development*, La Salle; Open Court, 1960).
Minkowski, H.: 1909, Raum und Zeit, *Physikalische Zeitschrift* **10**(3), 104–111 ('Space and Time', in H.A. Lorentz, A. Einstein, H. Minkowski & H. Weyl (1952), *The Principle of Relativity. A Collection of Original Memoirs on the Special and General Theory of Relativity* (with notes by A. Sommerfeld), Dover Publications, New York, pp. 73–96).
Misner, C.W., Thorne, K.S., Wheeler, J.A.: 1970, *Gravitation*, W.A. Freeman and Co., San Francisco.
Norton, J.: 1989, 'What was Einstein's Principle of Equivalence?', in D. Howard & J. Stachel (eds.), *Einstein and the History of General Relativity*, Einstein Studies, Vol. 1, Birkhäuser, Boston, pp. 5–47.
Norton, J.: 1992, 'The Physical Content of General Covariance', in J. Eisenstaedt & A.J. Kox (eds.), *Studies in the History of General Relativity*, Einstein Studies, Vol. 3, Birkhäuser, Boston, pp. 281–315.
Ortony, A. (ed.): 1979, *Metaphor and Thought*, Cambridge University Press, Cambridge.
Pais, A.: 1982, *Subtle is the Lord ..., The Science and the Life of Albert Einstein*, Oxford University Press, Oxford.
Pauli, W.: 1921, 'Relativitätstheorie', in A. Sommerfeld (ed.), *Encyklopädie der mathematishen Wissenshaften, mit Einschluss ihrer Anwendungen*, Vol. 5, Physik, part 2. B.G. Teubner, Leipzig, 1904–1922, pp. 539–775 (*Theory of Relativity*, Pergamon, London, 1958).
Poincaré, J.H.: 1902, *La Science et l'Hypothèse*, Flammarion, Paris.
Schutz, R.F.: 1985, *A First Course in General Relativity*, Cambridge University Press, Cambridge.
Sciama, D.W.:1953, 'On the Origin of Inertia', *Mon. Not. R.A.S.* **113**, 34–42.
Sciama, D.W.: 1959, *The Unity of the Universe*, Faber and Faber, London.
Sciama, D.W.: 1969, *The Physical Foundations of General Relativity*, Doubleday & Company, Inc.
Smolin, L. 1997: *The Life of the Cosmos*, Oxford University Press, New York, Oxford.
Speziali, P. (ed.): 1972, *Albert Einstein–Michele Besso. Correspondence 1903–1955*, Hermann, Paris.
Weinberg, S.: 1972, *Gravitation and Cosmology*, Wiley and Sons, New York.
Weinberg S.: 1992, *Dreams of a Final Theory*, Wintage.
Wheeler, J.A.: 1990, *A Journey into Gravity and Spacetime*, W.H. Freeman and Company, New York.

The Contribution of the History of Physics in Physics Education: A Review

FANNY SEROGLOU and PANAGIOTIS KOUMARAS
School of Education, Aristotle University of Thessaloniki, 54006 Thessaloniki, Greece

Abstract. Our research is focused on the selection, classification and comparative presentation of the various proposals concerning the contribution of the history of physics in physics education, that have been designed and/or carried out as part of either research or curriculum development during the last century. The framework of the classification is the result of the study of the aims of the teaching-learning of physics, as they have been presented since the 1960s, coupled with the current trends in science education, including discovery learning in the 1960s and constructivism nowadays. The study of the various proposals concerning the contribution of the history of physics in physics education revealed different points of view and led to the creation of new sub-categories of the initial framework. In an attempt to have a better overall view of the classification, we designed a chart so that the various proposals reported could be observed and commented on in a comparative way. The framework of the classification is presented on the vertical axis of a chart, while time from 1893, the year of the earliest reference available (Lodge 1893), till today is presented on the horizontal axis.

1. Introduction

Since 1893 (Lodge 1893), a significant number of ideas have been recorded concerning the fruitful use of the history of physics in physics education. For example, 2006 published papers are available from the Personal Library Software of WWW under the topic: History and Science and Education (December 1996). Such a large number of papers concerning the contribution of the history of physics in physics education requires a classification of all the proposals presented that would provide teachers and/or researchers with all the information they need as well as help them concentrate on the proposals touching their own project.

Selective collections of previous proposals on a certain subject are included in publications concerning curriculum development. For example, Nielsen and Thomsen present the reasons for using the history and philosophy of physics in physics education in order to change students' attitude towards the teaching-learning of physics. They also develop a curriculum that expresses a turning-point: Teaching of physics is no longer focused on the content of physics but on presenting physics as a human activity (Nielsen and Thomsen 1990).

Whiteley reports on eight proposals about the important role of the history of physics in physics education and argues that the history of physics may become the

F. Bevilacqua et al. (eds.), Science Education and Culture, 327–346.
© 2001 *Kluwer Academic Publishers. Printed in the Netherlands.*

keystone of a curriculum and also be used in the exams for evaluation (Whiteley 1993).

Sherratt makes a historical review of the evolution of science education during he first half of the 20th century and presents various trends, proposals, applications and problems about the use of the history of physics in the teaching of physics and in the curricula (Sherratt 1980, 1982).

A significant bibliography of publications, in quantity as well as in quality, about the use of the history of physics in physics education is presented in *Science Teaching: The Role of History and Philosophy of Science* (Matthews 1994). This book is a multidimensional report on the history of debate on the potentials of the rapprochement between history, philosophy and science education and highlights the problems of science education which may be alleviated by the beneficial contribution of the history and philosophy of science. It also presents curricula designed with the perspective of facilitating teaching – learning of science with the use of history of science and suggests ways in which the history and philosophy of science can be usefully included in teacher preparation programs.

We support the view that researchers, curriculum designers and teachers, working in this area, either at present or in the future, need a useful tool that will help them to get briefly acquainted with the different trends and the various proposals concerning the use of history of physics in physics education, as well as to define and confine their area of research and, furthermore, study the conditions and the perspectives of its touching on other areas of research. We make an attempt to cover this need with our classification. The framework of the classification is the result of the study of the changing and evolving aims, the current trends and the various factors influencing the teaching – learning of physics education during the years.

2. The Framework of the Classification

The framework of the classification is a result of the study of the aims of the teaching and learning of physics, as they have been presented since the 1960s, coupled with the current trends in science education including discovery learning in the 1960s and constructivism nowadays. The various trends introduce a number of factors that should be taken into account in order to accomplish a certain aim, i.e., in the context of constructivism, pupils' ideas become a very important factor influencing the aims of the teaching and learning of physics. In many cases, there is a very thin line between what we call an aim and what we define as a factor. That is a result of the fact that a parameter of the teaching and learning procedure may be defined as a factor in one decade and in the next decade, after the acknowledgement of its contribution to teaching and learning and expressing the recent research and curriculum developments, it may be defined as an aim. For example, in the 1980s students' metacognitive skills were considered to be an important factor which influenced learning, while now, in the 1990s the metacognitive aims of physics teaching are clearly defined in modern curricula (Gunstone and Northfield 1994; White and Gunstone 1989).

Recent research in science education points out that conceptual change, therefore learning, has not only a cognitive character but also a metacognitive and an emotional one (Duit 1994; Tyson et al. 1997). During the last 30 years, the objectives of physics teaching can be classified in four categories: cognitive, metacognitive, emotional (or affective) and practical objectives (Thijs and Bosch 1995). Practical objectives, which refer to developing skills in performing science investigations, analysing data, communicating and skills in working with others during labwork, are not included in the framework of our classification as proposals for the use of history of physics in this area of physics education have not been reported. Each one of the above category of aims expresses a different dimension of the teaching and learning procedure. Reflecting this view, the current framework of science education is three-dimensionally defined as cognitive, meta-cognitive and emotional. Each one of these dimensions includes a number of areas for study, which are all presented in brief in the following chapters.

2.1. THE COGNITIVE DIMENSION

The cognitive dimension includes the following categories: The teaching and learning of the content of physics, of the methodology of physics, of problem-solving skills and students' alternative ideas. Each one of these categories reflects a cognitive objective of the teaching – learning of physics.

The content of physics consists of the phenomena and concepts of physics, the theories and their models, the symbols and terminology of physics that students have to deal with during the teaching and learning process.

The teaching of physics methodology may either aim at the teaching of a certain methodology of physics, in the context of physics, or become a useful tool in facilitating the development of students' cognitive skills. A significant feature of the change in science curriculum during the past twenty-five years has been the shift away from the teaching of science as a body of established knowledge towards the experience of science as a method of generating and validating such knowledge (Hodson 1993).

Problem solving is concerned with the knowledge, comprehension, analysis and application of a set of exact concepts and/or relationships describing mostly verifiable phenomena of physics and demands definite cognitive strategies in which functional use is made of concepts and relationships (Mohopatra 1987). During the 1990s researchers propose that the concept of 'problem' in problem-solving processes should be extended to include the generation and development of bodies of knowledge such as models. For example, the differentiation between the use, the creation, the improvement and the revision of a scientific model may be characterised as a problem (Stewart and Hafner 1991).

According to constructivism, research on pupils' ideas about the phenomena and the concepts of physics coupled with the actual use of the information coming from the above research in instructional design may facilitate pupils' conceptual change. In the 1990s, students' alternative ideas have been recorded regarding the methodology of physics (Guillon 1995), furthermore, discussion has already started on the research of students' alternative ideas about problem-solving (Stewart and Hafner 1991).

2.2. THE METACOGNITIVE DIMENSION

The use of generative self-regulatory cognitive strategies that enable individuals to reflect on, construct meaning from and control their own activities, enhance the acquisition of knowledge by overseeing its use and by facilitating the transfer of knowledge to new situations (Glaser 1994; Duschl and Erduran 1996). Students are more likely to develop wide-ranging thinking skills if they are encouraged to think about their own thinking, to become aware of the strategies of their own thinking and actions (Nickerson et al. 1985; Perkins and Salomon 1989; Adey and Shayer 1993).

The metacognitive dimension includes the understanding of the nature of science and the understanding of the science-society interrelations which are both essential factors in the teaching and learning of science (Driver et al. 1996). The understanding of the nature of science involves the understanding of the nature of the content of science as well as the understanding of the nature of scientific methodology. The understanding of the nature of the content of science is reflected on the understanding of the philosophical dimension of explanation, of experiment, of model and on the recognition that physics is a functional model of interpreting, describing and predicting the evolution of phenomena. Studies on the understanding of the nature of science indicate that research should be focused on 'the understanding of the processes of scientific inquiry' (Durant et al. 1989; Robinson 1965). Some explicit reflection on the role of observation and experiment, the relationship between evidence and theory and their influence on the selection and application of a certain research methodology, is an essential component of the understanding of science (Driver et al. 1996).

Scientific and technological knowledge are, to a significant extent, culturally determined and reflect the social, religious, political, economic and environmental circumstances in which science and technology are practiced (Hodson 1993). Current debate on multicultural science education seeks to examine to what extent science is culturally determined and whether science transcends human differences, functioning as a hard-won vehicle for common engagement across cultures, religions and races (Matthews 1994). The interrelation of science and society may have a democratic, a utilitarian, a cultural and a moral character, in agreement with Thomas and Durant's overview of arguments in the literature for promoting public understanding of science (Thomas and Durant 1987).

Furthermore, as students present a number of alternative ideas about physics concepts and phenomenon at a cognitive level, they may also present alternative ideas about the nature of science (e.g., students consider scientific knowledge to be the 'absolute truth') and the interrelation between science and society (e.g., scientific and technological evolution cause an increasing number of environmental problems) at a metacognitive level. Recording such alternative ideas and developing ways and strategies on how to deal with those ideas in the classroom may encourage both learning and meta-learning.

2.3. THE EMOTIONAL DIMENSION

Until the 1980s, although educators and science education researchers used theories of motives in the development of teaching strategies, the influence of motivation was actually reduced to the aspect of providing the energy for cognitive development without having any influence on the cognitive structure itself. In the middle of the eighties psychological research rediscovered interest and motivation and since then many investigations have been made to throw light on its role. According to international meta-analysis about motivation and learning, motivational and emotional elements can be used to anticipate students' behaviour related to learning. Current research focuses on the correlation between the emotional/motivational and cognitive elements of learning (Fischer and Horstenhalt 1997).

The emotional dimension of the framework represents the attempts of researchers and curriculum designers in physics education to develop and apply (observable, measurable and improvable) methods in order to attract pupils to the world of physics. Efforts in this field vary from the awakening of pupils' interest, to their motivation, to the study of attitudes and finally to the study of behavioural intention.

2.4. BEYOND THE CURRENT FRAMEWORK OF SCIENCE EDUCATION

The framework of our classification also includes an area for proposals about the contribution of the history of physics in physics education, that we may come across during our research, and which cannot be classified within the current framework of science education. For example, if our classification was carried out in the 1960s, when the framework of science education was very different, some proposals inspired by the work of Thomas S. Kuhn would remain unclassified since these proposals refer either to students' alternative ideas or to the metacognitive dimension of teaching and learning. These very proposals influenced the evolution of the theories of conceptual change and constructivism in science education as well as the evolution of the framework of science education (Thagard 1992; Duschl 1994). Nevertheless, Kuhn's work had a direct influence on the development of Harvard Project Physics. However, at that time, educators and curriculum designers

Dimension	Category	Sub-category	Discipline
beyond the current framework			
emotional dimension	behavioural intention		science education
	attitudes		
	motives		
	interest		
metacognitive dimension	alternative ideas		science education
	science - society interrelation	moral	sociology & epistemology
		cultural	
		utilitarian	
		democratic	
	nature of science	methodology	epistemology
		content	philosophy
cognitive dimension	alternative ideas		science education
	problem-solving		science education
	methodology		science education, physics
	content		physics, science education, philosophy

Figure 1.

took into account only that part of his work that could fit in their contemporary framework of science education.

3. Classification of the Proposals for the Contribution of the History of Physics in Physics Education

During the study and the classification of the proposals different approaches and different perspectives of the same subject classified in the same area of the framework were indicated. This led to the creation of new sub-categories in the original framework. The original framework enriched with the new sub-categories is presented in Figure 1. Each sub-category is named after the discipline which mainly influences the perspective expressed by the proposals to be classified.

3.1. THE COGNITIVE DIMENSION

3.1.1. *The Teaching and Learning of the Content of Physics*

The proposals classified in this category express three different approaches regarding the use of the history of physics in the teaching and learning of the content of physics: the point of view of the philosopher, the point of view of the researcher in science education and the point of view of the physicist. Each one of these approaches to the teaching of the physics content similarly reflects the perspective of philosophy, the perspective of science education and the perspective of physics.

(I) *The Perspective of Philosophy*

In the first half of the 20th century, the philosophical extension of the theory of ontogeny-philogeny in biology, according to which ontogeny recapitulates philogeny, led science educators both to focus attention on and to increase interest in the history of physics and to include original material from the history of physics in the physics courses in the schools (Lodge 1893, 1905; Nunn 1919; Westaway 1929; Sherratt 1982). At that time educators supported the view that the same three successive phases of interest in the understanding of nature exist both in the history of physics as well as in the personal development of an individual. These three phases were wonder, utility and systematizing knowledge. These three phases were related to children of various ages. Children were supposed to respond 'most surely and actively to the direct appeal of striking and beautiful phenomena' at about the age of 11, while they were more interested in discovering the utility of knowledge and they were supposed to be ready to accept and study the systematization of knowledge after the age of 17 (Sherratt 1982).

In 1977 Piaget makes a different proposal indicating that we find stages in the history of science (before the 17th century) similar to the stages in a child's development. For example, in Aristotelian physics, the movement of a body is caused by the activation of its inner 'motor'. Furthermore, a body is supposed to move towards a certain destination. In the next stage, the inner 'motor' is replaced by an outer one. In the third stage, the movement of a body is studied using the concept of impetus and in the last stage before Newton, the movement of a body is interpreted by using the concept of acceleration. It has also been observed that children from the age of 4 until the age of 12 pass successively through all these four stages (Piaget 1977).

Nersessian commenting on the idea that 'ontogeny recapitulates phylogeny' indicates this recapitulation is neither feasible nor desirable, and although the changes in the ideas of students and scientists from one stage to the next are much alike, this does not mean that the processes that lead to these changes are necessarily the same (Nersessian 1994).

Many researchers have commented on the differences and the similarities that exist between students' reasoning today and scientists' reasoning in the history of physics. On the one hand, scientists generally master the most advanced levels of knowledge in their fields, whereas students do not (Wiser 1988). Scientists are meta-cognitively aware of their enterprise. They have an understanding of what they are out to investigate, the methods used to test their hypothesis, etc. In contrast, students generally construct knowledge about physics without being aware of the knowledge they have (Carey 1988). The world of the scientists in the history and of students nowadays are incredibly different and so must be their vastly different experiences in these worlds (Wiser 1988; McCloskey and Kargon 1988). The contemporary social background provides students with both knowledge and experience of instruments and phenomena, that scientists in the previous centuries actually ignored (Tselfes 1991). The concepts used by scientists in the history were invested with different meanings and interpretations than those of students today (e.g., the medieval explanations of projectile motion had gravity as a characteristic of the projectile itself, while students conceive of gravity as an external downward force that acts on the projectile) (Strauss 1988; McCloskey and Kargon 1988). Scientists constitute a community of individuals bent on solving problems in a particular domain. Students are individuals in a community to be sure, but not a community that shares the goals of problem solving for the particular concepts scientists have in mind. They are just trying to get the solution to the problem right (Carey 1988). There may be developing constraints on children's processing limitations and inferential capacities, whereas we do not imagine that these constraints are important for scientists in the history (Carey 1988). The development of cognitive schemata differs among individuals. The way a cognitive schema was developed by its creators does not reflect the actual way that students use nowadays to develop the same schema. Students and scientists have different starting-points in their thinking and different cognitive backgrounds (Tselfes 1991).

On the other hand, scientists in the history and students today are at the very beginning of conceptualizing aspects of their domains of knowledge. It may be that some early science grows out of individuals everyday experiences, which may be common to scientists 300 years ago and 20th century individuals (Strauss 1988). It is considered that both scientists in the history and students are at the same level in their struggle for crystallization of a paradigm: they define theoretical entities and attempt to relate them causally in a rough theory (McCloskey and Kargon 1988).

(II) *The Perspective of Science Education*

Since 1980, constructivism in physics education indicates the use of the history of physics as a source of inspiration for instructional design that would facilitate conceptual change. The history of physics may provide researchers with significant information about how a scientific concept has been constructed, changed and spread. The study of historical data provides researchers with similar proposals about students' learning processes in physics. For example, in the pre-*Principia* period, Newton himself dealt with a number of conceptual difficulties before coming to the formation of his laws. The way that Newton dealt with his conceptual difficulties may indicate a number of proposals about methods that would facilitate students to overcome their conceptual difficulties on the same subject (Steinberg et al. 1990; Izquierdo 1995).

According to another proposal, the teacher, by comparing the current scientific ideas with the historical ones in the classroom, helps his students to overcome their alternative ideas and accept the current scientific ideas (Wandersee 1985). For example, when the students attend or even participate in 'historical dialogues' dramatized in the classroom between Aristotle, Newton and Galileo, they recognize their own alternative ideas being expressed in the words of these three famous scientists and they are guided through discussion to the current scientific theory about movement (Conant 1957; Lochhead and Dufrence 1989; Solomon 1989).

(III) *The Perspective of Physics*
Since 1903, there have been many reports and studies indicating the history of physics as a source for experiments that clearly show the scientific principles. Furthermore, an anthology of experiments was recommended for instructional use (Sherratt 1980; Spurgin 1990). These proposals express the point of view of the physicist in search of the 'good' experiment.

3.1.2. *The Teaching and Learning of the Methodology of Physics*

(I) *The Perspective of Physics*
From the point of view of the physicist, students are facilitated, by their acquaintance with certain events and stories from the history of physics, in understanding the methodology of physics as they come to meet the ways and methods that famous physicists used in order to experiment and evolve their theories. The proposals classified in this sub-category, aim at the teaching and learning of a certain methodology of physics in the context of physics. Similar proposals were recorded towards the end of the previous century and the beginning of the current one, when the teaching of the methods of famous scientists was recommended by educators for the introduction of students to the methodology of physics (Cajori 1962; Sherratt 1982). Even though these proposals have such a long history, similar ones have been recorded until as recently as 1992.

(II) *The Perspective of Science Education*
Since 1965, there have been proposals influenced by research in science education, according to which the history of physics may provide teachers and researchers with instructional material to support the teaching-learning of the methodology of science as a tool for developing students' cognitive skills (Arons 1959, 1990; Brush 1969; Chambers 1989; Dunn 1993).

3.1.3. *The Teaching and Learning of Problem-Solving Skills*

During the last decade with the extension of the concept of 'problem' in the context of scientific research and model construction, proposals about the use of the history of physics in order to facilitate the development of students' problem-solving skills have been recorded. For example, the study and analysis of historical events that describe how scientists in the past dealt with such problems may start discussion and further consideration on the way students deal with similar problems in the classroom (Arons 1990).

According to the 'cognitive-historical' method, the problem-solving strategies scientists have invented and the representational practices they have developed over the course of the history of science are very sophisticated and refined outgrowths of ordinary reasoning and representational processes. Understood in this way, the cognitive activity of scientists becomes potentially quite relevant to learning, e.g., the study of the periods of transition before major conceptual 'revolutions' in physics indicates repeated use of specific heuristic procedures such as analogy, thought experiment, limiting case analysis and reasoning from imagistic representations. These techniques generate conceptual change in science and may help both teachers and students understand science. For example, when students, like Galileo and his followers, are introduced to Newtonian mechanics, they become familiar with constructing an abstract, mathematic representation of phenomena for the first time and, at the same time, adjusting such a representation to the natural world around them (Nersessian 1992).

3.1.4. *Research on Students' Alternative Ideas*

Since 1980, constructivism coupled with the research on students' alternative ideas has indicated history of physics as a source of fruitful information that may help and prepare researchers and/or teachers to expect students' alternative ideas (Benseghir 1989; Wandersee 1985). In the history of physics, in those cases where scientists presented contradicting arguments and theories, then in the same content area students' alternative ideas should be anticipated (Wiser and Carey 1983; Arons 1990; Sequeira and Leite 1991; Whiteley 1993).

Up to this point, proposals about the use of the history of science in the research on students' alternative ideas about the methodology of physics have not been recorded. However, it is our hypothesis that similar proposals may appear in the future.

3.2. THE METACOGNITIVE DIMENION

3.2.1. *The Nature of Science*

According to the proposals classified in this category, the history of physics may become a useful tool for presenting the nature of the content and of the methodology of physics in the classroom.

(I) *The Nature of the Content of Physics*

Since 1917, there have been proposals concerning the use of the history of physics in the classroom in order to emphasize the evolutionary nature of science and promote discussion on the traditional consideration that science presents an 'absolute truth' about the natural world (Sherratt 1982). In the early 1920s, Eric Holmyard suggested that the study of certain events from the history of science would encourage students to differentiate between the 'truth of science' and the 'truth of religion' (Holmyard 1923–4; Sherratt 1982). This trend may have been initiated by the previous revolutionary changes in science (e.g., Einstein's Theory of Relativity). In the last 30 years, an increasing number of proposals in this area reflects the idea that students will be facilitated in the operative use of the concepts of explanation, experiment and model if they understand the philosophical background of these concepts (Arons 1973, 1983, 1985, 1988; Brouwer and Singh 1983; Matthews 1988; Chambers 1989; King 1991). Previous theoretical models, ideas, experiments and further significant material from the history of physics that led to the evolution of these models, as well as of the scientific concepts involved, are presented in curricula, designed in the 1990s. The transition from particle to wave theory of light in optics, the transition from the caloric theory to the kinetic theory in thermodynamics and the transition from the Thomson to the Rutherford model in atomic physics are some paradigm shifts embedded within this curricula (Whiteley 1993).

(II) *The Nature of the Methodology of Physics*

According to proposals in this area, the history of physics may facilitate the understanding of different methodologies as students may compare such methodologies with the study of certain experiments from the history of physics (Arons 1959, 1984; Matthews 1988; Dunn 1993). We would like to point out that this metacognitive category deals with the comparative study of different methodologies of physics, whereas the cognitive category about methodology deals with the teaching and learning of one, and only one, methodology of physics. After 1989, a number of proposals suggest that the use of the history of physics in the classroom may facilitate students to realize the relation between observation and theory and may provide them with the opportunity to closely follow the way scientists deal with alternative theories and/or methods in order to choose the most appropriate one and thereby come to a conclusion (Brush 1989; Arons 1990, Whiteley 1993).

3.2.2. *Interrelation of Science and Society*

The proposals in this category refer to the study and comparative consideration of the various interrelations of science and society. These interrelations concern the inner structure of the scientific community as well as its operation in the wider social context. Current trends both in sociology and epistemology are reflected in the proposals of the four groups in this category: the democratic, the utilitarian, the cultural and moral interrelation of science and society.

(I) *Democratic Interrelation*
Proposals appeared in the last twenty years and in a way such proposals reflect the concept of democracy has come to a political maturity. Many people deal every day with or attend discussions about problems like the pollution of the environment or the aftermath of nuclear testing. The sometimes contradictory arguments, that scientists present when they give an account of these current universal problems, are part of the most recent history of science, which people should be informed about in order to be able to express their personal opinion when asked (NSTA 1982; Bybee et al. 1991).

(II) *Utilitarian Interrelation*
Proposals appeared after 1960 when a rapid evolution in technology was initiated and since then a number of technological objects and processes have become part of everyday life. Therefore, the way and the quality of life has changed, while it has become necessary for laymen to be technologically literate. For example, a course, coupled with a similar exhibition, about the story of the invention of telephone by Bell or of wireless by Marconi and about the evolution of communication to the present, may become a connecting link between school and every day technology (Barret and Stanyard 1979; NSTA 1982; Arons 1983; Hurd 1987; Bybee et al. 1991).

(III) *Cultural Interrelation*
The argument that the study of the history of science encourages people to appreciate science as a major cultural achievement was used, more than once, against the notion that the teaching of physics was too technocratic and culturally indifferent to the public, as for example, in the beginning of our century, when the importance of physics education as part of compulsory education was questioned (Westaway 1929). Since 1960 a number of proposals have been recorded about the recognition of the relation between physics and culture and of physics as part of our cultural heritage through the study of the history of physics. The theories of Copernicus and Galileo that influenced the way we perceive our position in the universe, Newton's theory that the same laws of motion apply both in Earth and in space, Einstein's Theory of Relativity are considered to be fundamental elements of culture (Arons 1965, 1988; Klopfer 1969; Sherratt 1982; Brouwer and Singh 1983; Jenkins 1989; Bybee et al. 1991; Dunn 1993).

(IV) *Moral Interrelation*
Certain incidents from the history of physics clearly show that a scientific discovery may either contribute to a better quality of life or become a destructive tool (Humby and James 1942; Brush 1969; Sherratt 1982; Brouwer and Singh 1983; Dunn 1993). The creation and use of the atomic bomb during the Second World War made scientists face their moral responsibility. Such proposals present a peak

in the years after the war trying to help science regain its lost morality due to the fact that science supported the war industry during the years of the conflict.

3.2.3. *Research on Student's Alternative Ideas*

Research on students' ideas about the nature of science and the interrelation of science and society, ideas which belong to the metacognitive dimension of learning, indicate the history of physics as a useful source of information for the interpretation of students' alternative ideas in this area.

3.3. THE EMOTIONAL DIMENSION

Research in this field changed focus from the awakening of pupils' interest, to their motivation, to the study of attitudes and finally to the study of behavioural intention, as has already been mentioned in previous paragraphs. As both focus and terminology changed, the proposals about the use of the history of physics in physics education classified in this area also followed these changes.

Since the beginning of the century, proposals have been recorded suggesting that the use of incidents from the history of physics in the classroom may become a source of admiration for the pioneers of science and strengthen students' interest (Lodge 1893, 1905; Westaway 1929; Brush 1969; Arons 1990). Proposals, recorded in the 1960s and 1970s, suggested that students become motivated to study physics by discussion arising from the presentation of historical incidents about the evolution of science (Klopfer 1969; Thomason 1992). Since 1970, proposals have suggested that the use of incidents from the history of physics or even dramatized presentations of 'historical dialogues' in the classroom may encourage students' positive attitude towards the teaching and learning of physics (Summers 1982; King 1991).

4. Comments on the Chart

In an attempt to have a better overall view of the classification, we designed a chart so that the various proposals reported could be observed and commented on in a comparative way. The framework of the classification is presented on the vertical axis of a chart, while time since 1893 (the year of the earliest reference available) till today is presented on its horizontal axis. If two more areas are added to the vertical axis of our chart, an area presenting the proposals applied in curricula and a second area presenting the proposals applied in teacher training courses, new opportunities are provided for further comment and comparative study on the proposals about the contribution of the history of physics in physics education and their application in curricula and teacher training. The chart is presented in Figure 2. The references corresponding to the numbers in Figure 2 are presented in the Appendix.

Area	Category	Subcategory	Discipline	References (1890–2000, left to right in time)
teacher training				5 46 59 48 49 39 28504751 60
curriculum				41 42 43 44 45 19 9 5335 47 13
beyond the current framework				
emotional dimension	behavioural intention		science education	
	attitudes			39 29
	motives			35 40
	interest			1 2 5 53 47
metacognitive dimension	alternative ideas		science education	
	science - society interrelation	moral	sociology & epistemology	37 38 53 56 21
		cultural		5 19 9 3512 1827563658 4732 21
		utilitarian		12 33 3127 34 4732
		democratic		31 32
	nature of science	methodology	epistemology	19 12 2752 283630472113
		content	philosophy	3 26 12 2756162836202913
cognitive dimension	alternative ideas		science education	2723 54 2447 2513
	problem-solving		science education	47 22
	methodology		science education	19 53 20 47 21
			physics	17 3 18
	content		physics	15 57
			science education	9 12 271654361110554713 14
			philosophy	1 2 3 4 5 6 7 8

Time axis: 1890 1900 1910 1920 1930 1940 1950 1960 1970 1980 1990 2000

Figure 2.

In Figure 2, among the first things we observe is a gap during a period of twenty years, in the 1940s and the 1950s, as there are not many proposals regarding the use of the history of physics in physics education recorded during the Second World War and the years that followed. Nevertheless, we should point out that although there are no proposals coming from European researchers, educators and curriculum designers during this period, a strong influence of history of science on the development of US curricula has been recorded (Conant 1947, 1957; Cohen 1950; Holton 1952). The proposals surrounding this period are those about the interrelation of science and society. The idea that 'scientific discoveries are social activities with social implications' is indicated in a study published in 1942. After 1965, there are a significant number of proposals about the use of the history of physics in physics education, aiming at the presentation of the interrelation of science and society with special emphasis given to the cultural and moral interrelation. The relation of science to culture provides science with a human face.

A second observation of the chart, presented in Figure 2, is the gradual shift in the focus of research interest from the cognitive to the metacognitive dimension since 1965. Such a shift is in agreement with the discussion on the teaching of the nature of science and the interrelation of science and society through the study of the history of physics, that has been going on during the last decade and aims at the development of students' metacognitive skills. For example, the teaching of the methodology of physics aimed to show the 'magnitude' of physics as a science in the beginning of the century, while since 1960 it has focused on the development of students' cognitive skills. However, methodology was presented as knowledge only up to 1980, since when it is also considered as meta-knowledge, as a tool for organizing and controlling reasoning and interpretation patterns used by the individual.

It is interesting to observe in the chart (Figure 2) that each curriculum expresses previous and/or contemporary proposals, like Harvard Project Physics and Whiteley's new Caribbean curriculum.

The application of these curricula brought to light a number of shortcomings and problems arising from the fact that teachers were not trained to deal with such curricula. At the same time proposals about the contribution of the history of physics for teacher training appeared. Current developments in both areas led to a number of application of proposals in teacher training.

For example, in the 1910s, 1920s and 1930s, a number of books were published recommending the use of the history of physics in physics teaching (Heath 1919; Taylor 1923; Wood 1925; Partington 1928; Roberts and Thomas 1934). In the middle of that period, studies on the problems teachers had to deal with as they used instructional material that included aspects from the history of physics appeared. Later, during the 1960s, the design of Harvard Project Physics was based on a number of previous proposals about the use of history of physics in physics education. In the following decades researches on teachers' problems in applying Harvard Project Physics in the classroom and on the causes of these problems were carried out (Brush 1969; Bileh and Malik 1977; Russell 1981). In the 1990s, curricula, which realized a number of proposals about the contribution of the history of physics in physics education, have been designed and applied in the classroom. This time, researches on teacher training in the history of physics and its use in physics teaching started about a decade before and have been going on during the application of curricula (Garrison and Bentley 1990; Arons 1990; Gallagher 1991; Whiteley 1993; Matthews 1994). Over the last few years, researchers in the area of teacher training and curriculum designers have co-operated in this field.

Appendix

Number	Reference	Number	Reference
1	Lodge 1893	31	NSTA 1982
2	Lodge 1905	32	Bybee et al. 1991
3	Sherratt 1982	33	Barret and Stanyard 1979
4	Nunn 1919	34	Hurd 1987
5	Westaway 1929	35	Klopfer 1969
6	Sherratt 1982	36	Arons 1988
7	Piaget 1977	37	Sherratt 1982
8	Nersessian 1994	38	Humby and James 1942
9	Arons 1965	39	Summers 1982
10	Lochhead and Dufrence 1989	40	Thomason 1992
11	Solomon 1989	41	Heath 1919
12	Arons 1973	42	Taylor 1923
13	Whiteley 1993	43	Wood 1925
14	Izquierdo 1994	44	Partington 1928
15	Fish 1903	45	Roberts and Thomas 1934
16	Arons 1985	46	Lauwerys 1935
17	Cajori 1962	47	Arons 1990
18	Sherratt 1982	48	Bileh and Malik 1977
19	Arons 1959	49	Russell 1981
20	Chambers 1989	50	Garrison and Bentley 1990
21	Dunn 1993	51	Gallagher 1991
22	Nersessian 1992	52	Arons 1984
23	Wiser and Carey 1983	53	Brush 1969
24	Benseghir 1989	54	Wandersee 1985
25	Sequeira and Leite 1991	55	Steinberg et al. 1990
26	Sherratt 1982	56	Brouwer and Singh 1983
27	Arons 1983	57	Spurgin 1990
28	Matthews 1988	58	Jenkins 1989
29	King 1991	59	Carey and Stauss 1968
30	Brush 1989	60	Matthews 1994

References

Adey, P. & Shayer, M.: 1993, 'An Exploration of Long-term Far-transfer Effects Following an Extended Intervention Programme in the High-school Science Curriculum', in D. Edwards, E. Scanlon & D. West (eds), *Teaching, Learning and Assessment in Science Education*, The Open University, London, pp. 190–220.

Arons, A. B.: 1959, 'Structure, Methods, and Objectives of the Required Freshman Calculus-Physics Course at Amherst College', *American Journal of Physics* **27**(9), 658–666.

Arons, A. B.: 1965, *Development of Concepts of Physics*, Addison-Wesley, Reading MA.

Arons, A. B.: 1973, 'Toward Wider Public Understanding of Science', *American Journal of Physics* **41**(6), 769–782.

Arons, A. B.: 1983, 'Achieving Wider Scientific Literacy', *Daedalus* **112**(2), 91–122.

Arons, A. B.: 1984, 'Education Through Science', *Journal of College Science Teaching* **13**, 210–220.

Arons, A. B.: 1985, 'Critical Thinking and the Baccalaureate Curriculum', *Liberal Education* **71**(2), 141–157.

Arons, A. B.: 1988, 'Historical and Philosophical Perspectives Attainable in Introductory Physics Courses', *Educational Philosophy and Theory* **20**(2), 13–23.

Arons, A. B.: 1990, *A Guide to Introductory Physics Teaching*, John Wiley, New York.

Barret, M. S. & Stanyard, T. N.: 1979, 'The Davy Bicentenary Event', *School Science Review* **61**, 203–213.

Benseghir, A.: 1989, *Transition Electrostatique – Electrocinetique: Point de Vue Historique et Analyse des Difficultes des Eleves*, Ph.D. Thesis, Universite Paris VII.

Bileh, V. & Malik, M. H.: 1977, 'Development and Application of a Test on Understanding the Nature of Science' *Science Education* **61**, 559–571.

Brouwer, W. & Singh, A.: 1983, 'The Historical Approach to Science Teaching', *The Physics Teacher*, April, 230–236.

Brush, S. G.: 1969, 'The Role of History in the Teaching of Physics', *The Physics Teacher* **7**(5), 271–280.

Brush, S.: 1989, 'History of Science and Science Education', in M. Shortlands & A. Warwich (eds), *Teaching the History of Science*, Blackwell, Oxford.

Bybee, R. W., Powell, J. C., Ellis, J. D., Giese, J. R., Parisi, L. & Singleton, L.: 1991, 'Integrating the History and Nature of Science and Technology in Science and Social Studies Curriculum', *Science Education* **75**(1), 143–155.

Cajori, F: 1962, *A History of Physics*, Dover Publications Inc., New York.

Carey, S.: 1988, 'Reorganization of Knowledge in the Course of Acquisition', in S. Strauss (ed.), *Ontogeny, Phylogeny and Historical Development*, Ablex Publishing Corporation, Norwood, NJ, pp. 1–27.

Carey, L. & Stauss, N. G.: 1968, 'An Analysis of the Understanding of the Nature of Science by Prospective Secondary Science Teachers', *Science Education* **52**, 358–363.

Chambers, J. H.: 1989, 'Alfred Russell Wallace. The Theory of Evolution and the Teaching of Science', *School Science Review* **71**, 143–147.

Cohen, I. B.: 1950, 'A Sense of History in Science', *American Journal of Physics* **18**, 343–359. Reprinted in *Science & Education* **2**(3), 1993.

Conant, J. B.: 1947, *On Understanding Science*, Yale University Press, New Haven.

Conant, J. B. (ed.): 1957, *Harvard Case Histories in Experimental Science*, 2 vols., Harvard University Press, Cambridge, MA.

Driver, R., Leach, J., Millar, R. & Scott, P.: 1996, *Young People's Images of Science*, Open University Press, Buckhingham/Philadelphia.

Duit, R.: 1994, 'Conceptual Change Approaches in Science Education'. Paper presented at the Symposium on Conceptual Change, Friedrich-Schiller-University of Jena, Germany. September 1–3, 1994.

Dunn, R.: 1993. 'Empires of Physics – A New Initiative in Science Education', *School Science Review* **75**, 135–137.

Durant, J., Evans, G. & Thomas, G.: 1989, 'The Public Understanding of Science', *Nature* **340**, 11–14.

Duschl, R. A.: 1994, 'Research on the History and Philosophy of Science', in D. L. Gabel (ed.), *Handbook of Research on Science Teaching and Learning*, Macmillan Publishing Company, New York.

Duschl, R. A. & Erduran, S.: 1996, 'Modelling the Growth of Scientific Knowledge', in G. Welford, J. Osborne & P. Scott (eds), *Research in Science Education in Europe – Current Issues and Themes*, Falmer Press, London.

Fischer, H. E. & Horstenhalt, M.: 1997, 'Motivation and Learning Physics'. Paper presented at ESERA Conference, Rome 1997.

Fish, A. H.: 1903, 'Science in a Liberal Education', *School World* **5**, 354.

Fleming, D.: 1967, 'Attitude: The History of a Concept', *Percept. Am. Hist.* **1**, 287–365.

Gallagher, J. J.: 1991, 'Prospective and Practicing Secondary School Science Teachers' Knowledge and Beliefs about the Philosophy of Science', *Science Education* **75**(1), 121–133.

Garrison, J. W. & Bentley, M.: 1990, 'Teaching Scientific Method: The Topic of Confirmation and Falsification', *School Science and Mathematics* **90**, 180– 197.

Glaser, R.: 1994, 'Application and Theory: Learning Theory and the Design of Learning Environment'. Paper presented at the 23rd International Congress of Applied Psychology, July 17–22, Madrid, Spain.

Guillon, A.: 1995, 'Scientific Processes and Laboratory Work in Physics', in D. Psillos (ed.), *European Research in Science Education-Proceeding of the Second Ph.D. Summer School*, Art of Text, Thessaloniki.

Gunstone, R. F. & Northfield, J.: 1994, 'Metacognition and Learning to Teach', *International Journal of Science Education* **16**(5), 523–537.

Heath, A. E.: 1919, 'The Philosophy of Science as a School Subject', *School Science Review* **1**(4), 131–134.

Hodson, D.: 1993, 'Teaching and Learning about Science: Considerations in the Philosophy and Sociology of Science', in D. Edwards, E. Scanlon & D. West (eds), *Teaching, Learning and Assessment in Science Education*, The Open University, London, pp. 5–32.

Holmyard, E. J.: 1923–4, 'The Historical Method of Teaching Chemistry', *School Science Review* **5**(20), 227–233.

Holton, G.: 1952, *Introduction to Concepts and Theories in Physical Science*, Addison-Welsey, New York.

Humby, S. R. & James, E. J. F.: 1942, *Science and Education*, Cambridge University Press, Cambridge.

Hurd, P. DeHart: 1987, 'A Nation Reflects: The Modernization of Science Education', *Bulletin of Science, Technology and Society* **7**, 9.

Izquierdo, M.: 1995, 'Cognitive Models of Science and the Teaching of Science, History of Science and Curriculum', in D. Psillos (ed.), *European Research in Science Education-Proceeding of the Second Ph.D. Summer School*, Art of Text, Thessaloniki.

Jenkins, E.: 1989, 'Why the History of Science?', in M. Shortland & A. Warwick (eds), *Teaching the History of Science*, Blackwell, Oxford.

King, B. B.: 1991, 'Beginning Teachers' Knowledge of and Attitudes Towards History and Philosophy of Science', *Science Education* **75**(1), 135–141.

Klopfer, L. E.: 1969, 'The Teaching of Science and the History of Sciences', *Journal of Research in Science Teaching* **6**, 87–95.

Lauwerys, J. A.: 1935, 'The Teaching of Physical Science', *School Science Review* **17**(66), 161–170.

Lochhead, J. & Dufrence, R.: 1989, 'Helping Students Understand Difficult Concepts Through the Use of Dialogues with History', in D. E. Herget (ed.), *The History and Philosophy of Science in Science Teaching*, Tallahassee, Florida State University.

Lodge, O.: 1893, *Pioneers of Science*, Macmillan, London.

Lodge, O.: 1905, *School Teaching and School Reform*, Williams and Norgate, London.

Matthews, M. R.: 1988, 'A Role for History and Philosophy in Science Teaching', *Educ. Phil. Theory* **20**, 67–81.

Matthews, M. R.: 1994, *Science Teaching. The Role of History and Philosophy of Science*, Routledge, New York.

McCloskey, M. & Kargon, R.: 1988, 'The Meaning and the Use of Historical Models in the Study of Intuitive Physics', in S. Strauss (ed.), *Ontogeny, Phylogeny and Historical Development*, Ablex Publishing Corporation, Norwood, NJ, pp. 49–67.

Mohopatra, J. K.: 1987, 'Can Problem-Solving in Physics Give an Indication of Pupils' Process Knowledge?', *International Journal of Science Education* **9**(1), 117–123.

National Science Teachers Association (NSTA): 1982, *Science-Technology-Society: Science Education for the 1980s*, NSTA, Washington, DC.

Nersessian, N.: 1992, 'Constructing and Instructing: The Role of "Abstraction Techniques" in Creating and Learning Physics', in R. A. Duschl & R. J. Hamilton (eds), *Philosophy of Science, Cognitive Psychology and Educational Theory and Practice*, State University of New York Press, New York.

Nersessian, N.: 1994, 'Conceptual Structure and Teaching: A Role for the History in Science Education', in V. Koulaidis (ed.), *Representations of the Natural World – A Cognitive, Epistemological and Didactical Approach*, Gutenberg, Athens.

Nickerson, R. S., Perkins, D. N. & Smith, E. E.: 1985, *The Teaching of Thinking*, Lawrence Erlbaum Associates, Hillsdale, NJ.

Nielsen, H. & Thomsen, P. V.: 1990, 'History and Philosophy of Science in Physics Education', *International Journal of Science Education* **12**(3), 308–316.

Nunn, T. P.: 1919, 'Science', in J. Adams (ed.), *The New Teaching*, Hodder and Stoughton, London.

Partington, R.: 1928, *The Composition of Water*, Bell and Sons, London.

Perkins, D. N. & Salomon, G.: 1989, 'Are Cognitive Skills Context-bound?', *Educational Researcher* **18**(1), 16–25.

Piaget, J.: 1977, *Conservations libre avec Jean Piaget*, Edition Robert Laffont, Paris.

Roberts, M. & Thomas, E. R.: 1934, *Newton and the Origin of Colours*, Bell and Sons, London.

Robinson, J. T.: 1965, 'Science Teaching and the Nature of Science', *Journal of Research in Science Teaching* **3**, 37–50. Reprinted in *Science & Education* **7**(6), 1998.

Russell, T. L.: 1981, 'What History of Science, How Much and Why?', *Science Education* **65**(1), 51–64.

Sequeira, M. & Leite, L.: 1991, 'Alternative Conceptions and History of Science in Physics Teacher Education', *Science Education* **75**(1), 45–56.

Sherratt, W. J. M.: 1980, *History of Science in Science Education*, Ph.D. Thesis, Leicester University.

Sherratt, W. J.: 1982, 'History of Science in the Science Curriculum: An Historical Perspective. Part I: Early Interest and Roles Avocated', *School Science Review* **64**, 225–236.

Shrigley, R., Koballa, T. Jr. & Simpson, R.: 1988, 'Defining Attitude for Science Educators', *Journal of Research in Science Teaching* **25**(8), 659–678.

Solomon, J.: 1989, 'The Retrial of Galileo', in D. E. Herget (ed.), *The History and Philosophy of Science in Science Teaching*, Florida State University, Tallahassee, pp. 332–338.

Spurgin, B.: 1990, 'Plus and Minus, Glass and Resin, Electrons and Cathode Rays', *School Science Review* **72**, 65–77.

Steinberg, M. S., Brown, D. E. & Clement, J.: 1990, 'Genius is not Immune to Persistent Misconceptions: Conceptual Difficulties Impeding Isaac Newton and Contemporary Physics Students', *International Journal of Science Education* **12**(3), 265–273.

Stewart, J. & Hafner, R.: 1991, 'Extending the Conception of 'Problem' in Problem-Solving Research', *Science Education* **75**(1), 105–120.

Strauss, S. (ed.): 1988, *Ontogeny, Phylogeny and Historical Development*, Ablex Publishing Corporation, Norwood, NJ.

Summers, M. K.: 1982, 'Philosophy of Science in the Science Teacher Education Curriculum', *European Journal of Science Education* **4**, 19–27.

Taylor, C. M.: 1923, *The Discovery of the Nature of the Air*, Bell and Sons, London.

Thagard, P.: 1992, *Conceptual Revolutions*, Princeton University Press, Princeton, NJ.

Thijs, G. D. & Bosch, G. M.: 1995, 'Cognitive Effects of Science Experiments Focusing on Students' Preconceptions of Force: a Comparison of Demonstrations and Small-Group Practicals', *International Journal of Science Education* **17**(3), 311–323.

Thomas, G. & Durant, J.: 1987, 'Why Should we Promote the Public Understanding of Science?', *Scientific Literacy Papers* **1**, 1–14.

Thomason, B.: 1992, 'Plant Sensitivity: A Historical Source', *School Science Review* **73**, 95–101.

Tselfes, V.: 1991, 'Teaching of Science in Compulsory Education - History of Science: What Possible Relation?'. Paper presented in the Symposium: *The Educational Development of the History of Science*, Thessaloniki, August 27–30, 1991.

Tyson, L. M., Venville, G. J., Harrison, A. G. & Treagust, D. F.: 1997, 'A Multidimensional Framework for Interpreting Conceptual Change Events in the Classroom', *Science Education* **81**, 387–404.

Wandersee, J. H.: 1985, 'Can the History of Science Help Educators Anticipate Students' Misconceptions?', *Journal of Research in Science Teaching* **23**(7), 581–597.

Westaway, F. W.: 1929, *Science Teaching*, Blackie and Son, London.

White, R. T. & Gunstone, R.F.: 1989, 'Metalearning and Conceptual Change', *International Journal of Science Education* **11**, 577–586.

Whiteley, P.: 1993, 'The History of Physics – Its Use in a Caribbean Physics Syllabus', *School Science Review* **75**, 123–127.

Wiser, M.: 1988, 'The Differentiation of Heat and Temperature: History of Science and Novice-Expert Shift', in S. Strauss (ed.), *Ontogeny, Phylogeny and Historical Development*, Ablex Publishing Corporation, Norwood, NJ, pp. 28–48.

Wiser, M. & Carey, S.: 1983, 'When Heat and Temperature Were One', in D. Gentner & A. L. Stevens (eds), *Mental Models*, Lawrence Erlbaum Associates Publishers, Hillsdale, NJ/London.

Wood, A.: 1925, *Joule and the Study of Energy*, Bell and Sons, London.

Contributors

Douglas Allchin studied biology and philosophy before completing his PhD at the University of Chicago. He has written, with Joel Hagen and Fred Singer, *Doing Biology* (Harper Collins 1996), and is editor of the Newsletter of the Sociology, History and Philosophy of Science in Science Teaching (SHIPS) group. His research interests are in the application of history and philosophy of science to science teaching, and the role of values in science.

Fabio Bevilacqua is an associate professor in the Physics Department, 'A. Volta', University of Pavia, Italy. He has undergraduate degrees in engineering and philosophy, and completed his PhD at Cambridge University's History and Philosophy of Science Department. He has published in the history of nineteenth century electrical theory, and has a particular interest in the work of Hermann Helmholtz. He is chairperson of the European Physical Society's History of Physics and Education Division. He is the director of a large research project at Pavia University which is developing and testing hypermedia materials for incorporating the history of physics into physics teaching.

Robert Carson is an associate professor of educational foundations at Montana State University. He teaches courses in philosophy, sociology, psychology and history of education. He reccived his PhD in philosophy of education from the University of Illinois. His research is in part concerned with science as a cultural influence, its role in liberal education, the meaning of liberal education in the context of a post-modernist, culturally pluralistic world, and the role that semiotic systems play in extending the possibilities of thought.

Alberto Cordero is professor of philosophy and history at Queen's College and the Graduate Centre, City University of New York; and Honorary Director of the Program for Scientific Thought, Universidad Peruana Cayetano Heredia, Lima, Peru. He has a PhD in philosophy from the University of Maryland; a MPhil degree in philosophy of science from Trinity College, Cambridge; and a MSc degree in nuclear physics from Worcester College, Oxford. Among his recent publications are: 'Rival Theories Without Observable Differences', in M. Pauri (ed.), *Observability, Unobservability and Their Impact on Scientific Realism,* Kluwer Academic Publishers, 2000, and 'Two Bad Arguments Against Naturalism', in J. Mosterín (ed.), *Current Issues in the Philosophy of Biology,* 1998.

Bo Dahlin is associate professor of Education at Karlstad University, Sweden. He took his BSc degree from the University of Stockholm 1971, and his PhD in Education from Gothenburg University 1989. Apart from Education, he has a background in Science and Philosophy. His main research interest is philosophy of education. His empirical research has been cross-cultural studies of conceptions of learning, based on a qualitative (phenomenographic) approach.

Jim Donnelly is senior lecturer in Science Education at the University of Leeds. He studied chemistry at University College, London. He has published in the fields of assessment, policy studies in science education and the history of technological education. In 2001 he will publish, with Edgar Jenkins, *Science Education: Policy, Professionalism and Change*, an analysis of the impact of recent policy initiatives on science teachers' professional situation during the last four deacdes.

Sibel Erduran is a research associate at King's College, University of London. Her BA degree is in biochemistry from Northwestern University, her MS in food chemistry from Cornell University, and her PhD in science education from Vanderbilt University. She has taught chemistry in an English-medium high school in Cyprus. Her research interests include the application of epistemological perspectives in chemistry education.

Igal Galili is a senior lecturer in the Science Teaching Center at the Hebrew University of Jerusalem, from where he obtained his BSc in physics, and MSc and PhD degrees in theoretical physics. His research areas are: conceptual knowledge in physics education; content and organization of students' knowledge of physics; history and philosophy of science and its implementation in science education.

Enrico Antonio Giannetto is a researcher in the history of physics at the Physics Department 'A. Volta', University of Pavia, Italy. He is a graduate of the University of Padova in theoretical, elementary particle physics. He studied the history of science at the Domus Galilaeana in Pisa, and obtained his doctorate in theoretical physics (a quantum-relativistic theory of condensed matter) at the University of Messina. His research interests cover the foundations of quantum-relativistic physics and cosmology, history of medieval and modern physics, and science education.

Ron Good is professor of physics and of science education at Louisiana State University. He has published widely in science education, has been a consultant for the National Science Foundation, and has served a term as editor of *The Journal of Research in Science Teaching*.

Amnon Hazan is coordinator of 'Science for All' program in the Science Teaching Center at the Hebrew University of Jerusalem. He obtained his BSc in physics from Bar-Ilan University, Ramat Gan, Israel, is MSc in science education from Tel-Aviv University and his PhD in science education from The Hebrew University of Jerusalem. His reseach is concerned with utilisation of history and philosophy of science in high school physics education, and it has appeared in recent issues of *International Journal in Science Education*, and the *American Journal of Physics*.

John Heilbron is a Senior Research Fellow at Worcester College Oxford, and Emeritious Professor of History at the University of California Berkeley, where he was formerly Vice-Chancellor. He is the author of numerous books including: *Elements of Early Modern Physics* (Uni. California Press, Berkeley, 1982); *The Dilemmas of an Upright Man. Max Planck as Spokesman for German Science* (Uni. California Press, Berkeley, 1986); *The Sun in the Church. Cathedrals as Solar Observatories* (Harvard Uni. Press, Cambridge, MA, 1999). He has published widely in physics and history of physics journals, as well as *Nature* journal.

Edgar Jenkins recently retired as Professor of Science Education Policy and Director of the Centre for Studies in Science and Mathematics Education at the University of Leeds, UK. He is the author of numerous articles and books concerned with the social history and politics of school science education and, from 1984 to 1997, was the Editor of the research review journal, *Studies in Science Education*. His most recent book is *Junior School Science Education in England and Wales Since 1900* (Woburn Press, 1998).

Nahum Kipnis was born and educated in the former USSR, where he received an MSc degree in physics and mathematics. He taught high school and college physics, and did research in experimental physics and the history of science. In 1984 he received a PhD degree in history of science from the University of Minnesota. The thesis was published as *History of the Principle of the Interference of Light* (Birkhaüser, Basel, 1991). He has published *Rediscovering Optics* (BENA Press, Minneapolis, 1992), a book that outlines a new technique for improving the learning of science by combining the history of science with investigative experiments.

Panagiotis Koumaras is a professor of educational material design and experimental teaching in the School of Education at the Aristotle University of Thessaloniki. He graduated from the physics department of the same university in 1975 and taught secondary school physics for ten years. In 1989 he graduated with a Ph.D. in Didactics of Physics from the physics department at Aristotle University. His research is concerned with the effectiveness of teaching materials, and with students' conceptual models.

Fritz Kubli has been a Gymnasium physics teacher since 1969. In 1970, he received his PhD at the Swiss Federal Institute of Technology in Zurich. His thesis allowed him to work with Louis de Broglie, the well-known pioneer of wave mechanics. Between 1973-1975, a joint project with the Genevean psychologist Jean Piaget gave him the opportunity to point out the implications of his development theory with regard to science teaching. Between 1995-1997, a research project financed by the Swiss National Foundation led to a study which investigated the function of narrative processes in science teaching. It has been published in German as a *Plädoyer für Erzählungen im Physikunterricht*.

Alexander Levine is an assistant professor of philosophy at Lehigh University, and holds a PhD in Philosophy from the University of California, San Diego. He prepared the English translation of *Reconstructing Scientific Revolutions: Thomas S. Kuhn's Philosophy of Science*, by Paul Hoyningen-Huene (University of Chicago Press, 1993), and has published articles in his two chief areas of research, the philosophy of science and philosophy of mind.

Olivia Levrini received her PhD from the Physics Department of the University of Bologna, Italy, with a dissertation thesis on "The possible interpretations of the formalism of General Relativity: Implications for teaching". Her current research work mainly concerns the design of teaching proposals in which history and philosophy of physics play a relevant role in improving physics understanding. The research involves the development of materials for pre-service teacher education and for teaching physics at university level.

Peter Machamer is a professor of history and philosophy of science at the University of Pittsburgh. He holds a BA degree from Columbia University, BA and MA degrees from Trinity College, Cambridge, and a PhD degree from the University of Chicago. He has published on Aristotle, Galileo, seventeenth century science, psychology, cognitive science, ethics and assorted other subjects. He is the editor of *The Cambridge Companion to Galileo* (Cambridge University Press, 1998).

Roberto de Andrade Martins is a professor at the Physics Institute 'Gleb Wataghin', State University of Campinas (UNICAMP), Brazil. He received a first degree in physics at Sao Paulo University, and PhD in logic and philosophy of science at UNICAMP. He researches topics in the history and philosophy of physics, and in physics education.

Michael R. Matthews is an associate professor in the School of Education at the University of New South Wales. He has degrees from the University of Sydney in science, philosophy, psychology, philosophy of science, and education. His recent books include *Challenging New Zealand Science Education* (Dunmore Press, 1995), *Science Teaching: The Role of History and Philosophy of Science* (Routledge, 1994) and *Time for Science Education* (Plenum Publishers, 2000). Additionally he has edited *Constructivism in Science Education: A Philosophical Examination* (Kluwer Academic Publishers, 1998).

Robert Nola is an associate professor of philosophy at the University of Auckland. He has degrees in science and philosophy. He has published in philosophy of science, 19th-century philosophy, and philosophy and science education. He is the editor of *Relativism and Realism in Science* (Kluwer Academic Publishers). His research areas are epistemology, metaphysics, and philosophy of science, especially issues to do with realism and relativism.

James Rutherford has recently retired from the directorship of the AAAS Project 2061. He obtained his AB degree from University of California, Berkeley, his MA from Stanford University and his EdD from Harvard University. He was a high school science teacher in California, an associate professor of education at Harvard, and professor of science education at New York University. In the 1970s he was director (with Gerald Holton & Fletcher Watson) of Harvard Project Physics.

William H. Schmidt received his undergraduate degree from Concordia College in River Forrest, Illinois, and his Masters and Doctoral degrees from the University of Chicago. He is the National Research Coordinator and Executive Director of the US National Center which oversees participation of the United States in the IEA sponsored Third International Mathematics and Science Study (TIMSS). He is widely published in numerous journals, has contributed chapters to many books, and most recently he has co-authored a number of books on the curriculum analysis for mathematics and science from the TIMSS.

Fanny Seroglou is a researcher in the School of Education in the Aristotle University of Thessaloniki, from where she graduated first in physics, then with a PhD in science education. Her research is concerned with the history of science in science education.

James Shymansky is E. Desmond Lee Professor of Science Education at the University of Missouri, St. Louis. He is a past editor of the *Journal of Reseach in Science Teaching,* and is currently directing a National Science Foundation project for K-6 teachers in rural districts of Missouri and Iowa. He has published extensively in science education journals.

Cibelle Celestino Silva is a PhD student at the Physics Institute 'Gleb Wataghin', State University of Campinas (UNICAMP), Brazil. She received a first degree in physics as Sao Paula University and an MSc degree in physics at UNICAMP, with a thesis on Newton's *Optics*.

HsingChi A Wang is a research assistant professor at the University of Southern California Center for Craniofacial Molecular Biology. She is also a research associate at Michigan State University, where she works with William Schmidt in examing the implications of the TIMSS research for American mathematics and science education. She had a Bachelor's degree in Physics from Taiwan, a Master's degree in Science Education, and a PhD in Curriculum and Instruction both from the University of Southern California. She has published in a number of science education journals.

Name Index

Subject Index